Bulges lie at the center of spiral galaxies. Until recently, they were thought to host uniquely old stellar populations and thus provide a key for understanding galaxy formation. Recent observations from the ground and space have drastically changed our view on the nature of bulges and shown that they can also contain dust, gas, and star-forming regions. This timely volume presents review articles by a panel of international experts who gathered at a conference at the Space Telescope Science Institute, Baltimore, to address several fundamental questions: What is a bulge? When and how did bulges form? And, on what timescales?

This volume provides a state-of-the-art picture of our new understanding of these fundamental building-blocks of galaxies, and a stimulating reference point for all those interested in galaxy formation.

C.M. Carollo is an Assistant Professor at Columbia University, New York.

H.C. Ferguson is an associate astronomer at the Space Telescope Science Institute, Baltimore.

R.F.G. Wyse is a Professor of Physics and Astronomy at Johns Hopkins University, Baltimore.

Globular Clusters

CAMBRIDGE CONTEMPORARY ASTROPHYSICS

Series editors
José Franco, Steven M. Kahn, Andrew R. King and Barry F. Madore

Titles available in this series
Gravitational Dynamics, *edited by O. Lahav, E. Terlevich*
and R. J. Terlevich (ISBN 0 521 56327 5)

High-sensitivity Radio Astronomy, *edited by N. Jackson and R. J. Davis*
(ISBN 0 521 57350 5)

Relativistic Astrophysics, *edited by B. J. T. Jones and D. Marković*
(ISBN 0 521 62113 5)

Advances in Stellar Evolution, *edited by R. T. Rood and A. Renzini*
(ISBN 0 521 59184 8)

Relativistic Gravitation and Gravitational Radiation,
edited by J.-A. Marck and J.-P. Lasota
(ISBN 0 521 59065 5)

Instrumentation for Large Telescopes,
edited by J. M. Rodríguez Espinosa, A. Herrero and F. Sánchez
(ISBN 0 521 58291 1)

Stellar Astrophysics for the Local Group,
edited by A. Aparicio, A. Herrero and F. Sánchez
(ISBN 0 521 63255 2)

Nuclear and Particle Astrophysics, *edited by J. G. Hirsch and D. Page*
(ISBN 0 521 63010 X)

Theory of Black Hole Accretion Discs,
edited by M. A. Abramowicz, G. Björnsson and J. E. Pringle
(ISBN 0 521 62362 6)

Interstellar Turbulence
edited by J. Franco and A. Carramiñana
(ISBN 0 521 65131 X)

Globular Clusters,
edited by C. Martínez Roger, I. Pérez Fournón and F. Sánchez
(ISBN 0 521 77058 0)

The Formation of Galactic Bulges

Edited by

C. M. CAROLLO
Columbia University, New York

H. C. FERGUSON
Space Telescope Science Institute, Baltimore

R. F. G. WYSE
Johns Hopkins University, Baltimore

PUBLISHED BY THE PRESS SYNDICATE OF THE UNIVERSITY OF CAMBRIDGE
The Pitt Building, Trumpington Street, Cambridge, United Kingdom

CAMBRIDGE UNIVERSITY PRESS
The Edinburgh Building, Cambridge CB2 2RU, UK   www.cup.cam.ac.uk
40 West 20th Street, New York, NY 10011-4211, USA   www.cup.org
10 Stamford Road, Oakleigh, Melbourne 3166, Australia

© Cambridge University Press 1999

First published 1999

Printed in the United States of America

10/12 pt. Typeset in LaTeX by the author

*A catalog record for this book is available from
the British Library.*

Library of Congress Cataloging-in-Publication Data is available

ISBN 0 521 66334-2   hardback

Contents

CONTENTS

Part 5. Bulge Phenomenology

Part 6. Conference Summary

Preface

The innermost, denser regions of galaxies, i.e., the 'bulges', are a fundamental component of galaxies whose properties *define* the entire Hubble sequence. Understanding the origin of bulges is thus a required step toward understanding how such a sequence has come to place, i.e., toward deciphering how stars and galaxies condensed from the diffuse material in space into the structure that we observe today. Several decades of exploration of the Milky Way and Local Group bulges, and of nearby bulges external to the Local Group, have slowly built the orthodox view that bulges as a family should be reasonably old isotropic rotators with near-solar mean chemical abundances (although with a very wide abundance distribution function), i.e., nothing more than low-luminosity ellipticals. However, some major breakthroughs in the last few years concerning bulges in the local and early universe suggest that the time is ripe to perhaps reconsider this orthodoxy. The new picture that emerges from the most recent Hubble Space Telescope (HST) and 10m-class ground-based telescopes studies challenges the canonical beliefs about what bulges really are, how and when they form, and about the physical mechanisms that are important in determining their fundamental properties. Basic, and yet fundamental questions still need an answer: *(i)* Are bulges a one-parameter or a multi-parameter family? What are the average properties of bulges in terms of stellar populations and dynamics? What are the deviations from these averages? Are there more families of bulges with different origins and fates? *(ii)* What is the connection between bulges and host galaxies? What are the physical connections at the present epoch between bulges and starburst nuclei, starburst rings, and the inner disks? How do the properties of bulges depend on the properties of the surrounding galaxy (e.g. surface brightness, spiral structure)? Are starbursts/winds/mergers essential to bulge formation? Is central activity related to the secular evolution of bulges, and how? How is angular momentum transferred and/or dissipated in the formation of bulges? *(iii)* Is there a universal bulge-bar connection? Are bars universal precursors of bulges? What are the threshold conditions for triggering the growth and disruption of bars? Is the bar-driven growth of a bulge associated with the growth of a central (dark) object? What is the role of the central black hole's growth in bulge formation/evolution? What are the constraints on the initial conditions if bulges condense without bars? What role does bar/bulge formation play in hierarchical models of galaxy formation? *(iv)* What is the interplay of physical processes along the cosmic history of bulges? What are the plausible initial conditions for the formation of bulges within self-consistent cosmological scenarios? What local constraints do we have on bulge formation as a function of time? How do those constraints compare with the available data on high-redshift objects?

These proceedings are the written memory of a mini-workshop on *When and how do bulges form and evolve?* held at the Space Telescope Science Institute (STScI),

Baltimore, on October 5-7, 1998; they document our state-of-the-art understanding of this field, and the currently forseen lines of future research on it. The aim of the STScI mini-workshop was to create the framework for re-examining preconceptions about the nature and origin of bulges, and for critically interpreting the latest exciting results that the NASA Space Science Program, in unison with the latest ground-based programs, have been providing. More generally, the workshop was envisaged to set, at a very timely moment, the context for a discussion on the structure and physical properties of bulges in the early and present-day universe, and for a focused debate on the formation and cosmological evolution of bulges and their host galaxies. The meeting was organized to have a real 'workshop' format, with plenty of time for discussion, and to bring together scientists who explore this fundamental issue of modern cosmology from different perspectives. Only a fragmentary picture of 'bulge formation' is built in fact when separately studing the few bulges of the Local Group in great detail, when probing the integrated properties of a large sample of external nearby bulges at the expense of detailed information, or when interpreting the faint patches of light from the very early universe: all the pieces of the puzzle need to be merged together in order to build a self-consistent picture of how these complex systems formed.

Starting from an operative definition of 'bulge', suggested to be *the central concentration of mass in excess of the inward extrapolation of the outer, constant scalelength, exponential disk'*, the speakers were asked to attempt to answer the following specific questions:

1. What really is a bulge?

2. When did bulges form? i.e.: When was 1/2 of the mass assembled, when did 1/2 of the stars form, what is the spread from galaxy to galaxy?

3. What are the typical timescales of bulge formation (compared e.g. to the free-fall time)? i.e.: What is the star formation timescale, what is the dynamical timescale, what is the age spread within one galaxy?

4. What physical processes/properties determine the bulge-to-disk ratio? i.e.: What is the role of winds, starbursts, and nuclear activity, what is the role of gas dynamics versus stellar dynamics, what is the role of bars?

Within this framework of themes, the workshop stimulated a critical discussion of the latest observational studies of the properties of our own Milky Way bulge, of nearby bulges, and of distant bulge-like structures, and a confrontation between theoretical expectations versus observational constraints. In Parts 1 to 4 of this book, the invited and contributed papers are collected according to the above themes and to the general philosophy of the meeting. A few contributed papers describing some recent developments from 'phenomenological' studies of bulges are collected in Part 5. Part 6 concisely summarizes where we stand in our current understanding of the formation and cosmic evolution of (disk-embedded) spheroidal stellar systems.

We are grateful to Mike Fall for suggesting such a format for the meeting (and book), to the other members of the Scientific Organizing Committee, Andy Fruchter, Laura Fullton, Paul Goudfrooij, Hashima Hasan, Tim Heckman, Patricia Knezek, Piero Madau, Crystal Martin, Colin Norman, Massimo Stiavelli for input and help, to the Space Telescope Science Institute for support and hospitality, and to NASA for co-sponsoring this initiative and supporting the realization of this book. Indeed, the motivations and aims of the workshop and its proceedings align with one of NASA's most exciting current research fields and future cornerstone scientific goals: understanding the origin of galaxies. Some of the major breakthroughs of the past couple of years in this field have been made with telescopes and instruments developed within NASA's Space Science Program, e.g., HST with WFPC2, NICMOS and STIS, and also ASCA, ISO; more breakthroughs are expected to come with future missions such as the Next Generation Space Telescope. The STScI mini-workshop on bulge formation has contributed to putting the latest spectacular results together into a general framework for discourse on this fundamental topic of modern astronomy, thereby building a firm basis for future research into the challenging search for our own origins.

It was with sadness that we learned right at the beginning of the workshop that Olin Eggen, one of the legendary figures of 20th-century Astronomy and one of the pioneers in the study of Galactic structure, had just died. Following the suggestion of Mike Rich, we dedicate to him these proceedings, to commemorate his life in, and impact on, Astronomy, as a tribute to his leading and inspirational role in the past five decades of astronomical research.

C. Marcella Carollo, Henry Ferguson & Rosemary Wyse
Baltimore, Maryland
May, 1999

In Memory of Olin Eggen

Olin Eggen, one of the great figures of modern optical astronomy, died in Canberra Hospital on October 2, 1998, just a few days before the beginning of this workshop. He was 79.

Eggen was born on a farm in Wisconsin in 1919, grew up with a younger sister and brother in the town of Orfordville, WI, and worked his way through a science degree at the University of Wisconsin in Madison, bartending and playing the piano in nightclubs. He spent part of the war in occupied Europe as a courier for the US Office of Strategic Services, posing as a Swedish salesman for a ball-bearing company.

After the war, Eggen returned to the University of Wisconsin and took his PhD in astrophysics. At that time, Joel Stebbins and Albert Whitford in Madison had refined the technique of photoelectric photometry to measure the brightnesses and colors of stars. This was a great step forward in observational astrophysics. In 1946, the estimated size of the universe was about ten times smaller than we now believe it to be, and very little was known about how stars form, evolve and die. The precision and versatility of photoelectric photometry was vitally important in bringing our knowledge up to its present state. Eggen began his research in astronomy using photoelectric photometry to study variable stars and star clusters, and he continued with this technique throughout his long career.

After his PhD, Eggen was hired at Lick Observatory in California where he worked until 1956, and in this period he made a couple of extended observing visits to Mount Stromlo. He then moved to the Royal Greenwich Observatory in Sussex, as chief assistant to the Astronomer Royal, returning to California for five years as Professor of Astronomy at Caltech. When Bart Bok left Mount Stromlo in 1966, Eggen succeeded him as director.

Astrophysics was Eggen's life. His driving issues were how the Galaxy works and how the stars evolve. He worked about 15 hours a day, never took holidays, spent a week each month gathering data from photoelectric observations at the 40-inch telescope at Siding Spring, and otherwise lived a rather private and monastic life. In his recent memoir he wrote "What a glorious 50 years it has been ... a life on the dome floor, in the dark". As director of the observatories, astrophysics was always the guiding priority. He facilitated the construction of new instruments and was much involved in the Anglo Australian Telescope project. He also found time to produce 98 research papers during his period as director of Mount Stromlo (1966-1978).

His scientific achievements include a spectacular paper on how our Galaxy formed. This paper, written in 1962 with his colleagues Lynden-Bell and Sandage, is known as ELS and is one of the most influential astrophysics papers ever written. Eggen is also famous for his work on the evolution of stars and on the properties of large loose groups of stars (the Eggen moving groups) that have a common origin and then move together through our Galaxy. Although most of his groups are part of the disk of the Galaxy, he argued strongly for the reality of some apparent moving groups in the galactic halo. This was not widely accepted at the time, but recent work shows that halo groups do indeed exist.

Underlying his private and sometimes gruff exterior, was a person of charm and wit and insight. In social moments, Eggen was a great story teller but the boundary between fact and fantasy was sometimes blurred. From England, he brought a red and black Austin Healey which became a familiar item on the Canberra roads in the late 1960s. He insisted that this car had three carburettors. This was not at all true, and his more outrageous stories were called 'three carburettor stories' around the observatory.

After his directorship at Mount Stromlo, Eggen became staff astronomer at the Cerro Tololo Inter-American Observatory in Chile, where he remained until he died. But he regarded Australia as home, and returned to Mount Stromlo each year for a visit. It seemed fitting that he should have come back to Canberra to die. We shall miss him.

Ken Freeman
Canberra, Australia
May 1999

Part 1

INTRODUCTION

What are Galactic Bulges?

By GERARD GILMORE

Institute of Astronomy, Madingley Road, Cambridge CB3 0HA, UK

What are Galactic Bulges? I provide a brief overview of the observations and their interpretations, concluding that remarkably little is robust, remarkably little is well-defined, and remarkably little is well-explained. Galactic bulges are a subject ripe for HST and the large telescopes.

1. Definitions

What is a galactic bulge? A crucial pre-requisite to answering this question is to define the terminology. 'Galactic' seems innocuous, but is not. In discussion of bulges many people decree bulges to be synonymous with (small) ellipticals. Others note that a bulge is defined only relative to a disk. Are bulges simply small ellipticals? Did they form in the same way? Is the existence of a disk irrelevant to the history of a bulge?

Which leaves us to define a 'bulge'. One must beware definitions which are self-fulfilling as much as one must beware definitions which are not restrictive. A consideration in a valuable definition must be the utility of bulges in defining the Hubble sequence. The invaluable Carnegie Atlas of Galaxies (Sandage & Bedke 1994) raises both a working definition and a clue to interpretation:

"One of the three classification criteria along the spiral sequence is the size of the central amorphous bulge compared with the size of the disk. The bulge size, seen best in nearly edge-on galaxies, decreases progressively, while the current star-formation rate and the geometrical entropy of the arm pattern increases, from early Sa to Sd, Sm and Im types".

This raises a definition 'central, amorphous', and describes what has recently been rediscovered, that the properties of bulges are very tightly correlated with the present-day properties of their disks. Since almost all galaxies can be classified in the Hubble-type system, this definition of a bulge clearly has some merit. It also has some interesting implications, which might be also considered as self-fulfilling by definition. The first is 'central'. Why central? After all, everything in a galaxy which is not in a violent interaction is centered on the center? What this means is really 'more centrally concentrated than the disk': bulges are smaller than disks. Why then are they visible at all? Bulges must of course be of higher surface brightness than the inner regions of disks to be visible, i.e., they have higher phase-space density. And presumably formed from more highly dissipated gas. This emphasises another important point: bulges are extra light, they are not misshapen disks.

Might there be bulges which are more extended than disks, and/or of lower surface brightness? Such things would be hard to see, and would probably be called a (stellar) halo. It is interesting to consider if stellar halos are any more than stellar bulges which are excluded by definition rather than nature.

The second key parameter is 'amorphous'. This requires steady-state, with no significant recent star formation or dust lanes. Of course, as noted, bulges are 'seen best in nearly edge-on galaxies', so that the central bulge and disk are excluded from consideration. That is, the 'amorphous' parameter is restricted to the outer parts of bulges. This allows, though does not require, that bulges are still forming stars in their centers, with later diffusion of orbits, perhaps in the same way as generates the age-velocity dispersion relation for stars in the Solar neighbourhood.

The third important factor is the correlation between spiral structure and bulge:disk ratio. This emphasises that bulges and disks coexist in the same potential. That is, bulges are sufficiently compact and massive that their gravitational potential significantly affects disk stability. It allows the possibility, though does not require, that bulges and disks mutually interact gravitationally, and that secular evolution from disks into bulges may happen.

Finally, as with any definition, one must recognise limitations: is it sensible to consider everything which is excess light above the inner disk to be 'bulge'? Should one allow for nuclear components of the disk, or in addition to both disk and bulge? For the present, it is perhaps most efficient to proceed considering representative properties, leaving minority effects aside.

2. Photometry: What, Roughly, Do We Know?

There has been a rapid increase in the quality and quantity of bulge photometry in recent years. The introduction of large area digital cameras in both the optical (CCD) and near-infrared has allowed measurements limited by the complexity of galaxies, rather than the quality of the data. This allows full 2-D data modelling (Shaw 1987) and the ability to subtract reliably a bulge model from the data, to investigate underlying disk and nuclear structure. Possibly the single most important result of such analyses is that disks do continue under bulges. That is, bulges are extra light, added onto a 'normal' disk. This is a key constraint on formation and evolution models. It also substantially confuses observations, especially in later-type galaxies, where the disk contributes a lot of light. The important general features of bulges are now however tolerably consistently described, from the several excellent recent photometric studies.

2.1. *Diversity*

Starting with the most obvious, not all galaxies have similar bulges. The most interesting and best studied case is the Local Group galaxy M33. Here Bothun (1992), following several earlier studies referenced therein, showed there is a central luminosity excess containing 2% of the light. Thus, formally, M33 has bulge:disk ratio 0.02. The inner profile does not follow an $r^{1/4}$ law. Minniti *et al.* (1993) showed the central luminosity excess is made of high-mass young stars. Thus, does M33 have a bulge, in which case bulges are forming today for the first time? Or does it have a nuclear disk structure, in which case studies of more distant late-type galaxies are confusing bulge and inner disk? Is it even meaningfull in such cases to make the distinction? Clearly in this case the definition of smooth and amorphous is violated, so one must beware comparing different things. But one must also beware ignoring age ranges by preconception.

This single example may seem simply regrettable. However, the work reported by Trager at this meeting, and Peletier & de Grijs (1998), shows that apparently young central luminosity excesses in late-type spirals are the norm. One might of course calculate the luminosity this central star-forming excess will have when it is old and amorphous, and see if M33 will ever have a 'bulge' in that sense. Such a calculation however requires knowledge of future star formation, which is unavailable. Is M33 the clue that bulges of late-type galaxies are just starting to form? Is there evidence that this inner star-forming event will ever produce a 'bulge'? Schweitzer has for some time supported the case for continuing bulge formation, with NGC5102 and NGC7252 being especially convincing cases (e.g. Schweizer 1990). Further examples are possibly too rare, if this is to be the normal path of bulge formation.

2.2. *Similarity*

There are several recent comprehensive studies of bulge photometric properties, primarily Dutch PhD theses, which establish the general properties of bulges of early type spirals. While these general features are perhaps well-established, considerable caution remains. It is salutary to look at the plotted component decompositions in one of the most recent works, Andredakis, Peletier & Balcells (1995, their figure A1). The quite erratic behaviour of the inner disk profiles in those decompositions suggests caution in acceptng the full range of parametrisations as being well-established.

The general results are that bulge luminosity profiles are not all the same. There is a marginal correlation between the slope of the luminosity profile and bulge:disk ratio, though this is dominated by a few extreme points. More robustly, especially from the photometry of de Jong (1995), we know that bulges are a minority component, with bulge:disk ratio being in the range 0.2 to 0.02 for galaxy types 2 to 8. The bulges are also physically small: a typical bulge has scale length about 0.3kpc, compared to a typical disk which has scale length a few kpc. Statistically, bulges are 10% of the size of their disks. That is, the scale length of a bulge is similar to the thickness of a disk. Differences in kinematic support, to say nothing of formation history, might be expected. Wyse (this meeting) has emphasised that the Milky Way follows these correlations rather well.

The small sizes of bulges make observations difficult: even at quite bright magnitudes, bulges are smaller than 1-2 arcseconds in apparent size. We await with interest the outcome of the present extensive HST multi-wavelength studies to improve the data, given this situation.

One of the other fundamental results from the photometry is that bulges are more like their partner disks than they are like each other: there is a rather good correlation between bulge color and disk color. Although extant kinematic and spectroscopic data are more limited, it seems bulges also follow the elliptical galaxy color-linestrength relation, and the elliptical galaxy fundamental plane.

Does this imply a common evolutionary history between bulges and their disks, between bulges and similar luminosity ellipticals, neither, or both? Can one blame it all on the depth of the relevant potential well, or is a common star formation history implied?

3. Kinematics

Rather few kinematic studies of bulges are available. The earlier work, establishing that bulges are kinematically similar to ellipticals of the same luminosity, is discussed by Wyse (this meeting).

One possibly interesting constraint derives from the very small scatter in the Tully-Fisher relation. An elegant study of spirals in the Ursa Major group by Verheijen (1997 thesis: see Tully & Verheijen 1997 for the relevant result) reaffirms that the scatter in this relation is very small. However, Tully-Fisher compares the depth of the potential well to the total luminosity, bulge plus disk. How does the disk know the amount by which its luminosity should be reduced to compensate for the bulge? This relation is so tight over such a very wide range of absolute magnitudes, and hence bulge:disk ratios, it does require that the baryon:total mass fraction be similar everywhere, and that the formation histories of bulges not be too dissimilar. One cannot mix very young and very old bulges without changing mass:light ratios; a result which is apparently not seen.

Thus, while systematically different bulge formation histories as a function of Hubble type are consistent with all the data, a diversity of histories at a single type is not.

Fortunately the available information is about to be revolutionised. Building on the success of the TIGER integral field spectrograph, (Bacon *et al.* 1995) 2-D imaging spectrographs are at present being provided for almost every major telescope. Complemented by the spatial resolution of HST, we can be confident that central kinematics, line strength distributions, and dust maps will be available for a large sample of spiral galaxies in the near future. It will be interesting to see what part of what we think we now understand survives closer examination.

4. Ages

Age determinations are difficult in astronomy, at both an absolute (how old is a globular cluster?) and a relative (what is the globular cluster system age range?) scale. Determination of ages of mixed, unresolved stellar populations is a notoriously difficult problem. Considerable efforts have been made to build plausible stellar population mixture models, for comparison with observations. With careful selection of a mix of colors and absorption-line indices some progress is being made (e.g. Kuntschner & Davies 1998; Trager this meeting). Nonetheless, it remains true that it is extremely difficult to distinguish the effects of age and metallicity, especially when combined with reddening, and noting that a very broad range of metallicities, and perhaps ages, will contribute light to any pixel. In the inner parts of intermediate and late-type spirals, where significant contributions must be provided by inner disk star formation, reliable determinations of bulge ages are extremely problematic.

The basic data set is broad-band color data, which, as noted, shows bulges to be similar in color to inner disks. It remains unknown if the metallicity distribution functions of the two populations are similar. If they are, then similar ages are implied. But what is the luminosity-weighted age of an inner disk?

In so far as there is agreement between analyses, the concensus supports old ages for (most of) the stars in (most of) the bulges of (most) galaxies earlier than about Sbc type, with a rapid lurch to near zero-age for later type bulges. This extreme dependance on type seem so implausible that one looks forward to the forthcoming data to establish the subject more soundly.

Somewhat better data are available for the Milky Way, where direct HST and ground color-magnitude data below the main-sequence turn-off exist, and have been interpreted to imply old ages (e.g. Renzini, this meeting; Rich this meeting). Even here, however, problems with disentangling the effects of metallicity, reddening, and foreground disk 'contamination' are considerable, allowing only the statement that the bulk of the bulge is predominately quite old. A complexity is that foreground disk stars appear in the color-magnitude diagram at the same place as do young bulge stars. Thus, while one can simply discuss the bulk of the population, that is avoiding the most interesting question. It is worth remembering how hard this analysis is, before one blithely adopts low spatial resolution integrated light results on other galaxies!

Recently, Feltzing & Gilmore (1999) have used HST archive data for a set of fields towards the bulge, and between 2° and 8° latitude, to consider the spatial distributions of the stars above and below the turnoff. They show that indeed stars above the turnoff are distributed more like disk stars than are the stars below the turnoff. Thus one may safely conclude that effectively all the apparently young stars are foreground disk. The bulge is indeed predominately old, where studied. Just how old is another problem: the available isochrones are not a superb match to the data, the metallicity, its range, and the alpha-element enhancements are unknown, and extinction remains a problem. One can say robustly that the field bulge stars above 1-2 scale heights are comparable in age to the inner globular clusters, unless their abundances are very different.

It must be emphasised that this tells us nothing about the ages of the very metal-rich stars: it is often assumed they are as old as are the dominant half-solar metallicity stars, but we have no relevant information either way.

A further possible complexity is that the center of the Milky Way is a site of very active star formation. Sgr B is of course one of, if not the, most extreme star forming complexes in the Galaxy. There are three massive clusters of young stars within 50pc of the Galactic Center. One among these, the quintuplet has age about 4Myr, and mass similar to the young globular clusters in the LMC (Figer *et al.* 1999). Closer in, Genzel *et al.* (1996) have discovered not only a population of O-stars, but that these stars form a counter-rotating system.

Extensive surveys for OH/IR stars, a subset of which are young, intermediate mass stars, have been carried out by the Leiden group (see Sevenster 1997). These surveys are interesting in both strengthening evidence for continuing star formation in the inner Galaxy, and for providing dynamical tracers (see below). The mean star formation rate over the few Myr can be derived from these studies, and corresponds to about 1 solar mass per year. Recall that the mass of the inner (COBE) bulge, while poorly known, is of order 10^{10} solar masses.

The combination of evidence is overwhelming: continuing star formation in the inner disk is making a mass of stars comparable to the inner bulge mass, if the recent star formation rate is typical of that over the galaxy lifetime. Where are the relevant intermediate-age stars? Are they the inner bulge super-metal-rich stars, too metal-rich and so too red to be noticed in our color-magnitude data? Are they still in the disk? What is the relation between the inner disk and inner bulge?

Studies of the central regions of other disk galaxies frequently show continuing central star formation, and central gas disks, but then frequently a minimum in the star formation and molecular gas mass at larger radii. There are gas disks in early-type galaxies, seen by HST (e.g. Cen A, Schreier *et al.* 1998): continuing central star formation is the norm.

This implies late infall, by accretion or asymmetry in the potential. More importantly, it raises the question of the dynamical fate of these young stars. No understanding of bulges is possible independantly of an understanding of the evolution of inner disks. One is perhaps best advised to fall back on the Sandage/Bedke definition, and note that bulges are *seen best in nearly edge-on galaxies*: the centers are multi-mixture systems.

5. Chemical Abundances

Chemical element ratios are the best clock available to calibrate early star formation rates. The well known 'excess' of the $[\alpha/\mathrm{Fe}]$ ratio in stars which formed rapidly in fairly un-enriched gas is a key discriminant for rapid star formation models. Rapid in this context means on a timescale short enough that Type I supernovae are not substantial contributors to chemical evolution. This time is poorly known, but is perhaps 1Gyr. An absence of α-enrichment does not mean slower star formation, as the punctuated-burst model of the LMC illustrates (Gilmore & Wyse 1991). Nonetheless, the presence of α-enrichment is a clear signal.

Chemical abundances in unresolved galaxies can be studied only through the mix of absorption line-strength indices. Such data as are available (e.g. Idiart *et al.* 1996) suggest that bulges follow the same line-strength velocity-dispersion relation as do elliptical galaxies. This at least reduces the dimensionality of the problem, even if not solving it.

More complete data can be obtained in the Local Group. For M31 the HST photometry suggests a rather metal rich (near solar) bulge even at very large distances from the center. Note that a very extended bulge would violate all the scaling and correlation relations noted above, so it is not in the least obvious that one is really talking about a 'bulge' at all in this case. Future observations of the center of M31 will be more interesting.

For the Milky Way, extensive studies of the outer bulge (Ibata & Gilmore 1995), and regions about two scale heights from the center (e.g. Sadler, Rich & Terndrup 1996) agree in general: the metallicity distribution function is extremely broad, ranging from -2dex to well above solar. The broad maximum peaks near one-half solar, similar to the solar neighbourhood. There is very recent evidence (Frogel, this meeting) from IR photometry for an abundance gradient in the inner two scale heights, or few hundred parsecs. These studies have important implications for formation models (see below). Confirmation, and extension to include the distribution function of α-elements, is of extreme importance for understanding bulges.

Direct spectroscopic analysis of single stars in the two bulges of the Local Group (MWG, M31) should resolve this at least in those cases. At present, the situation for the Milky Way remains startlingly poor: one wonders why so few spectra and analyses have been attempted, and looks forward to the solution of this problem by the VLT.

6. Bulge Formation

There is only one model of bulge formation which is not fundamentally at variance with any fundamental observation discussed above, that of high-redshift rapid formation of a small stellar system. During or after star formation, a large number of secondary processes may, or may not, have contributed to formation of the present spatial distribution. These include: initial lumpy conditions; early mergers; early dynamical instabilities in a disk; and scattering of old stars by a central massive black hole (Gerhard and Binney 1985). None of this of course implies that this model is correct; just that we do not (yet) know enough to disprove it.

The abundance data for the Milky Way become relevant here. These studies provide two crucial constraints on bulge formation models. First the metallicity distribution function is quite unlike that of the outer disk, where the distribution function is narrow, or of the globular cluster system. Thus, models which form bulges from disks, either by mergers or by secular evolution, must also create extreme radial gradients in the width of the disk abundance distribution function with galactocentric radius. It would be interesting to know the present inner disk abundance distribution function in the Milky Way, to test directly any disk-bulge relationship. Second, abundance gradients on small scales, as reported by Frogel, imply dissipation, or a mixture of two discrete populations with different scale heights. It is extremely difficult to see how such a gradient could be consistent with available secular evolution models.

Of course, it might also be true that the the Milky Way is not exactly similar to every other bulge in the Universe. Thus, other formation models remain of interest.

7. Bars, Nuclei, Secular Evolution

7.1. *Secular Evolution*

A second class of formation model is currently topical, secular evolution of a disk into a bulge. This is related to the general question of bulges and bars.

Very dramatic progress in numerical simulations of cold stellar disks has quantified bar instability growth, and the process of bulge formation from vertical instabilities and scattering (Pfenniger, this meeting). The movies certainly seem impressive. The more quantitative studies however demonstrate serious limitations; it remains to be seen if these are fundamental, or merely require some further development of the models. Presently however, the models predict dynamical evolution which is too fast, and too extreme. The dynamical models match observations of OH/IR stars in the central galaxy tolerably after only a few Gyr of evolution, and destroy any resemblance to a real galaxy after a few Gyr. Perhaps realism might help: the models have a limited at best treatment of gas, accretion, etc.

A plausible secular evolution model must start from realistic initial conditions, and show an ability to reproduce the galaxy distribution function on relevant timescales, before it can be considered a fair alternative to more successful models. Nonetheless, the numerical studies of disk evolution are basically just gravity, and must contain some truth. Quantifying the true importance of secular effects remains a challenge, but an important one. We clearly have much to learn here.

One interesting, though as yet preliminary result is a correlation between bulge mass and the mass of the galactic central massive black hole. If this survives better data, it directly links bulges and nuclei. This is not necessarily bizarre: recent studies confirm the old suspicion that central massive black holes stabilise disks against bar growth (e.g. Sellwood & Moore 1999). Massive black holes can also scatter a dense inner disk into a bulge population, growing a bulge of old stars over time.

7.2. *Bars vs Bulges*

Galactic bars, like Irish ones, attract crowds. Irish bars, and perhaps galactic ones, come with a health warning. There are very many recent studies, reviews, and conferences available, so that a detailed introduction to galactic (or Irish) bars is unnecessary here. There is however one important point of direct relevance here which is worth a lot more careful attention than is always given: What is a Bar? We noted above the complexity involved in defining a bulge. Bars seem even harder. This might seem unrelated to the present discussion, except that a description of a bulge as a bar, in preference to a description as a triaxial spheroid, has significant dynamical and formation implications. Is the Galactic bulge a bar? Is any bulge a bar?

It is common to suggest that a very large fraction of all disk galaxies are barred. Usually this attribution is based on the shape of isophotes rather than kinematics. The first important point to note is that not everything mis-shapen is a bar. For example, the LMC is clearly barred. However, the LMC bar is not a dynamical bar. It is offset from the kinematic center, and is better described as a distortion. In fact, asymmetries which might easily be confused with bars are very common, perhaps the norm. Inspection of good-quality images of galaxies shows that many galaxies have distortions, $m = 1, ...,$ at all levels. Many things which one might consider bars are perhaps no more than inner spiral arms. So to the health warning: not all non-circular shapes are bars. Many are perhaps simply non-circular. It remains unclear if any dynamical inference is valid from a description of some oval luminosity distribution as a bar. A helpful working basis would be to restrict 'bar' to a description of specific kinematic and morphological features in a thin disk. Perhaps in future it will be proven such things can be thick, like spheroids: with a few very special possible exceptions, that proof is lacking.

Now to the question: is the Galactic bulge a bar? Evidence in favour is basically the inner gas rotation curve. Evidence against is the inability to find any signs of bar effects more than one disk scale height out of the Plane. The OH/IR star analysis of Fux *et al.* (see Pfenniger, this volume) sees effects within about 0.1kpc of the Plane. The gas dynamical effects are similarly constrained. At larger distance stellar kinematics (Ibata

& Gilmore 1995) are consistent with Kent isotropic oblate models. A recent analysis of the COBE flux maps (Binney, Gerhard & Spergel 1997) shows a rather mildly triaxial bulge. A small amplitude inner triaxiality is also required by direct high-resolution inner bulge studies (Unavane & Gilmore 1998), as well as by several studies at larger distances. But is triaxiality synonymous with a bar? Or is the bar in the disk, and the bulge is something else?

While this may seem a hair-splitting point, it does have significant implications for formation models, and the growth of pre-conceptions, and hence deserves more careful use of terminology than has become the norm.

8. Conclusions

Galactic bulges are a key to the definition of galaxy types, and perhaps are a key to the formation of galaxy types. A brief overview is provided by Wyse, Gilmore & Franx (1997). That review noted *Bulges are diverse and heterogeneous, and although their properties vary systematically, sometimes they are reminiscent of disks, sometimes of ellipticals. The extant observational data are however limited.* To that we add that, while remarkably little is firmly established, a remarkable amount is being done: galactic bulges are coming of age, and will soon be accurately definable. It remains to be seen if the term relates to a genus or a species.

REFERENCES

ANDREDAKIS, Y., PELETIER, R., BALCELLS, M. 1995 *MNRAS*, **275**, 874

BACON, R., ET AL. 1995 *A&AS*, **113**, 347

BOTHUN, G. 1992 *AJ*, **103**, 104

BINNEY, J., GERHARD, O., SPERGEL, D. 1997 *MNRAS*, **288**, 365

FELTZING, S., GILMORE, G. 1999 *A&A*, in press

FIGER, D.F., MCLEAN, I., MORRIS, M. 1999 *ApJ*, **514**, 202

GENZEL, R., ET AL. 1996 *ApJ*, **472**, 153

GERHARD, O., BINNEY, J. 1985 *MNRAS*, **216**, 467

GILMORE, G., WYSE, R.F.G 1991 *ApJ*, **367**, L55

IBATA, R., GILMORE, G. 1995 *MNRAS*, **275**, 605

DE JONG, R. 1995, Ph. D. Thesis, Groningen

KUNTSCHNER H., DAVIES, R.L. 1998 *MNRAS*, **295**, L29

MINNITI, D., OLSZWESKI, E., RIEKE, G. 1993 *ApJ*, **410**, L79

PELETIER, R., DE GRIJS, R. 1998, preprint (astro-ph/9808232)

SADLER, E., RICH, M., TERNDRUP, D. 1996 *AJ*, **112**, 171

SANDAGE, A., BEDKE, J. 1994 *The Carnegie Atlas of Galaxies*. (Carnegie Institute of Washington, Washington DC). (CAG)

SCHREIER, E., ET AL. 1998 *ApJ*, **499**, L143

SCHWEIZER, F. 1990 in *Dynamics & Interactions of Galaxies* (ed. R. Weilen) p60. (Springer-Verlag, New York)

SELLWOOD, J., MOORE, E. 1999 *ApJ*, **510**, in press

SEVENSTER, M. 1997, Ph. D. thesis, Leiden

SHAW, M. 1987 *MNRAS*, **229**, 691

TULLY, B., VERHEIJEN, M. 1997 *ApJ*, **484**, 145

UNAVANE, M., GILMORE, G. 1998 *MNRAS*, **295**, 145

WYSE, R.F.G., GILMORE, G., FRANX, M. 1997 *ARA&A*, **35**, 637

Part 2

THE EPOCH OF BULGE FORMATION

PART 2: THE EPOCH OF BULGE FORMATION

This section focuses on the question of *when* bulges form, with emphasis on the more observationally-accessible ages of the stars. Many of the speakers were careful to remind us that the epoch at which the bulk of the bulge stars formed could be very different from the epoch of their assembly.

Renzini advocates that the vast majority of bulge stars are old, based mostly on the fossil evidence of low-mass stars in local resolved bulges. He proposes further that bulges are so closely related to elliptical galaxies that one may treat them together, as 'spheroids' (note that much of the later discussions would tend to restrict this to bulges of early-type disk galaxies). Renzini favors redshifts higher than 3 for the 'epoch' of spheroid star formation, based on a combination of the observed redshifts of Lyman-break galaxies (perhaps bulges in formation), together with the ages derived from stellar evolution considerations.

Lilly *et al.* adopt the opposite, but complementary, approach to Renzini, by studying direct signatures of high-redshift bulge formation. They present evidence that the millimeter sources detected by SCUBA are distant analogs of local ultra-luminous infra-red galaxies, and are spheroids forming in a starburst. The identified counterparts to the SCUBA sources would lead to a redshift of 'spheroid formation' of perhaps as high as 2, but more recent than that favored by Renzini (although the resulting stellar age difference, which depends on assumed cosmology, may not be large).

Shorter contributions discuss in more detail what can be inferred from the Milky Way bulge and from studing the colors and absorption line-strengths of the integrated stellar populations in external disk galaxies. Frogel concludes from an IR analysis of the inner regions of the Milky Way bulge that the bulk of its stars are metal-rich and old. Trager *et al.* present a (preliminary) study of various line-strength indices in the bulges of selected galaxies, and suggest that the stars in bulges of early-type spirals are older than those in bulges of late-type disk galaxies, perhaps a similar trend to that of the mean ages of the disk stars. Peletier & Davies present HST IR–optical colors for the bulges of selected, mostly early-type disk galaxies, and also interpret the observed colors in terms of an old stellar age for these bulges.

A theoretical perspective, based on monolithic collapse models, is given by van den Bosch, who proposes that bulges form early, from the inner, lower angular-momentum regions of a proto-galaxy. This model has both assembly and star formation in bulges at high redshift, although exactly how 'high' depends on the over-density of the proto-galaxy as a whole.

Origin of Bulges

By A L V I O   R E N Z I N I

European Southern Observatory, Karl-Schwarzschildstr. 2, D-85748 Garching b. München,
Germany

Insight into the origin of bulges is sought in this review only from the properties of their stellar populations. Evidence concerning the age of the Galactic bulge stellar population is reviewed first, then the case of the bulge of M31 is discussed. The similarity of bulges and ellipticals is then illustrated, inferring that the problems of the origin of bulges and of the origin of ellipticals may well be one and the same: i.e. the origin of galactic spheroids. In this mood, the current evidence concerning the age of the dominant stellar populations of early-type galaxies is then reviewed, both for low- as well as high-redshift galaxies, and both for cluster as well as field ellipticals. All reported evidence argues for the bulk of the stars in galactic spheroids having formed at high redshift, with only minor late additions and a small dependence on environment. An attempt is made to evaluate how current formation scenarios can account for this observational evidence. The role of spheroids in the cosmic star formation and metal enrichment history is also briefly discussed. Finally, some critical questions are asked, answers to which may help our further understanding of the formation and evolution of galactic spheroids.

1. Introduction

Much on our speculations on *how* bulges originated depends on what we believe about *when* they formed. Some scenarios prefer bulges to be young, or middle age, late comers anyway. Others prefer a rapid, early build up of bulges, and push back to very early times the epoch of their formation. For this reason I will mostly concentrate on reviewing evidence on ages, leaving the last section to speculations on origins. While they may provide additional clues, some morphological and dynamical properties – such as bars, ripples, or peanut shapes – are largely ignored in this review.

Section 2 focuses on the Galactic bulge, the one we can study best, and in all details. Next closer bulge to us is that of M31, to which Section 3 is dedicated. No other prominent bulge exists in the Local Group (M33 does not really have a bulge), and Section 4 emphasizes that most bulges of spirals are quite similar to ellipticals, so the problem of bulge ages merges with that of dating ellipticals, and becomes the more general problem of dating *spheroids*. This is the subject of Section 5, i.e. dating spheroids at low, as well as high redshifts. In Section 6 cluster and field early-type galaxies are compared to each other, and Section 7, on speculations, is last.

Overall, a wide body of observational evidences is presented showing that the bulk of stellar populations in galactic spheroids are very old. This is true all the way from the bulge of our own Galaxy to high redshift cluster ellipticals. The main issue that remains open is whether star formation and assembly of spheroids were concomitant events, or whether the bulk of stars formed in smaller entities that then hierarchically coalesced, with this process extending over much of the cosmological time.

2. The Age of the Galactic Bulge Relative to the Halo

Dating of bulge stars is complicated by several factors, such as crowding, depth effects, variable reddening, metallicity dispersion, and contamination by foreground disk stars. In an attempt to circumvent some of these limitations Ortolani *et al.* (1995) have selected the bulge globular clusters NGC 6528 and NGC 6553 for HST study. These clusters are

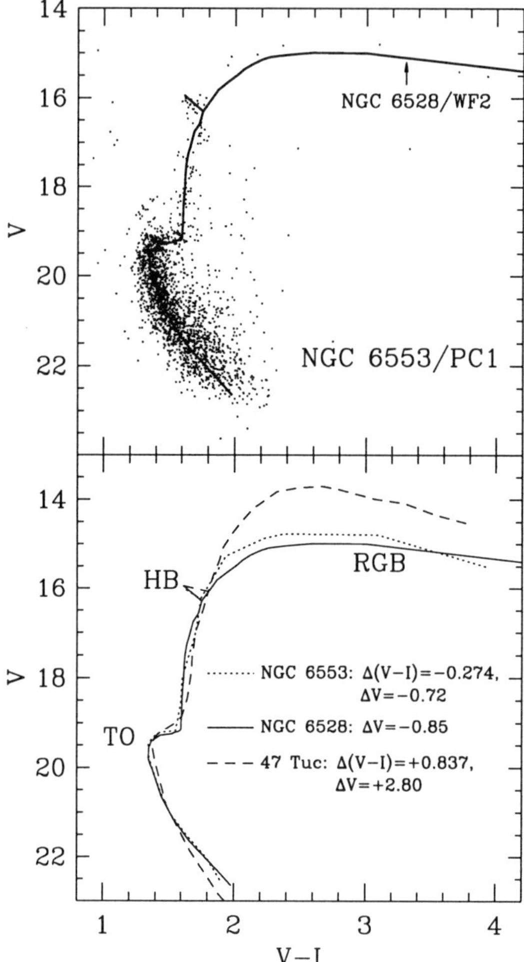

FIGURE 1. Upper panel: the color-magnitude diagram of the bulge globular NGC 6553 for stars in the PC field of WFPC2, with superimposed the mean locus of the cluster NGC 5528, as sampled by chip #2 of WFPC2. Lower panel: The mean loci of NGC 6528, NGC 6553, and 47 Tuc. Each locus has been shifted as indicated, in order to bring into coincidence the end of the HB (from Paper II).

respectively located at $\sim 4°$ and $6°$ from the galactic center, and their overall metallicity [M/Fe] is about solar (Barbuy et al. 1999), close to the average for stars in Baade's Window (McWilliam & Rich 1994). Like most other clusters within ~ 3 kpc from the Galactic center, they belong to the population of Bulge globular clusters, having the same kinematical properties and metallicity distribution of Bulge stars (e.g. Minniti 1995). (To qualify these clusters as *disk clusters* is clearly a misnomer.)

The upper panel in Figure 1 shows the CMD of NGC 6553 as sampled by the PC1 chip of WFPC2, that combines good statistics with relatively low differential reddening. Superimposed on it is the mean locus of the CMD of NGC 6528 as sampled by the WF2 chip of WFPC2. The data points of NGC 6553 have been dereddened as indicated in the lower panel so to make its turnoff color equal to that of NGC 6528. Then the CMD of this latter cluster has been shifted vertically to bring its horizontal branch (HB) to coincide with that of NGC 6553. Note that the mean locus of NGC 6528 provides an excellent fit to the NGC 6553 data, from the MS all the way to the tip of the RGB. The virtual identity of the CMD of the two clusters is further demonstrated in the lower panel of Figure 1, where the mean loci of the two clusters are compared to each other.

Also shown in the lower panel of Figure 1 is the mean locus of the inner halo globular cluster 47 Tuc ([Fe/H]=−0.7), which has been shifted in color and magnitude in order to bring into coincidence its HB with that of the two bulge clusters. As can be seen in this figure, the luminosity difference between the HB and the main sequence turnoff of the the Bulge clusters is the same as (or even slightly larger than) that of 47 Tuc. This comparison demonstrates that the two Bulge clusters are as old as the halo clusters (to within ± ∼ 2 Gyr), and therefore the bulge underwent rapid chemical enrichment to solar abundance and beyond, very early in the evolution of our Galaxy. Due to the *relative* nature of the dating procedure, this conclusion is independent of uncertainties in reddening, distance, and *absolute* age determinations.

The next step in the Ortolani *et al.* study is represented by an attempt at dating the Bulge *field* stellar population itself, still in a relative fashion with respect to the clusters. Figure 2 shows that the MS luminosity function of the cluster NGC 6528 is indistinguishable from that of the stars in Baade's Window (the low-reddening bulge field at ∼ 4o from the Galaxy center), that was obtained from observations with the ESO NTT with superb seeing ($0''.4$). From this comparison Ortolani *et al.* infer that the whole Bulge formed quickly, some 15 Gyr ago (if this is the age of the halo clusters), and set an upper limit of ∼ 10% by number to any intermediate age population in the Bulge. Indeed, a larger proportion of intermediate age stars would have resulted in a shallower fall off of the bulge luminosity function around TO (i.e., for $20.5 \gtrsim V \gtrsim 19.5$), where instead it coincides with that of the cluster.

Further insight on the formation time scale of the Bulge comes from the detailed abundance studies of its stars. In 12 field K giants McWilliam & Rich (1994) find a moderate α-element enhancement ([Mg/Fe]$\simeq$[Ti/Fe]$\simeq$+0.3, but with [Ca/Fe]$\simeq$[Si/Fe]$\simeq$ 0), moderate r-process element enhancement, while s-process elements appear solar with respect to iron. Barbuy *et al.* (1999) have analyzed 2 stars in NGC 6553 finding somewhat more enhanced α-element overabundance, with [Na/Fe]$\simeq$[Al/Fe]$\simeq$[Ti/Fe]$\simeq$+0.6 and [O/Fe]$\simeq$[Mg/Fe]$\simeq$[Si/Fe]$\simeq$+0.3. Note that different systematics may go a long way towards explaining the differences between these two studies. General consensus exists on the interpretation of α-element and r-process element enhancements as due to a *quick* star formation and metal enrichment, with elements produced by Type II supernovae being incorporated into new stars before the bulk of iron from Type Ia SNs is produced. However, how quick is *quick* remains uncertain. Basically, the bulk of stars must form before the explosion of most SNIa's, but the actual *distribution* of SNIa explosion times following a burst of star formation remains empirically indeterminate and theoretically very model dependent (cf. Greggio 1996). According to general wisdom it takes at least ∼ 1 Gyr for a fair fraction of SNIa to release their iron. If so, at least 90% of the Bulge stars formed within the first Gyr of the object that we now call the Milky Way. In conclusion, the *fossil* evidence tells us that the whole Galactic spheroid is pretty old indeed, and formed on a rather short timescale.

There are important lessons to draw from these conclusions. Our Milky Way is a rather late-type spiral galaxy in a very loose group that is located rather away from major density peaks in the distribution of galaxies. Nevertheless, her whole spheroidal component looks ∼ one Hubble time old, from the halo globular clusters all the way to the inner bulge. With a mass of ∼ $2 \times 10^{10} M_\odot$, the old age for the bulk of the spheroid

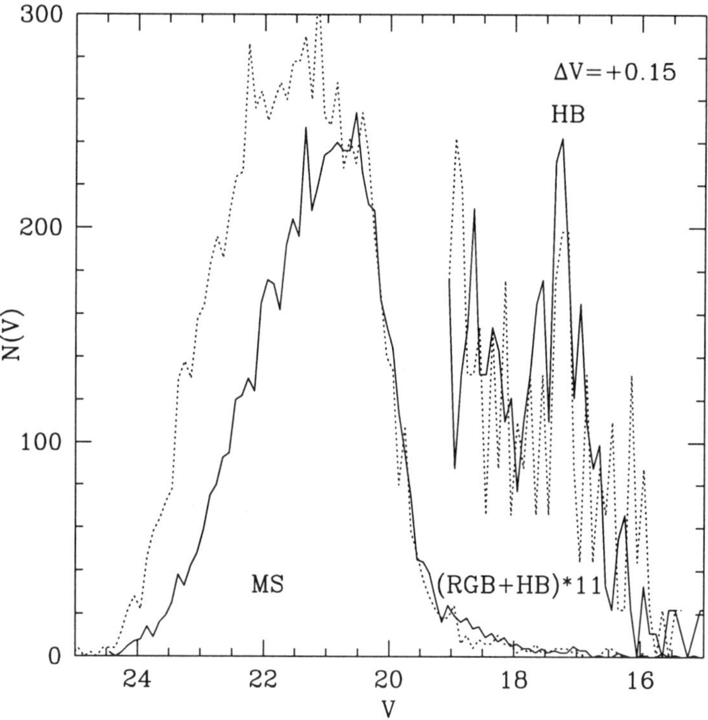

FIGURE 2. The luminosity function (LF) of main sequence (MS) and red giant (RGB+HB) stars in NGC 6528 (WF2 field, dotted line) and in Baade's Window (BW, solid lines). The cluster LF has been shifted by $\Delta V = +0.15$ so as to bring into coincidence its HB peak (marked on the figure) with that of BW, and multiplied by a factor of 2 so as to normalize the two distributions at $V = 20.45$, where both are reasonably complete, or to the same number of RGB+HB stars brighter than V=19.45 in this figure. Note that below $V \simeq 21$ the bulge LF is progressively more incomplete compared to that of the cluster. The cluster LF has been suitably broadened with a Montecarlo simulation to mimic the depth effect present in the BW field. For this display, the RGB+HB LFs have been multiplied by a factor 11, in order to avoid overlap with the LF of the disk foreground stars (from Ortolani *et al.* 1995).

population implies an average star formation rate $\sim 20\,M_{\odot}\mathrm{yr}^{-1}$ at the epoch of spheroid formation, some 14-15 Gyr ago (having assumed $\sim 10^9$ yr for the duration of the star formation process). This value is as small as the smallest star formation rates of $z \gtrsim 3$ galaxies (Steidel *et al.* 1998). Such galaxies have also effective radii of 1-3 kpc (typical of galactic bulges, cf. Giavalisco *et al.* 1996), and it is rather tempting to speculate that with Lyman-break galaxies one may have caught bulge formation in the action. With the Galactic spheroid accounting for $\sim 20\%$ of the stellar mass of the Milky Way, one can conclude that $\gtrsim 20\%$ of all stars in our Galaxy have formed 'at fairly high redshift'.

3. The Next Bulge: M31

In ground based and pre-COSTAR HST studies the suspicion had been advanced for the presence in the bulge of M31 of a major intermediate-age component, as suggested by the detection of putative bright AGB stars (e.g. Rich & Mould 1991; Rich, Mould, & Graham 1993; Rich & Mighell 1995; Davidge *et al.* 1997). However, bright AGB stars ($M_{\mathrm{bol}} \simeq -5$) are also produced by old, metal rich globular clusters, such as the

Bulge clusters discussed in the previous section (Frogel & Elias 1988; Guarnieri, Renzini, & Ortolani 1997). Moreover, with insufficient angular resolution blends of RGB stars can be mistaken for bright AGB stars (e.g. Renzini 1998b), and the presence of an intermediate age population in the bulge of M31 could not be unquestionably proven with such data.

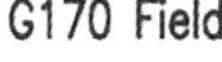

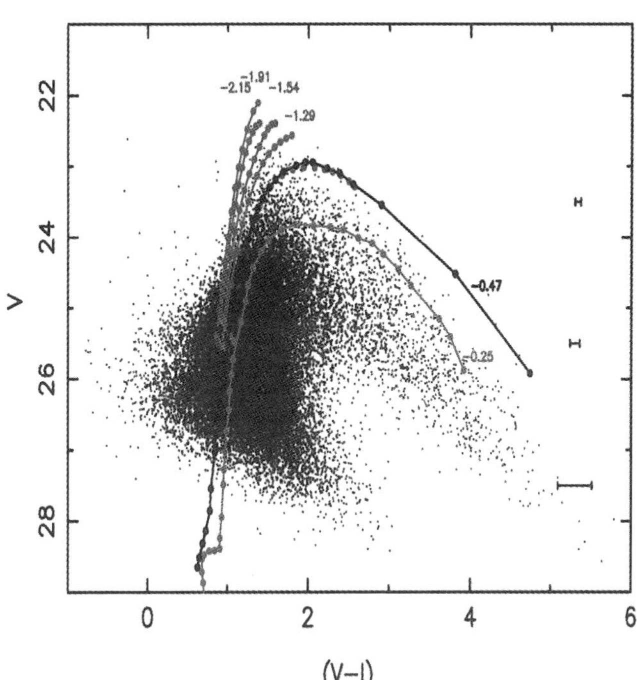

FIGURE 3. The WFPC2 color-magnitude diagram of the field near the globular cluster G170, located at a projected distance of $\sim$ 1.55 kpc from the center of the bulge (from Jablonka *et al.* 1999). Also shown is the red giant branch loci for several metallicities [Fe/H], as indicated.

WFPC2 observations of the bulge of M31 are now becoming available. Jablonka *et al.* (1999) have obtained deep CMDs for various locations in the bulge of M31, confirming that what on low resolution images appeared as *bright AGB stars* are indeed blends of fainter stars. Figure 3 shows one of such CMDs, relative to the field in the vicinity of the very metal rich globular cluster G170, located at a projected distance of 1.55 kpc from the center of M31. The CMD of the field near the cluster G177 (at 0.8 kpc from the center) is virtually identical. Perhaps the most striking aspect of this CMD is the predominance of a fairly homogeneous metal rich population, with the upper RGB bending down in the $V - (V - I)$ CMD due to strong TiO blanketing as typical of metal rich globular clusters (see Figure 1). Very few metal poor star appear to be present, while the bulk of stars are more metal rich than [Fe/H]$=-0.25$.

In conclusion, there is no evidence for an intermediate age population in the bulge of M31. Its almost uniformly metal rich population points to the presence of a 'G-dwarf Problem', which may be a general characteristics of (at least) the inner regions of galactic spheroids (e.g. Greggio 1997). The metallicity distribution of the M31 bulge may provide important insight for understanding the formation process. It rises two intriguing questions: 1) Where are the stars that produced the metals now locked in the bulge stars we see? and 2) Where have the metals produced by this bulge stellar population gone? The tentative answer to the first question is 'they are out in the halo of M31', which could be tested extending deep HST imaging to larger galactocentric distances (but see Rich, Mighell, & Neill 1996). The tentative answer to the second question is 'they have been ejected out in the IGM by an early galactic wind'. If these are the correct answers, then the even more tentative conclusion is that the bulge formed outside-in by dissipative merging and collapse of mostly gaseous pregalactic lumps (see the contributions by Carlberg and by Elmegreen in this volume), with the resulting starburst then ejecting the residual gas and a lot of metals along with it.

4. Bulges vs Ellipticals

The properties of bulges are extensively reviewed at this meeting, and there is no point trying to summarize them here. In this section I would like to emphasize only one aspect: the close similarity of the bulges of spiral galaxies with elliptical galaxies. While also this aspect is further illustrated by others at this meeting, Figure 4 gives a very direct impression of the extent to which bulges are similar to ellipticals (from Jablonka, Martin, & Arimoto 1996), and therefore may share a common origin. The bulk of bulges appear to follow precisely the same $Mg_2 - M_r$ relation of ellipticals, with just a minority of them (i.e. 5 out of 26 in the Jablonka *et al.* sample) having Mg_2 values appreciably lower than those of ellipticals of similar luminosity. The same similarity also exists between the $Mg_2 - \sigma$ relations of bulges and ellipticals (Jablonka et al. 1996).

As well known, the Mg_2 index depends on both age and metallicity; actually on both the age and metallicity *distributions*. Therefore, the close similarity of the $Mg_2 - M_r$ relations argues for spiral bulges and ellipticals sharing a similar star formation history and chemical enrichment. One may argue that origin and evolution have been very different, but differences in age distribution are precisely compensated by differences in the metallicity distributions. This may be difficult to disprove, but I tend to reject this alternative on aesthetic grounds. It requires an unattractive cosmic conspiracy, and I would rather leave to others the burden of defending such a scenario.

In conclusion, it appears legitimate to look at bulges as ellipticals that happen to have a prominent disk around them, or to ellipticals as bulges that for some reason have missed the opportunity to acquire or maintain a prominent disk. Therefore, we can legitimately refer to *spheroids* as the class of objects that includes ellipticals and the bulge+halo component of spirals. In this mood, the problem of the origin of bulges becomes the problem of the origin of spheroids.

5. The Epoch of Spheroid Formation

Great progress has been made in recent years towards charting and modeling galaxy formation and evolution. Yet, the origin of the galaxy morphologies, as illustrated by the Hubble sequence, has so far defied a generally accepted explanation. This is also the case for spheroids, i.e. bulges and ellipticals alike, with two quite different scenarios still confronting each other. In one scenario spheroids come from the destruction of pre-existing disks or part of them. In the case of ellipticals, by merging spirals, a widely entertained notion since the original proposal by Toomre (1977). In the case of bulges, by some bar instability randomizing the orbits of stars originally in the inner part of a disk (e.g. Combes *et al.* 1990; Raha et al. 1991; Hasan, Pfenninger, & Norman 1993), or by being merger remnant ellipticals that managed to re-acquire a new disk. This

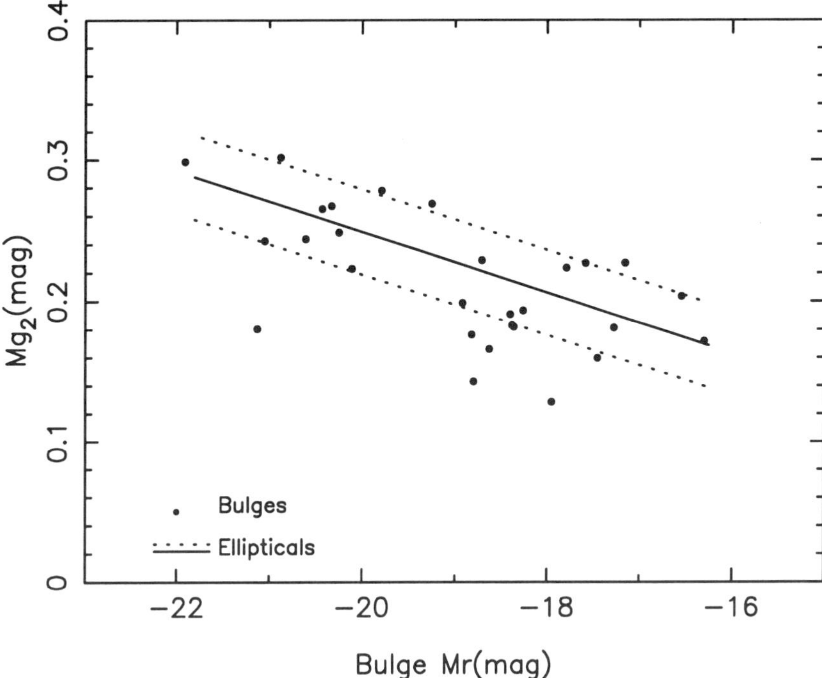

FIGURE 4. The Mg$_2$ − M_r relation for a sample of bulges. The solid line is the mean relation for elliptical galaxies, and the dotted lines limit the area occupied by ellipticals (from Jablonka *et al.* 1996).

latter scenario is now motivated by hierarchical clustering cosmologies, and ellipticals are modeled to form through a series of merging events (between spirals) taking place over a major fraction of the cosmological time (e.g. Baugh, Cole, & Frenk 1996; Kauffmann 1996).

The other scenario assumes instead the whole baryonic mass of the galaxy being already assembled at early times in gaseous form, and for this reason it is sometimes qualified as *monolithic*. The original idea can be traced back to the Milky Way collapse model of Eggen, Lynden-Bell, & Sandage (1962), with early examples including the models of Larson (1974) and Arimoto & Yoshii (1987). In this case, the disk of spirals is a late comer, somehow acquired later by a more ancient spheroid.

Through the 1980's much of the debate focused on the age of ellipticals as derived from the integrated spectrum of their stellar populations. In general, advocates of the merger model favored an intermediate age for the bulk of stars in ellipticals, but the matter remained controversial given the well know age-metallicity degeneracy and the crudeness of stellar population models of the time (for opposite views see O'Connell 1986, and Renzini 1986).

A first breakthrough came from noting the tightness of the color-σ relation of ellipticals in the Virgo and Coma clusters (Bower, Lucey, & Ellis 1992). This demands a high degree of *synchronicity* in the star formation history of ellipticals, that is most naturally accounted for by pushing back to early times most of the star formation. Making minimal use of stellar population models, this approach provided for the first time a *robust*

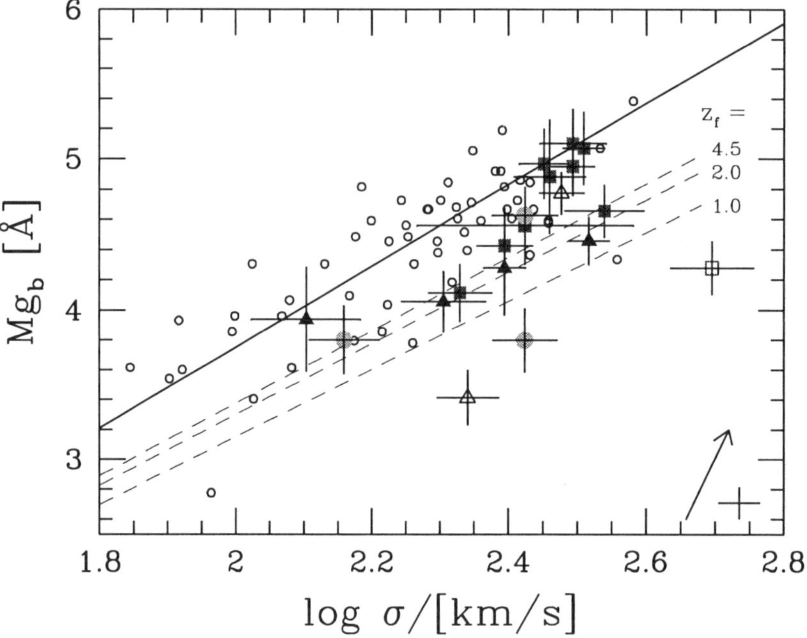

FIGURE 5. The $\mathrm{Mg_b} - \sigma$ relation for a sample of ellipticals in two clusters at $z \simeq 0.37$ (symbols with error bars) is compared to the same relation for a sample of galaxies in the Virgo and Coma clusters (from Bender *et al.* 1997). The dashed lines represent the expected location of single burst, passively evolving galaxies for various formation redshifts (with $H_\mathrm{o} = 50$, $q_\mathrm{o} = 0.5$). The aperture correction is shown near the lower/right corner.

demonstration that at least *cluster* ellipticals are made of very old stars, with the bulk of them having formed at $z \gtrsim 2$.

The main lines of the Bower *et al.* argument are as follows. The observed color scatter of cluster ellipticals is related to the age dispersion among them by the relation:

$$\delta(U - V) = \frac{\partial(U - V)}{\partial t}(t_\mathrm{H} - t_\mathrm{F}) \tag{5.1}$$

where t_H and t_F are the age of the 'oldest' and 'youngest' galaxies, respectively. Here by age one intends the luminosity-weighted age of the stellar populations that constitute such galaxies. The time derivative of the color is obtained from evolutionary population synthesis models, which give $\partial(U-V)/\partial t \simeq 0.02 \,\mathrm{mag/Gyr}$ for $t \simeq 10$. The observed color scatter is $\delta(U - V) \simeq 0.04$ mag, consistent with pure observational errors. Hence, one gets $t_\mathrm{H} - t_\mathrm{F} \lesssim 0.04/0.02 = 2$ Gyr, and if the oldest galaxies are 15 Gyr old, the youngest ones ought to be older than 13 Gyr, from which Bower *et al.* conclude they had to form at $z \gtrsim 2$. If the oldest galaxies were instead as young as, say 5 Gyr, then the youngest should be older than at least 3 Gyr, which would require a high degree of synchronicity in their formation, which seems unlikely.

Evidence in support of the Bower *et al.* conclusion has greatly expanded through the 1990's, and is now compelling. This came from the tightness of the fundamental plane relation for ellipticals in local clusters (Renzini & Ciotti 1993), from the tightness of the color-magnitude relation for ellipticals in clusters up to $z \sim 1$ (e.g., Aragon-Salamanca *et al.* 1993; Taylor *et al.* 1998; Kodama et al. 1998; Stanford, Eisenhardt, & Dickinson 1998), and from the modest shift with increasing redshift in the zero-point of the fun-

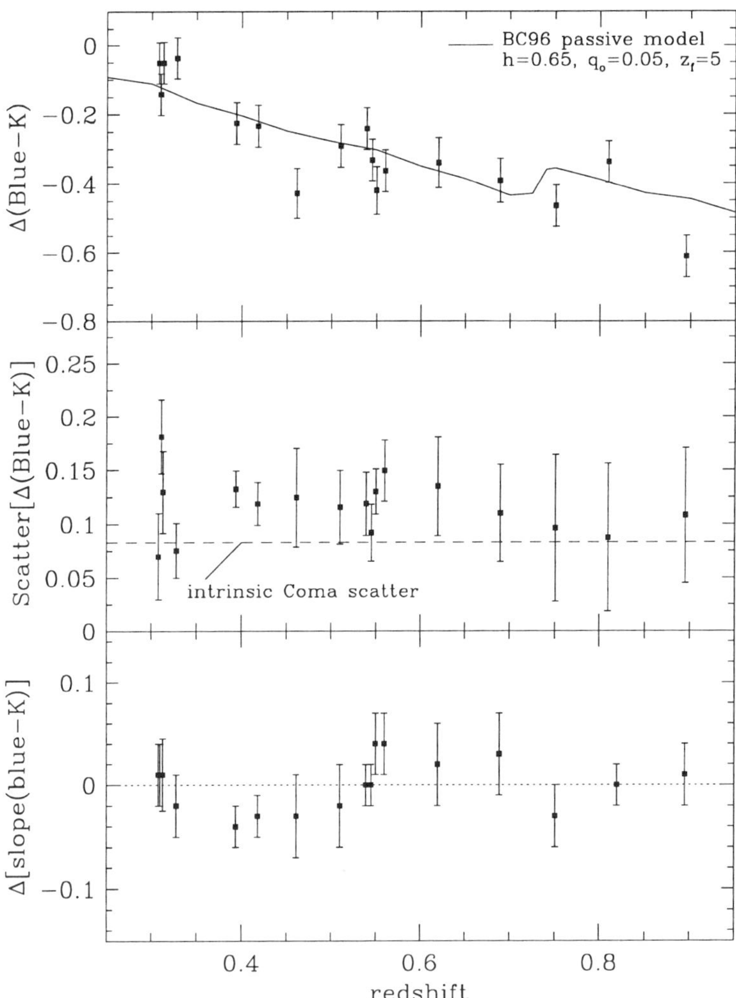

FIGURE 6. The color evolution of early-type galaxies in clusters out to $z \simeq 0.9$ (Stanford, Eisenhardt, & Dickinson 1997; Dickinson 1997). The 'blue' band is tuned for each cluster to approximately sample the rest frame U-band, while the K band is always in the observed frame. Top panel: the redshift evolution of the blue$-K$ color relative to the Coma cluster. A purely passive evolution models is also shown. Middle panel: the intrinsic color scatter, having removed the mean slope of the color-magnitude relation in each cluster and the contribution of photometric errors. The intrinsic scatter of Coma galaxies is shown for reference. Bottom panel: the redshift evolution of the slope of the (blue$-K$) $- K$ color-mag diagram, modulo the slope for galaxies in Coma.

damental plane, Mg$-\sigma$, and color-magnitude relations of cluster ellipticals (e.g., Bender *et al.* 1997; Dickinson 1995; Ellis *et al.* 1997; van Dokkum et al. 1998; Pahre, Djorgovski, & de Carvalho 1997; Stanford, Eisenhardt, & Dickinson 1998; Kodama *et al.* 1998). All these studies agree in concluding that most stars in ellipticals formed at $z \gtrsim 3$, though the precise value depends on the adopted cosmology. Figure 5 illustrates the case of the Mg$-\sigma$ relation for ellipticals in two clusters at $z \simeq 0.37$, while Figure 6 documents the constancy of the color disperion of cluster ellipticals all the way to $z \sim 1$.

It is worth emphasizing that all these studies follow the methodological approach pioneered by Bower *et al.* (1992). They focus indeed on the tightness of some correlation among the global properties of cluster ellipticals, which sets a robust constraint on their age dispersion as opposed to an attempt to date individual galaxies. Moreover, the move to high redshift offers two fundamental advantages. The first advantage is that looking at high z provides the best possible way (I should say *the* way) of removing the age-metallicity degeneracy. If spheroids are made of intermediate-age, metal rich stars, they should become rapidly bluer and then disappear already at moderate redshift (e.g. Kodama & Arimoto 1997). The observational opportunity of studing galaxies at large lookback times makes quite obsolete attempts at finding combinations of spectral indeces that may distinguish between age and metallicity effects in nearby galaxies. The second advantage is that at high redshift one gains more *leverage*: for given dispersion in some observable one can set tighter and tighter limits to the age dispersion. This comes from the color time derivatives being larger the younger the population. For example, the derivative $\partial(U - V)/\partial t$ is ~ 7 times larger at $t = 2.5$ Gyr than it is at $t = 12.5$ Gyr (e.g. Maraston 1998), and therefore a given dispersion in this rest-frame color translates into a ~ 7 times tighter constraint on age and therefore on formation redshift. This is further illustrated also by the case of isolated high redshift ellipticals. For example, Spinrad *et al.* (1997) found a *fossil* (i.e. passively evolving) elliptical at $z = 1.55$ for which they infer an age of at least 3.5 Gyr, hence a formation redshift in excess of ~ 5. An even much higher formation redshift may be appropriate for the extremely red galaxy in the NICMOS field of the HDF-South, whose spectral energy distribution is best accounted for by an old, passively evolving population at $z \simeq 2$ (Stiavelli *et al.* 1999).

6. Cluster vs Field Spheroids

Much of the evidence discussed in the previous Section is restricted to cluster ellipticals. In hierarchical models, clusters form out of the highest peaks in the primordial density fluctuations, and cluster ellipticals completing most of their star formation at high redshifts could be accommodated in the models (e.g. Kauffmann 1996; Kauffmann & Charlot 1998a). However, in lower density, *field* environments, both star formation and merging are appreciably delayed to later times (Kauffmann 1996), which offers the opportunity for an observational test of the hierarchical merger paradigm.

The notion of field ellipticals being a less homogeneous family compared to their cluster counterparts has been widely entertained, though the direct evidence has been only rarely discussed. Visvanathan & Sandage (1977) found cluster and field ellipticals to follow the same color-magnitude relation, but Larson, Tinsley, & Caldwell (1980) – using the same database – concluded that the scatter about the mean relation is larger in the field than in clusters. More recently, a larger scatter in field versus cluster ellipticals was also found for the fundamental plane relations by de Carvalho & Djorgovski (1992). However, at least part of the larger scatter among field ellipticals certainly comes from their distances being more uncertain than for clusters.

Taking advantage of a large sample (~ 1000) of early-type galaxies with homogenously determined Mg_2 index and central velocity dispersion, Bernardi *et al.* (1998) have recently compared the $Mg_2 - \sigma$ relations (which are distance independent!) of cluster and field galaxies, and the result is shown in Figure 8. As it is evident from the figure, field, group, and cluster ellipticals all follow basically the same relation. The zero-point offset between cluster and field galaxies is 0.007 ± 0.002 mag, with field galaxies having lower values of Mg_2, a statistically significant, yet very small difference. This is in excellent agreement with the offset of 0.009 ± 0.002 mag, obtained by Jorgensen (1997) using 100 field and 143 cluster galaxies.

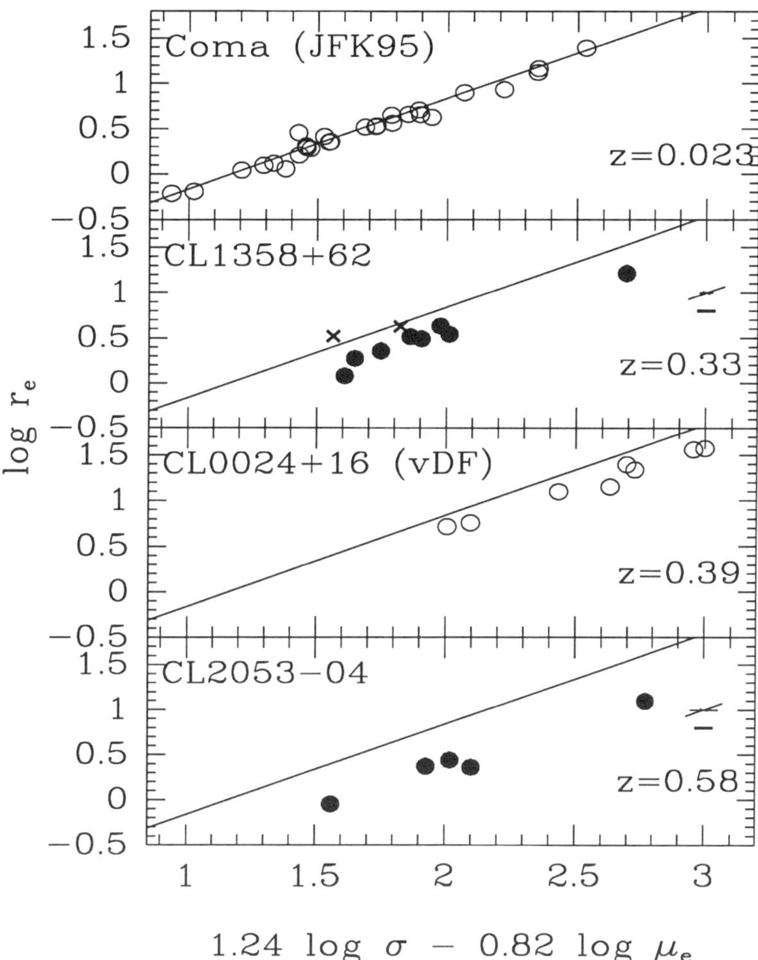

FIGURE 7. The fundamental plane relations of clusters at increasing redshifts (Franx *et al.* 1997). Note that the slope of the fundamental plane remains constant. This is also the case when including the cluster MS1054–03 at $z = 0.83$ (van Dokkum *et al.* 1998).

Using the time derivative of the Mg_2 index from synthetic stellar populations, Bernardi *et al.* conclude that the age difference between the stellar populations of cluster and field early-type galaxies is at most ~ 1 Gyr. The actual difference in the mass-weighted age (as opposed to the luminosity-weighted age) could be significantly smaller that this. It suffices that a few galaxies have undergone a minor star formation event some Gyr ago, with this having taken place preferentially among field galaxies.

The comparison between these empirical findings and the theoretical simulations is somehow complicated by the rather loose way in which cluster, group, and field environment are defined in the observational studies on the one hand, and in the theoretical simulations on the other. For example, in the models of Kauffmann (1996) there is a ~ 4 Gyr age difference between model ellipticals now residing in a $10^{15} M_\odot$ dark matter halo and those residing in a $10^{12} - 10^{13} M_\odot$ halo. This age difference would correspond to

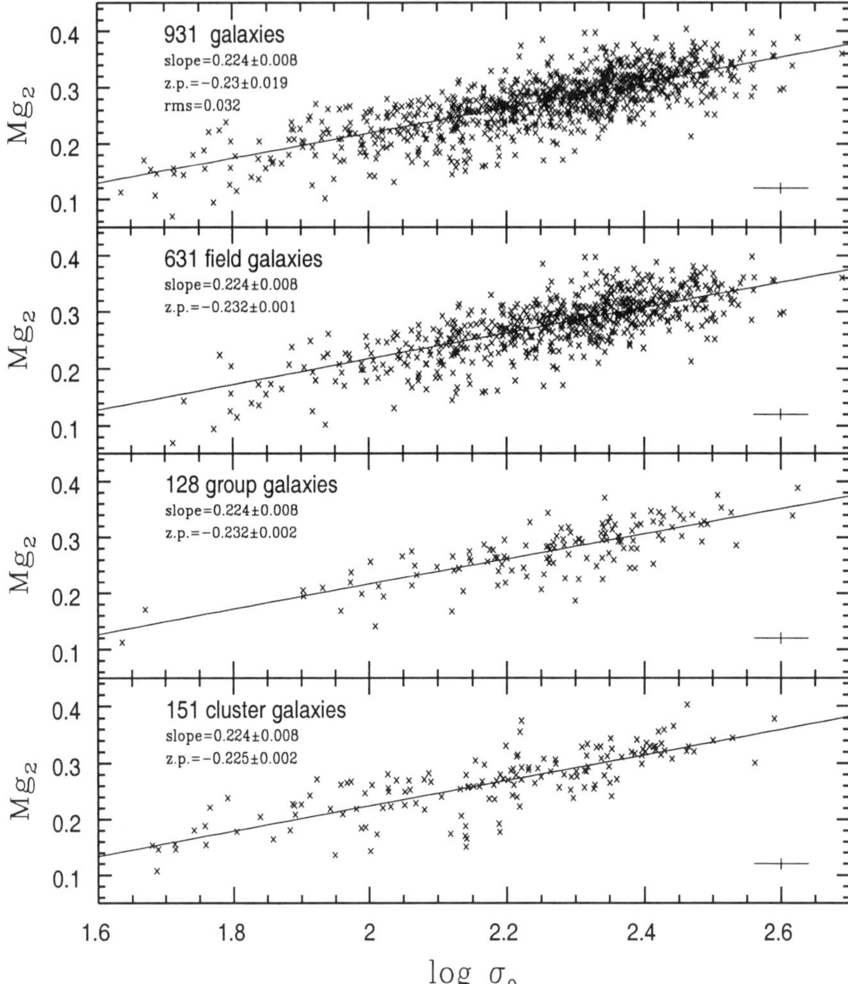

FIGURE 8. The Mg_2–σ relation for a sample of early-type galaxies (upper panel), as well as for the field, group and cluster subsamples (lower panels), from Bernardi *et al.* (1998). The corresponding number of objects, the slope, and the zero-point (z.p.) are shown in the upper/left corner of each panel. The least squares fits to the Mg_2–σ_0 relation are also shown as solid lines. For the three subsamples the slope as derived for the total sample was retained, and only the zero-point was determined. The typical error bar is shown in the lower/right corner.

a difference $\Delta Mg_2 \simeq 0.023$ mag, which Bernardi *et al.* data exclude at the $\sim 4.5\sigma$ level. However, in Kauffmann & Charlot (1998a) cluster and field are defined as those dark matter halos with circular velocity $V_c = 1000$ km s^{-1} and < 600 km s^{-1}, respectively, and the age difference between model ellipticals in such two environments is greatly reduced. Empirically, Bernardi *et al.* have assigned to the field those galaxies that do not belong to known clusters or groups, and one does not know what the average circular velocity is in such environments. Clearly, the problem is to find a common language between observers and model makers, i.e. a common way of defining cluster and field environments before comparing to each other data and simulations.

7. Discussion

As documented in the previous sections, compelling evidence now exists for the bulk of stars in galactic spheroids being very old, i.e. formed at redshifts beyond ~ 3, and possibly even much beyond this value. This applies to ellipticals and bulges alike, in clusters as well as in the lower density regions still inhabited by spheroids, including the bulge of our own Galaxy. This is what was expected (actually postulated) in the monolithic collapse scenario, while it appears to be quite at variance with most realizations of the hierarchical merging scenario.

7.1. *Monolithic vs Hierarchical*

The fact that spheroids are made of old stars does not necessarily invalidate the hierarchical merging paradigm, which actually offers a still unique description on how large galaxies could have been assembled. One possibility to comply with the observations may be to tune hierarchical models to mimic the monolithic model, by pushing most of the action back to an earlier cosmological epoch. With most of the merging taking place at high redshifts, among still mostly gaseous components, merging itself would promote widespread starburst activity. The natural observational counterparts of these events may be represented by the Lyman-break galaxies at $z\gtrsim3$ (Steidel *et al.* 1996), where star formation rates could be as high as $\sim 1000\ M_\odot\mathrm{yr}^{-1}$ (Dickinson 1998). It remains to be explored whether such tuning of algorithms and parameters of the hierarchical model could produce model universes fulfilling all other observational constraints. Alternatively, stars now in spheroids do indeed form at very high redshifts, but they are assembled into big spheroids only at much later times (as favored, e.g. by Kauffman 1996).

One testable prediction of the hierarchical merging model is that – obviously – bigger galaxies form later by assembling smaller pieces, and their stellar populations are appreciably younger than those of smaller galaxies. Therefore, intrinsically brighter galaxies should get bluer at a faster rate with increasing z, compared to fainter ones. As a consequence, the color-magnitude, color-σ, Mg-σ and fundamental plane relations should *flatten* with increasing redshift (lookback time). No such effect has been detected yet: the slope of the color-magnitude relation appears to be the same all the way to at least $z \simeq 1$ (see bottom panel in Figure 6, and Figure 7). The predicted flattening is actually a consequence of the postulate that ellipticals are made by merging spirals, with the gas in the disks being turned into stars when the two dark matter halos merge. Hence, in this frame late merging implies late star formation as well. On the other hand, it remains to be seen whether dissipationless merging of gas-free galaxies can produce the spheroids we see at low redshift, with their very high phase-space density. If so, the color-mag and similar relations should actually get *steeper* with increasing redshift.

The other prediction of the hierarchical model is that big galaxies should progressively disappear with increasing redshift, and several claims have been made *pro* and *con* the actual disappearance of ellipticals in various redshift surveys. Unfortunately, this approach is less conclusive than it may appear at first sight: when ellipticals are selected following either color or morphological criteria a small residual star formation should suffice to let otherwise old galaxies drop out of the selected samples, even if their main (spheroidal) body is already in place.

To overcome the intrinsic weakness of this approach, Kauffmann & Charlot (1998b) avoid using either color or morphology criteria, and adopt a pure $K-$band magnitude limited selection criterion. In this way the number evolution of massive galaxies is followed, independently of morphology or trace star formation, hence providing a more fundamental test of the models. Comparing to a $K < 19$ sample of galaxies with measured redshift, they conclude that their pure luminosity evolution (PLE) models are excluded by a large margin. Such models would predict $\sim 50\%$ of the galaxies in the sample to be at $z > 1$, while only $\sim 10\%$ is observed, hence they argue for number evolution due to merging being at work. The same test can be attempted on the somewhat bigger K−band magnitude limited sample of Cohen *et al.* (1998), which includes 195 objects down to $K = 20$. Among these objects, 24 turned out to be stars and for 34 objects no redshift could be determined. Among the residual 137 objects, 21 have $z > 1$. The vast majority of objects without a measured redshift are likely to be galaxies at $z > 1$, whose strong spectral features have moved out of the range of the optical spectrograph. If so, the sample would have $\sim 21 + 34 = 55$ out $\sim 137 + 34 = 171$ galaxies at $z > 1$, or $\sim 32\%$. Interpolating on Figure 4 in Kauffmann & Charlot (1998b) one can roughly estimate that their PLE model predicts $\sim 60\%$ of galaxies in a $K \leq 20$ sample to be at $z > 1$, while their hierarchical model predicts $\sim 10\%$. So, the Cohen *et al.* sample suggests a value that is just midway between the predictions of the two models. Clearly, existing samples are still too small for reaching any firm conclusion, especially when considering that large fluctuations may take place between one pencil beam survey and another due to fluctuations in the sampled large scale structures. For example, Cohen *et al.* (1998) emphasize that approximately half of the galaxies in their sample lie in five 'redshift peaks', likely due to clustering. Therefore, Poisson statistics may be more profitably applied to the number of sampled 'structures', rather than to that of galaxies.

7.2. *The Role of Spheroids in the Cosmic History of Star Formation*

With spheroids containing at least 30% of all stars in the local universe (Schechter & Dressler 1987; Persic & Salucci 1992) or even more (Fukujita, Hogan, & Peebles 1998), one can conclude that at least 30% of all stars – hence $\sim 30\%$ of metals – have formed at $z \gtrsim 3$ (Renzini 1998a; Dressler & Gunn 1990). This is several times more than suggested by a conservative interpretation of the early attempt at tracing the cosmic history of star formation, either empirically (Madau *et al.* 1996) or with theoretical simulations (e.g. Baugh *et al.* 1996). Yet, it is in fine agreement with the recent direct estimates from the spectroscopy of Lyman-break galaxies (Steidel *et al.* 1998), as well as with sub-mm observations (Hughes et al. 1998), where the cosmic SFR runs flat for $z \gtrsim 1$, as in one of the options offered by the models of Madau, Pozzetti, & Dickinson (1998).

7.3. *The Role of Spheroids in the Metal Enrichment of the Early Universe*

The global metallicity of the present day universe is best estimated in clusters of galaxies, where it is $\sim 1/3$ solar. This can be taken as representative of the overall metallicity since clusters and *field* have converted into star and galaxies nearly the same fraction of baryons (Renzini 1997). With $\sim 30\%$ of all stars having formed at $z \gtrsim 3$, and the metallicity of the $z = 0$ universe being $\sim 1/3$ solar, it is straightforward to conclude that the global metallicity of the $z = 3$ universe had to be at least $\sim 1/3 \times 1/3 \sim 1/10$ (Renzini 1998a,c). Damped Ly_α systems (DLA) may offer an opportunity to check this prediction, though they may provide a vision of the early universe that is biased in favor of cold, metal-poor gas that has been only marginally affected by star formation and metal pollution. Metal-rich objects that may exist at high redshift, such as giant starbursts that would be dust-obscured, metal-rich passively-evolving spheroids, and the hot ICM/IGM, obviously do not exist among DLAs. Still, these objects may contain much of the metals in the $z \sim 3$ universe as they do in the present day universe. In spite of these limitations the average metallicity of the DLAs at $z = 3$ appears to be $\sim 1/20$ solar (Pettini *et al.* 1997, see their Figure 4), just a factor of 2 below the expected value from the *fossil evidence*. However, this is still much higher than the extreme lower limit $Z \simeq 10^{-3} Z_\odot$ at $z = 3$ as inferred from Ly_α forest observations (Songaila 1997). Ly_α forest material is believed to contain a major fraction of cosmic baryons at high z, hence (perhaps) of metals. There is therefore a potential conflict with the estimated global metallicity at $z \simeq 3$, and the notion of Ly_α forest metallicity being representative of the the universe metallicity at this redshift. Scaling down from the cluster yield, such low metallicity was achieved when only $\sim 0.3\%$ of stars had formed, which may be largely insufficient to ionize the universe and keep it ionized up to this redshift (Madau 1998, but see Gnedin & Ostriker 1997). This suggests that Ly_α forest may not trace the mass-averaged metallicity of high redshift universe, and that the universe was very inhomogeneous at that epoch. The bulk of metals would be partly locked into stars in the young spheroids, partly would reside in a yet undetected hot IGM, a phase hotter than the Ly_α forest phase.

7.4. *Open Questions*

Several questions remain open at this stage. Some of them can soon get answers from observations, others from new theoretical simulations, or from extracting more information from old ones. Of course, the list of interesting questions could actually be much longer, and include e.g. the origin(s) of all those structural and morphological aspects that have been set deliberately aside in this review.

- How can hierarchical models be tuned to produce the uniform age of stars in the Galactic bulge?
- and the uniformity of stellar metallicity in the bulge of M31?
- What fraction of *'ellipticals'* would belong to clusters, groups, and field in simulations of galaxy formation?
- How much number evolution of spheroids has taken place between $z = 1$ and $z = 0$?
- What is the redshift distribution of a fair and complete sample of $K \le 20$ galaxies?
- Is the fraction of spheroids formed by merging *spirals* very large or very small?
- At which redshift do color-magnitude (and analogous) relations for ellipticals begin to flatten? Do they flatten at all?
- At $z \simeq 1$ do global relations for ellipticals in the field differ from those of galaxies in clusters, and if so by how much?
- Are Lyman-break galaxies spheroids in formation? What is their mass?
- What is the global metallicity of the universe at $z = 3$?
- Does an early assembly of bulges help forming the right disks?
- Is the early universe re-ionized and maintained ionized by forming spheroids?

It is my feeling that it will not take much before having fairly secure answers to most of these questions.

I would like to thank Ralf Bender, Marc Dickinson, Marijn Franx and Pascale Jablonka for their kind permission to reproduce here some of the figures from their papers. I would also like to thank the Space Telescope Science Institute for its hospitality during the meeting.

REFERENCES

ARAGON-SALAMANCA, A., ELLIS, R.S., COUCH, W.J., CARTER, D. 1993 *MNRAS*, **262**, 764

ARIMOTO, N., YOSHII, Y. 1987 *A&A*, **173** 23

BARBUY, B., RENZINI, A., ORTOLANI, S., BICA, E., GUARNIERI, M.D. 1999 *A&A*, **341**, 539

BAUGH, C.M., COLE, S., FRENK, C.S. 1996 *MNRAS*, **283**, 1361

BENDER, R., SAGLIA, R.P., ZIEGLER, B., BELLONI, P., GREGGIO, L., HOPP, U., BRUZUAL, G.A. 1997 *ApJ*, **493**, 529

BERNARDI, M., RENZINI, A., DA COSTA, L.N., WEGENER, G., ET AL. 1998 *ApJ*, **508**, L43

BOWER, R.G., LUCEY, J.R., ELLIS, R.S. 1992 *MNRAS*, **254**, 613

COHEN, J.G. ET AL. 1998, preprint (astro-ph/9809067)

COMBES, F., DEBBASCH, FRIEDLI, D., PFENNIGER, D. 1990 *A&A*, **233**, 82

DAVIDGE, T.J., ET AL. 1997 *AJ*, **113**, 2094

DE CARVALHO, R.R., DJORGOVSKI, S. 1992 *ApJ*, **389**, L49

DICKINSON, M. 1995, in *Fresh Views of Elliptical Galaxies* (ed. A. Buzzoni, A. Renzini & A. Serrano), ASP Conf. Ser. **86**, p283. (ASP)

DICKINSON, M. 1997, in *Galaxy Scaling Relations* (ed. L.N. da Costa & A. Renzini), p215. (Springer-Verlag, New York)

DICKINSON, M. 1998, in *The Hubble Deep Field* (ed. M. Livio, S.M. Fall & P. Madau), p219. (Cambridge)

DOMINGUEZ-TENREIRO, R., TISSERA, P.B. & SAIZ, A. 1998 *ApJ*, **508**, L123

DRESSLER, A., GUNN, J. E. 1990, in *Evolution of the Universe of galaxies. The Edwin Hubble Centennial Symposium* (ed. R.G. Kron), ASP. Conf. Ser. **10**, p200. (ASP)

EGGEN, O.J., LYNDEN-BELL, D., SANDAGE, A. 1962 *ApJ*, **136**, 748

ELLIS, R.S., SMAIL, I., DRESSLER, A., COUCH, W.J., OEMLER, A. JR., BUTCHER, H., SHARPLES, R.M. 1997 *ApJ*, **483**, 582

FRANX, M., KELSON, D., VAN DOKKUM, P., ILLINGWORTH, G., FABRICANT, D. 1997, in *The Nature of Elliptical Galaxies* (ed. M. Arnaboldi, G.S. Da Costa & P. Saha), ASP Conf. Ser. **116**, p512. (ASP)

FROGEL, J.A., ELIAS, J.H. 1988 *ApJ*, **324**, 823

FUKUJITA, M., HOGAN, C.J., PEEBLES, P.J.E. 1998 *ApJ*, **503**, 518

GIAVALISCO, M., STEIDEL, C.C., MACCHETTO, F.D. 1996 *ApJ*, **470**, 189

GNEDIN, N., OSTRIKER, J.P. 1997 *ApJ*, **486**, 581

GREGGIO, L. 1997, in *The Interplay between Massive Star Formation, the ISM and Galaxy Evolution* (ed. D. Kunth *et al.*), p89. (Edition Frontieres)

GREGGIO, L. 1997 *MNRAS*, **285**, 151

GUARNIERI, M.D., RENZINI, A., ORTOLANI, S. 1997 *ApJ*, **477**, L21

HASAN, H., PFENNIGER, D., NORMAN, C. 1993 *ApJ*, **409**, 91

HUGHES, D., SERJEANT, S., DUNLOP, J., ROWAN-ROBINSON, M., ET AL. 1998 *Nature*, **394**, 241

JABLONKA, P., MARTIN, P., ARIMOTO, N. 1996 *AJ*, **112**, 1415

JABLONKA, P., ET AL. 1999 *ApJL*, in press

JORGENSEN, I., FRANX, M., KJAERGAARD, P. 1996 *MNRAS*, **280**, 167

JORGENSEN, I. 1997 *MNRAS*, **288**, 161

KAUFFMANN, G. 1996 *MNRAS*, **281**, 487

KAUFFMANN, G., CHARLOT, S. 1998a *MNRAS*, **294**, 705

KAUFFMANN, G., CHARLOT, S. 1998b *MNRAS*, **297**, L23

KODAMA, T., ARIMOTO, N. 1997 *A&A*, **320**, 41

KODAMA, T., ARIMOTO, N., BARGER, A.J., ARAGON-SALAMANCA, A. 1998 *A&A*, **334**, 99

LARSON, R.B. 1974 *MNRAS*, **173**, 671

LARSON, R.B., TINSLEY, B.M., CALDWELL, C.N. 1980 *ApJ*, **237**, 692

MADAU, P. 1998, preprint (astro-ph/9807200)

MADAU, P., FERGUSON, H.C., DICKINSON, M.E., GIAVALISCO, M., STEIDEL, C.C., FRUCHTER, A. 1996 *MNRAS*, **283**, 1388

MADAU, P., POZZETTI, L., DICKINSON, M. 1998 *ApJ*, **498**, 106

MARASTON, C. 1998 *MNRAS*, **300**, 872

MCWILLIAM, A., RICH, R.M. 1994 *ApJS*, **91**, 794

MINNITI, D. 1995 *AJ*, **109**, 1663

NAVARRO, J.F., STEINMETZ, M. 1997 *ApJ*, **438**, 13

O'CONNELL, R.W. 1986, in *Stellar Populations* (ed. C. Norman, A. Renzini & M. Tosi), p167. (Cambridge)

ORTOLANI, S. RENZINI, A., GILMOZZI, R., MARCONI, G., BARBUY, B., BICA, E., RICH, R.M. 1995 *Nature*, **377**, 701

PAHRE, M.A., DJORGOVSKI, S.G., DE CARVALHO, R.R. 1997, in *Galaxy Scaling Relations: Origins, Evolution and Applications* (ed. L. da Costa & A. Renzini), p197. (Springer-Verlag, New York)

PERSIC, M., SALUCCI, P. 1992 *MNRAS*, **258**, 14

PETTINI, M., SMITH, L.J., KING, D.L., HUNSTEAD, R.W. 1997 *ApJ*, **486**, 665

RAHA, N., SELLWOOD, J.A., JAMES, R.A. KAHN, F.D. 1991 *Nature*, **352**, 411

RENZINI, A. 1986, in *Stellar Populations* (ed. C. Norman, A. Renzini & M. Tosi), p213. (Cambridge)

RENZINI, A. 1997 *ApJ*, **488**, 35

RENZINI, A. 1998a, in *The Young Universe* (ed. S. D'Odorico, A. Fontana, E. Giallongo), ASP Conf. Ser. **146**, p298. (ASP)

RENZINI, A. 1998b *AJ*, **115**, 2459

RENZINI, A. 1998c, preprint (astro-ph/9810304)

RENZINI, A., CIOTTI, L. 1993 *ApJ*, **416**, L49

RICH, R.M., MIGHELL, K. 1995 *ApJ*, **439**, 145

RICH, R.M., MIGHELL, K., NEILL, J.D. 1996, in *Formation of the Galactic Halo – Inside and Out* (ed. H.L. Morrison & A. Sarajedini), ASP Conf. Ser. **92**, p544. (ASP)

RICH, R.M., MOULD, J.R. 1991 *AJ*, **101**, 1286

RICH, R.M., MOULD, J.R., GRAHAM, J. 1993 *AJ*, **106**, 2252

SCHECHTER, P.L., DRESSLER, A. 1987 *AJ*, **94**, 563

SONGAILA, A. 1997 *ApJ*, **490**, L1

SPINRAD, H., DEY, A., STERN, D., DUNLOP, J., PEACOCK, J., JIMENEZ, R., WINDHORST, R. 1997 *ApJ*, **484**, 581

STANFORD, S.A., EISENHARDT, P.R., DICKINSON, M. 1998 *ApJ*, **492**, 461

STEIDEL, C.C., GIAVALISCO, M., DICKINSON, M., ADELBERGER, K.L. 1996 *AJ*, **112**, 352

STEIDEL, C.C., GIAVALISCO, M., PETTINI, M., DICKINSON, M., ADELBERGER, K.L. 1996 *ApJ*, **462**, L17

STEIDEL, C.C., ADELBERGER, K.L., GIAVALISCO, M., DICKINSON, M., PETTINI, M. 1998, preprint (astro-ph/9811399)

STIAVELLI, M., TREU, T., CAROLLO, C.M., ROSATI, P., VIEZZER, R., CASERTANO, S., ET AL. 1999 *A&A*, **343**, L25

TAYLOR, A.N., DYE, S., BROADHURST, T.J., BENITEZ, N., VAN KENPEN, E. 1998 *ApJ*, **501**, 539

TOOMRE, A. 1977, in *The Evolution of Galaxies and Stellar Populations* (ed. B.E. Tinsley & R. Larson), p401. (Yale)

VAN DOKKUM, P.G., FRANX, M., KELSON, D.D., ILLINGWORTH, G.D. 1998 *ApJ*, **504**, L17

VISVANATHAN, N., SANDAGE, A. 1977 *ApJ*, **216**, 214

Deep sub-mm Surveys: High Redshift ULIRGs and the Formation of the Metal-Rich Spheroids

By SIMON J. LILLY[1], STEPHEN A. EALES[2], WALTER K. GEAR[3], TRACY M. WEBB[1], J. RICHARD BOND[4],

AND

LORETTA DUNNE[2]

[1]Department of Astronomy, University of Toronto, 60 St. George Street, Toronto, Ontario M5S 3H8, Canada

[2]Department of Physics and Astronomy, Cardiff University, P.O. Box 913, Cardiff CF2 3YB, UK

[3]Mullard Space Science Laboratory, University College London, Holmbury St. Mary, Dorking, Surrey RH5 6NT, UK

[4]Canadian Institute for Theoretical Astrophysics, University of Toronto, 60 St. George Street, Toronto, Ontario M5S 3H8, Canada

Deep surveys of the sky at millimeter wavelengths have revealed a population of ultra-luminous infrared galaxies (ULIRGs) at high redshifts. These appear similar to local objects of similar luminosities (such as Arp220) but are much more 'important' at high redshift than at low resdhift, in the sense that they represent a much larger fraction of the total luminous output of the distant Universe than they do locally. In fact the ULIRGs at high redshift are producing a significant fraction ($\geq$ 15%) of the total luminous output of the Universe averaged over all wavelengths and all epochs. The high-z ULIRGs could plausibly be responsible for producing the metal-rich spheroidal components of galaxies, including the bulges of spiral galaxies. In this case we would infer from the redshift distribution of the sources that much of this activity is probably happening relatively recently at $z \leq 2$.

1. Introduction

Despite a great deal of progress in recent years, there still remain major uncertainties in our observational picture of the formation and evolution of galaxies in the high redshift Universe. Not least, the relationship between the star formation activity seen at high redshift and the present-day morphological components of the galaxy population, including the bulges that are the subject of this conference, remains unclear. The origin of the stars in the metal-rich spheroidal components of present-day galaxies, which constitute a half to two-thirds of all stars in the Universe (see Fukugita *et al.* 1998), is thus an unsolved observational question. The formation of the bulk of metal-rich spheroid stars in highly dissipational mergers of gas-rich systems at high redshifts is an attractive scenario, except for the absence (hitherto) of a substantial population of luminous star forming galaxies at high redshifts with the high star formation rates (several $10^2 - 10^3$ $M_\odot \mathrm{yr}^{-1}$) that would be required to produce substantial spheroidal components of galaxies on typical dynamical timescales of 10^8 yr.

Several papers at this conference have highlighted the evidence in the present-day Universe that the spheroidal populations probably formed within the first 1/3 of the history of the Universe, i.e. at $z \geq 1$. Certainly, the evolution seen in the optically-selected galaxy population out to $z \sim 1$ appears to be primarily due to relatively small galaxies with irregular morphologies and to the disk components of larger galaxies (see e.g.

Brinchmann *et al.* 1998 and Lilly *et al.* 1998a, also Guzman *et al.* 1997, Mallen-Ornelas *et al.*, in preparation, and references therein) and it is thus likely that the spheroids were to a large degree in place by $z \sim 1$. The nature of the ultraviolet-selected 'Lyman-break' galaxies seen at $z > 3$ (Steidel *et al.* 1996) and their relationship to present-day galaxies is still quite uncertain (see e.g. Dickinson 1999, Trager *et al.* 1997 and references therein), and very little is really known about the nature of galaxies in the crucial intermediate redshift range $1.5 < z < 3$.

However, it is very clear that the observational picture of the high redshift Universe that has been gained from optical and near-infrared observations must be seriously incomplete. The νI_ν energy content of far-IR/sub-mm background detected by the FIRAS and DIRBE instruments on COBE (Puget *et al.* 1996, Hauser *et al.* 1998, Fixsen *et al.* 1998) is at least as large (see e.g. Dwek *et al.* 1998) as that of the optical/near-IR background that is obtained by integrating the galaxy number counts (e.g. Pozzetti *et al.* 1998). While some of the far-IR background may result from AGN activity, it is likely that of order a half of the energy from stellar nucleosynthesis at cosmological redshifts emerges as re-processed radiation in the far-IR. Indeed, in terms of the energy from recent star formation activity, the balance may be tipped even further in favor of the far-IR because we know that a significant fraction of the optical background will be coming from old stars - the energy of the optical/near-IR background is already three times higher at K than at U, see Pozzetti *et al.* (1998).

Determining the nature and redshifts of the sources responsible for the far-IR/sub-mm background is therefore vital to our understanding of galaxy evolution. Several groups (e.g. Smail *et al.* 1997, 1998, Hughes *et al.* 1998, Barger *et al.* 1998 and ourselves) are pursuing deep surveys in the sub-millimetre waveband at 850 μm with the new SCUBA bolometer array (Holland *et al.* 1999; Gear *et al.* in preparation) on the 15m James Clerk Maxwell Telescope (JCMT) located on Mauna Kea. Working at 850 μm has a number of rather interesting features since it is well beyond the peak of the far-IR background ($100-200\mu$m). Not least, the k-corrections at 850 μm are extremely beneficial as the rest-wavelength moves up with redshift towards the peak of thermal dust emission around 100 μm. In consequence, a typical starburst galaxy (i.e. with an effective dust temperature of around 30K and effective emissivity $\propto \nu^{1.5}$) has a roughly constant observed flux density at 850 μm over the entire $0.5 < z < 5$ redshift range, especially if $\Omega = 1$ (see Figure 3 of Lilly *et al.* 1999) and observations at 850 μm are thus as sensitive to obscured star formation at very high redshifts, $z \sim 5$ as they are at $z \sim 0.5$! This remarkable fact has a number of interesting consequences. First, 'flux-density limited' samples will approximate 'luminosity limited' (or 'volume limited') samples; secondly, the redshift distribution is likely to be only a weak function of flux density; thirdly, the knowledge of precise redshifts is not critical for determining bolometric luminosities; and finally, one finds that the intuition of optical observers towards quantities such as the redshift distribution sometimes requires modification!

In this paper, we review what is currently known about the sources responsible for the 850 μm background. We take $H_0 = 50h_{50}$ kms^{-1}Mpc^{-1} and for simplicity generally assume a matter-dominated $\Omega = 1$ cosmology.

2. Resolving the sub-mm Background into Discrete Sources

In the last six months, four independent groups have published first results from deep surveys at 850 μm. Smail *et al.* (1997, 1998) have undertaken an ingenious survey using the gravitational lensing effect of moderate redshift clusters of galaxies to amplify background sub-mm sources and now have a sample of 17 sources at $S_{850} > 6$ mJy (3σ).

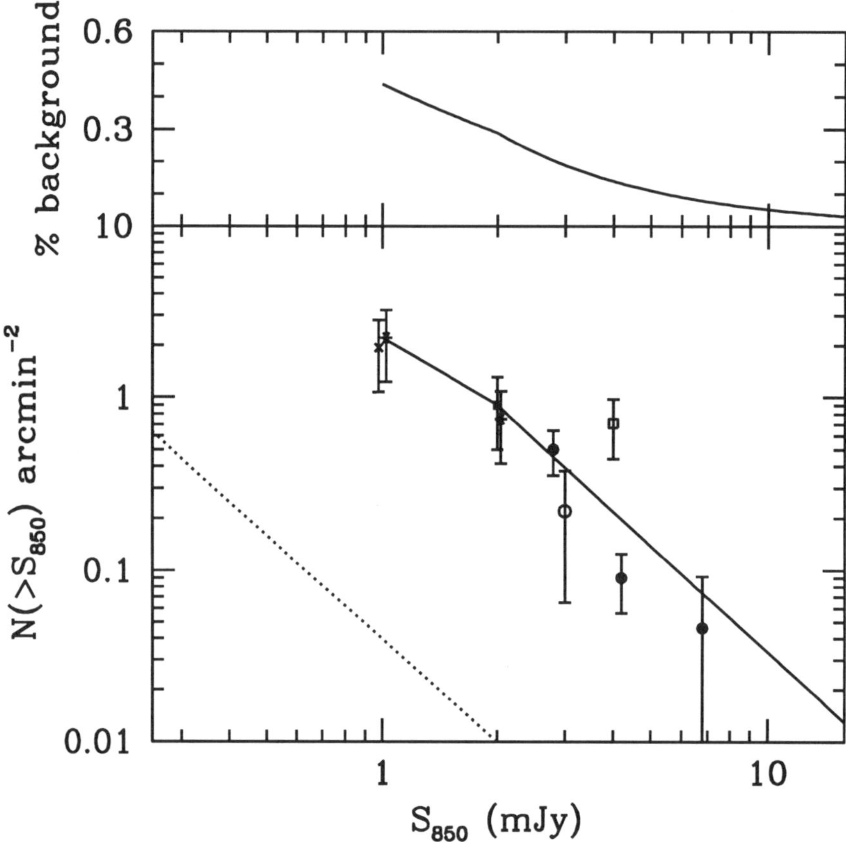

FIGURE 1. Lower panel: the cumulative number counts at 850 μm from recent published surveys compared with a 'no-evolution' model prediction (dotted line) from Eales *et al.* (1999) (adapted from Blain *et al.* 1999b). Sources of data are: Solid dots (Eales *et al.* 1999), open square (Smail *et al.* 1997), open circle (Barger *et al.* 1998), stars (Hughest *et al.* 1998), cross (Blain *et al.* 1999b). Upper panel: the derived cumulative fraction of the 850 μm background (Fixsen *et al.* 1998) that is produced by these sources assuming the solid curve in the lower panel.

The remaining surveys have been 'field' surveys. Hughes *et al.* (1998) published a single very deep image of the HDF that revealed 5 sources at $S_{850} > 2$ mJy (4σ), Barger *et al.* (1998) had 2 sources at $S_{850} > 3$ mJy (3σ) and our own program (Eales *et al.* 1999) has 12 published sources with $S_{850} > 3$ mJy (3σ) with another 20 or so sources at various stages of identification and analysis - the properties of these appear consistent with the first 12, but will not be discussed here.

Given the small numbers involved, the number counts of sources from these surveys are consistent (see Figure 1 - adapted from Blain *et al.* 1999b) and indicate substantial excesses over the number of sources predicted in 'no-evolution' replications of the local IRAS 60 mm luminosity function (see Smail *et al.* 1997, Eales *et al.* 1999). The direct counts at $S_{850} > 2$ mJy have been extended to about 1 mJy with a $P(D)$ analysis in the HDF (Hughes *et al.* 1998) and by a lens inversion analysis by Smail *et al.* (1999).

While many of the sources have been detected at low S/N ratios, the chopping and nodding employed in sub-mm observations lend themselves to a number of straightforward statistical tests (e.g. searching for negative images at the same level of significance) and the great majority of the claimed sources are probably real. It should be noted that all of the blank-field surveys are approaching, or have reached, the confusion limit. For instance, at the $S_{850} \sim 3$ mJy 3σ limit of our own survey there are already about 40 beams per source, the conventional point at which confusion effects become significant.

The upper panel in Figure 1 shows the cumulative fraction of the 850 μm background (taken to be 12 mJy arcmin^{-2} , Fixsen *et al.* 1998) that is produced by these detected sources. It should be noted that already by 3 mJy we are accounting for 20% of the background, a fraction that increases to 50% at the faintest limit, $S_{850} \sim 1$ mJy, probed by the statistical studies.

3. Identifications of the sub-mm Sources

3.1. *How Reliable are the Identifications?*

The SCUBA beam at 850 μm is 15 arcsec FWHM, necessitating a probabilistic approach to identifications on deep optical or near- infrared images. Some of the sub-mm sources are μJy radio sources enabling more accurate, arcsec-level, positions to be determined and these can be identified relatively unambiguously. The fraction of sources that are detectable as faint radio sources is not well determined at this point. In the HDF, Richards (1999) claimed 3 of 5 of the Hughes *et al.* sub-mm sources were detected at $S_{8.5GHz} > 10$ mJy (although this required a quite large and controversial offset of 6 arcsec with respect to the Hughes *et al.* (1998) sub-mm astrometric reference frame). In our own CFRS-14 sample, for which the radio catalogue extends to $S_{5GHz} \sim 16$ mJy (Fomalont *et al.* 1992), we find about 33% of sub-mm sources to be radio sources (and also, since they have similar surface densities, a similar fraction of radio sources to be sub-mm sources). In the future, millimetre wavelength interferometry may produce better positions for the remainder.

All of the survey programs have searched for identifications with extragalactic objects. It is possible to compute the probability that the nearest member of a population of candidate identifications (i.e. optical galaxies) with surface density n is located within a distance d from a random position on the sky, $P = e^{-\pi n d^2}$ (e.g. Downes *et al.* 1986) and this P statistic has been used by many workers in the identification of sub-mm sources (e.g. Hughes *et al.* 1998, Smail *et al.* 1998). There is already a subtlety in the use of P, in that if the density of sources n used is based on the magnitude of the candidate identifications, i.e. $n(< m)$, then P will suffer an *a posteriori* bias, but this can be (and has been) dealt with either analytically or through Monte Carlo simulations. The P statistic represents a starting point, but is not what is really required, which is rather the probability that a particular claimed identification is, in fact, correct. The quantity P tells us the fraction of sources in a sample of size N that would be expected to have an incorrect candidate identification lying within this distance d, i.e. $N_{spurious}(< P) = NP$. Thus, a low value of P for any individual source is not, on its own, enough to make an identification secure. Rather, one has to look at the sample as a whole and determine the number of identifications in the sample (with a certain value of P) relative to the number of spurious identifications (with that same P) that would have been expected if the two populations were completely unrelated. Only if this ratio is high can a particular individual source with that value of P be regarded as securely identified. This is illustrated in Figure 2, which shows the distribution of (corrected) P values for the identifications in the three main published programs (the solid histograms)

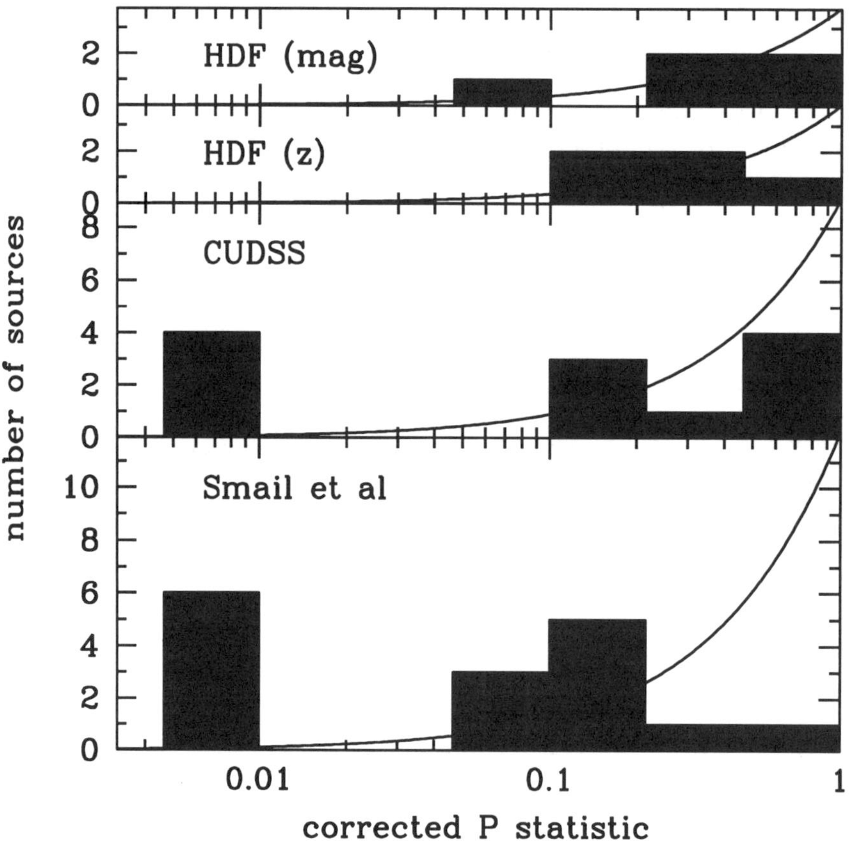

FIGURE 2. The distribution of P values in the three main published surveys (solid histograms - upper two panels Hughes *et al.* (1998) based on magnitudes and redshifts, then our own sample and the Smail *et al.* (1998) lensing sample). The best measure of whether individual identifications are correct is given by comparing the number of identifications in the whole sample with a particular value of P with the number, NP, that would have been expected by chance (solid line). This suggests that about 50% of the sources in all the samples have been correctly identified.

compared with the distribution of P expected if the optical and sub-mm populations were completely unrelated (the smooth line). Statistically, sources lying 'below' the smooth line cannot therefore be regarded as identified, regardless of their value of P, since that number of objects would have been expected by chance!

Our conclusion from Figure 2 is that in all of the deep sub-mm samples studied to date we have reliable identifications for only about half (40-60%) of the sub-mm sources, regardless of whether identifications for the remainder have been claimed or not. This is a handicap, but as we will see below, it is not as serious as one might suppose. Furthermore, the statistics of the identifications already enable us to make an important statement: *At least half (and quite possibly all) of the high latitude 850 μm sources must be extragalactic in nature.*

3.2. *The Redshift Distribution of the Identifications*

Hughes *et al.* in their HDF sample emphasized the high redshifts of their identifications. In our own program (Lilly *et al./* 1998, 1999), we found that many of the sub-mm identifications had already been cataloged in the CFRS program and that in fact three of the first 12 sources had spectroscopic redshift measurements at $z < 1$ (at $z = 0.074$, 0.55 and 0.66). For the remainder, we have estimated redshifts on the basis of the optical-infrared $UVIK$ colors. We concluded that all the initial eight identifications (of which at least six may be regarded as secure) were optically luminous galaxies (comparable to present-day L_*) spanning a broad range of redshifts $0.08 < z < 3$, with four at $z < 1$. The upper limit at $z \sim 3$ comes from detections in the U-band. With the present rather limited data, the observed properties of the four unidentified empty field sources in our sample would be broadly consistent with those of the identified galaxies if they were placed anywhere over a rather wide range of redshifts, $2 < z < 10$. Redshifts as low as $z \sim 1$ however would not be excluded by the present data but would require a higher extinction, as in VIIZw031 (see Trentham *et al.* 1999). A reasonable guess for the median redshift is $< z > \sim 2$.

As discussed in Lilly *et al.* (1999 - see their Figure 10ab), these results appear to be broadly similar to those of the other surveys, especially if the Richards (1999) modification of the HDF identifications are adopted. The lensed sample of Smail *et al.* (1998) does not at present have redshift estimates (except for constraints based on detection in B or V) but appears to have a similar distribution in I_{AB} magnitude especially when an average lens amplification of a factor of 2.5 is taken into account.

3.3. *The Nature of the sub-mm Sources*

As noted above, any source detected at $S_{850} \geq 3$ mJy that lies at $z > 0.5$ must have a luminosity above that of Arp 220, i.e. $L \geq 3 \times 10^{12} h_{50}^{-2} L_\odot$. Assuming the energy comes from star formation as opposed to black-hole accretion, this luminosity corresponds to a substantial star formation rate of $\geq 600 h_{50}^{-2} M_\odot$ yr^{-1}.

The broad-band spectral energy distributions of the identifications in our own sample, as defined from the optical through the far-IR component to the radio, from measurements or limits at 0.8 μm, 15 μm, 450 μm, 850 μm and at 5 GHz, are consistent with the measured/estimated redshifts of the identifications and a rest-frame SED that broadly matches that of Arp 220.

The galaxies have a range of optical colors, but are on average a little redder in $(V - I)$ than typical field galaxies, consistent with what is known about the ultraviolet properties of local ULIRGs (Trentham *et al.* 1999). The $(V-I)$, I distribution for our identifications and for those in the Smail *et al.* program match nicely the expectations based on local ULIRGs (Trentham *et al.* 1999). The HST morphologies of the $z > 0.5$ identifications in our sample range from relatively normal-looking galaxies to clear examples of mergers, but nearly all show some sign of peculiarity in the form of secondary nuclei or asymmetrical outer isophotes. Little is known about the ultraviolet morphologies of ULIRGs at low redshift, but the Trentham *et al.* (1999) study shows considerable diversity and substantial differences from the optical morphologies.

In summary, *in essentially all respects that can presently be studied, the $z > 0.5$ sources in our sample appear to be very similar to local ULIRG prototypes such as Arp220.*

4. The Significance of ULIRGs at High Redshift

The results outlined above lead robustly to a very important conclusion: *ULIRGs as a class are a much more important component of the galaxy population (in that they produce a much higher fraction of the total luminous output) at high redshift than at low redshift.* In the local Universe, ULIRGs of luminosities greater or equal to that of Arp 220 (i.e. $2 \times 10^{12} h_{50}^{-2} L_\odot$) contribute only about 1% of the far-IR luminous output of the

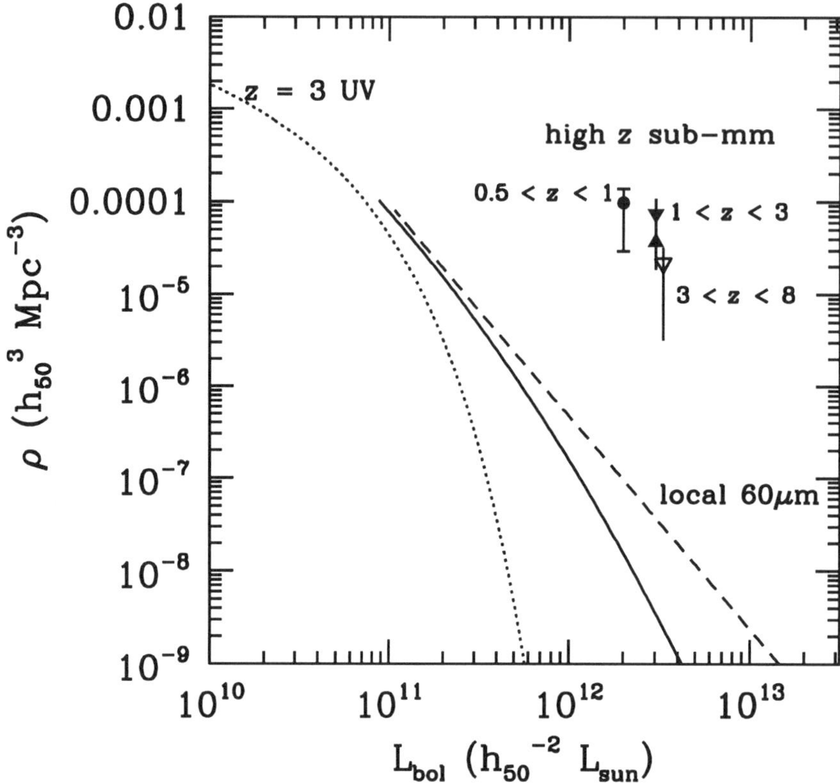

FIGURE 3. Estimate of the cumulative bolometric luminosity function of the sub-mm population from our own sample (points), the local IRAS 60μm population, and the $z \sim 3$ ultraviolet-selected 'Lyman-break' population (uncorrected for extinction). The $1 < z < 3$ sub-mm points show minimum and maximum values according to whether the four 'empty field' sources are at $z > 3$ or $1 < z < 3$. Likewise, the $3 < z < 8$ point assumes that the 'empty fields' are at $z > 3$.

galaxy population (Soifer *et al.* 1987, Saunders *et al.* 1990, Sanders and Mirabel 1996). In contrast, at high redshifts ($z > 1$), similar objects must produce at least 30% of the far-IR/sub-mm background (for $\Omega = 1$, more for low Ω since Arp220 would lie further down the $N(S)$ distribution), which we have seen is at least equal in energy content to the optical/near-IR background. Thus, high-luminosity obscured objects are much more common, relatively, at high redshift, and are in fact producing a substantial fraction (15%) of the total luminous output of the Universe averaged over all epochs and all wavebands.

The cumulative bolometric luminosity function in the far-IR derived from our own sample is shown in Figure 3 compared with the local IRAS luminosity function and the ultraviolet luminosity function for the Lyman-break galaxies constructed by Dickinson (1999).

5. Interpretation of the Redshift Distributions

The relatively small number of sources (no more than 50% of the sample) that can possibly be at very high redshifts ($z > 3$) already sets quite strong constraints on the amount of high-luminosity obscured star formation that can take place at these redshifts. This is because, as pointed out in previous papers (Lilly *et al.* 1998, 1999), the beneficial k-corrections produce a strong weighting of high redshift star formation activity in the production (in redshift space) of the 850 μm background relative to the production of stars (in redshift space). This weighting is simply $f_\nu(\nu_{em})/f_\nu(\nu_{obs})$, or $(1 + z)^{3.5}$ over much of the redshift range of interest $0 < z < 6$. In Figure 4 (from Lilly *et al.* 1999), we have computed the redshift distribution of the 850 μm background light for a number of different star formation histories, assuming that the energy of this star formation emerges with the spectral energy distribution of an obscured starburst, like Arp 220. It can be seen that galaxy formation/evolution scenarios in which 50% of all dust-enshrouded star formation in the Universe occurred prior to $z = 3$ predict that 85% of the 850 μm background had been produced at $z > 3$. The distribution of observed light at $S_{850} > 2.8$ mJy in our identified source sample is also shown. This does not reach unity because the unidentified sources have been omitted since their redshifts are unconstrained (but it is assumed for this purpose that they have $z > 2.5$) and the contribution from the two less securely-identified galaxies estimated to lie at around $z \sim 2.5$ is shown as a dotted line.

Even if we make no assumption at all about the redshifts of the fainter sources with $S_{850} < 3$ mJy, our observations would already appear to require that at least 15% of the 850 μm background must be produced at $z < 3$, which is only barely consistent with a scenario in which 50% of obscured star formation takes place at $z \geq 3$.

If we speculatively assume that the redshift distribution of fainter sources with $S_{850} < 3$ mJy follows that at $S_{850} > 3$ mJy (a plausible, but not watertight, assumption given the flatness of the 850 μm flux density-redshift relation, see also the models of Blain *et al.* 1999a) then our results suggest that the great bulk of obscured star formation in the Universe occurred at redshifts $z < 3.0$. While this analysis can not be regarded as conclusive until we penetrate deeper in to the background, Figure 4 suggests that the cumulative production of the 850 μm background appears to follow well the expectations of models in which the luminosity density in the far-IR (at least in high luminosity obscured objects) peaks in the $1.2 < z < 2$ range and falls thereafter. Interestingly, initial indications for a decline in the ultraviolet luminosity density of the Universe at high redshifts (Madau 1996, 1997) have not been borne out by more recent work (Steidel *et al.* 1999).

6. The Relationship to the Lyman-Break Ultraviolet-Selected Galaxy Population

As shown in Figure 12 of Lilly *et al.* (1999) the bolometric output in the far-IR of the high luminosity ($L \geq 3 \times 10^{12} h_{50}^{-2} L_\odot$) high z ULIRG population already matches that in the ultraviolet of the whole 'Lyman-break' population of galaxies, even though the former only comprise the 'top' 20% of the 850 μm background.

Estimates of the reddening of the Lyman-break galaxies (LBG) based on the observed ultraviolet continuum slope suggest that for typical LBG the far-IR luminosity would be between $2 - 7$ times that seen in the ultraviolet (Dickinson 1999, see also Pettini *et al.* 1998) for SMC and Calzetti (1997) extinction curves, with higher values being claimed by Meurer *et al.* (1997). These typical objects, with 'corrected' star formation rates of

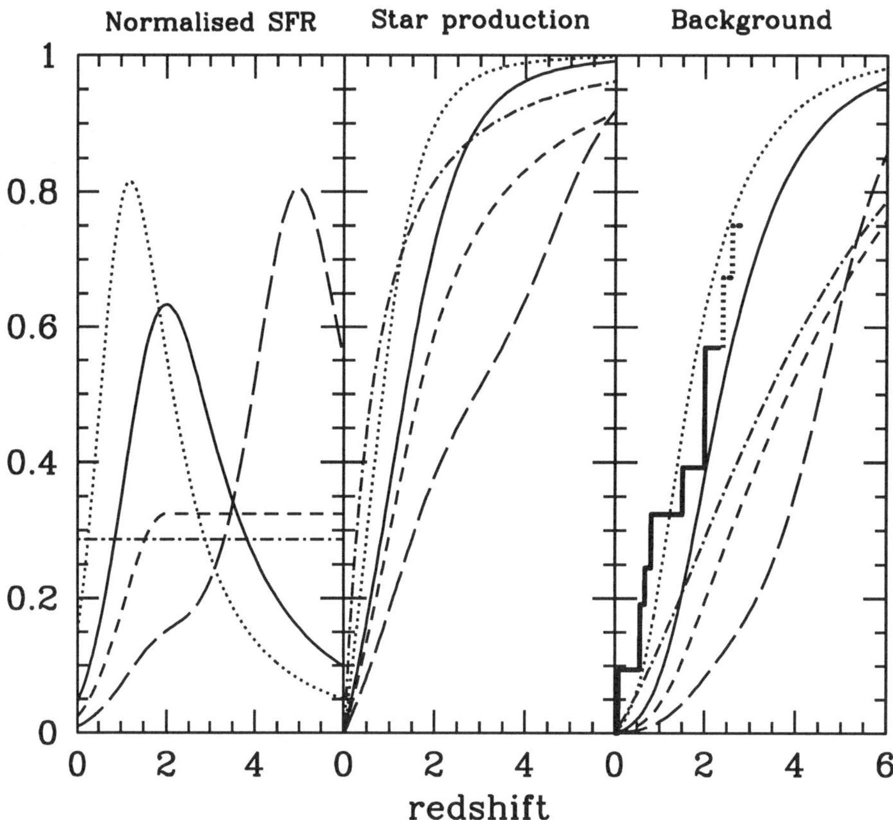

FIGURE 4. Production of the 850 μm background from different star formation histories: The left hand panel shows five different heuristic star formation histories. For each model, the cumulative production of stars is shown in the center panel, and the cumulative distribution of light in the 850 μm background is shown in the right-hand panel. Because of the highly beneficial k-corrections at 850 μm, the light in the background is heavily weighted in favor of high redshift star formation. Models in which half the obscured star formation in the Universe occurred prior to $z \sim 3$ predict that only 15% of the background should come from $z < 3$. We can rule this out unless all of the sources with $S_{850} < 3$ mJy have $z > 3$ - a highly unlikely situation. The irregular line in the left-most panel shows the distribution in redshift of the background produced by the observed sources - illustrating the effect of assuming that the sources below the limit of the survey in fact have the same redshift distribution. This would require a falling luminosity density at high redshifts, and would imply that most stars formed in these obscured objects did so at $z \leq 2$.

up to $30 - 100 h_{50}^{-2} M_{\odot} \mathrm{yr}^{-1}$ would be undetectable with SCUBA at present, but would have to be responsible for a significant fraction of the background.

Obviously the estimation of bolometric luminosities on the basis of extinction in the ultraviolet is highly uncertain requiring an assumed extinction curve that largely reflects the geometrical distribution of stars and dust. This is especially true in the high extinction regime (see the three examples of local ULIRGs in Trentham *et al.* 1999). Nevertheless, preliminary indications (Steidel, private communication) are that the number of very highly extinguished LBG with 'corrected' Arp220-level luminosities (i.e. after correction with the Calzetti reddening curve) are roughly consistent with the number directly observed in the sub-mm surveys described here, which have $\phi \sim 10^{-4} h_{50}^3$ Mpc^{-3}, Figure 3). This agreement is encouraging, though possibly fortuitous given the uncertainties in the reddening correction applied to the optical sample. The sub-mm sample may also contain some ULIRGs that are so heavily obscured as to be completely absent from the present LBG samples.

7. The Nature of the Obscured Energy Sources

A difficult question concerns the fraction of the far-IR energy that comes from hidden active galactic nuclei. In the local Universe, the evidence from mid-IR emission features (Genzel *et al.* 1998) is that AGN provide a significant but not dominant (25% - 50%) contribution to ULIRGs at these luminosities and this seems a reasonable first guess as to the situation at high redshifts. The ultraviolet spectrum of the highly luminous sub-mm source SMM02399-0136 (Ivison *et al.* 1998) shows spectroscopic indications for an AGN but this same source also exhibits strong, starburst like, CO detections (Frayer *et al.* 1998, see also Frayer *et al.* 1999). For all but the brightest sources (which may be tackled by SIRTF) these mid-IR diagnostics may be unobservable until NGST flies - and even then only if it has a mid-IR spectroscopic capability.

Ascribing a dominant fraction of the energy output of this population to AGN would require a major upwards revision in the total energy output of AGN. On the other hand, several authors have stressed the inadequacy of the 'known' quasar population to produce the required mass of black holes (integrated over the population). Using the Magorrian *et al.* (1996) relationship between black hole mass and stellar bulge mass ($M_{BH} \sim 0.006 M_*$) in local galaxies and assuming a radiative efficiency $\epsilon \sim 0.1$ for black hole accretion and an energy release of $0.016 M_Z c^2$ for the return of m_Z of metals (Songaila *et al.* 1990) it is easy to show that the bolometric light output associated with the production of the black holes and stellar metals in typical spheroids should be comparable:

$$\frac{L_{AGN}}{L_{star}} \sim \frac{\epsilon M_{BH} c^2}{0.016 M_Z c^2} \sim \frac{M_{BH}}{M_{star}} \frac{1}{0.0032} \sim 2. \tag{7.1}$$

On the other hand, the approximate consistency with the extinction-corrected properties of the LBG population noted above suggests that much of the far-IR background is indeed coming from stars. Observational resolution of this important question, at better than the factor of two level, will be challenging.

8. The Formation of Spheroids?

The identification, in the sub-mm, of a population of galaxies at high redshift that are producing a substantial fraction of present day stars in high luminosity systems is important because it is then attractive to identify these as producing the metal-rich spheroidal components of galaxies, including the bulges of present-day spiral galaxies. Local ultra-luminous IR galaxies have long been proposed as being triggered by major mergers and resulting in the production of massive spheroids (see Sanders and Mirabel 1997 and references therein). The high individual luminosities ($\geq 3 \times 10^{12} h_{50}^{-2} L_\odot$) and implied star formation rates ($\geq 600 h_{50}^{-2} M_\odot \mathrm{yr}^{-1}$) are consistent with making substantial stellar populations on dynamical timescales.

It should be stressed that the ULIRG population revealed in the sub-mm surveys at high redshift has a sufficiently high number density (more than 100 times higher than in the present-day Universe - Figure 3) that they could be responsible for producing a significant fraction of all stars that have been formed in the Universe, since they are responsible for producing, in the far-IR, a significant fraction of the entire bolometric extragalactic background light. In more absolute terms, a star formation rate of $600h_{50}^{-2}M_\odot\mathrm{yr}^{-1}$ maintained for $4h_{50}^{-1}$ Gyr at a number density of $\phi \sim 10^{-4}h_{50}^3$ Mpc^{-3} would yield a stellar mass density of $2.4 \times 10^8 M_\odot$ Mpc^{-3}. It should be recalled (e.g. Fukugita *et al.* 1998) that the spheroids contain a half to two-thirds of all stars in the Universe or about $0.7 - 2.0 \times 10^8 h_{50} M_\odot$ Mpc^{-3}. So the numbers are in the right ball-park.

The combination of the high integrated production of stars, the high star formation rates, the incidence of merger-like morphologies and the obvious presence of substantial amounts of dust, make it attractive, though still speculative, to associate these galaxies with the production of the metal-rich spheroid component of galaxies. In this case, these first data from our survey (see Figure 4) suggest that much of this activity, conservatively at least 50%, and probably much more, has happened at relatively recent epochs, i.e. $z < 3$.

The research of SJL and JRB is supported by the Natural Sciences Engineering Research Council of Canada and by the Canadian Institute for Advanced Research. The Research of SAE and WKG is supported by the Particle Physics and Astronomy Research Council in the United Kingdom. The support of all of these agencies is gratefully acknowledged.

REFERENCES

BARGER, A., COWIE, L., SANDERS, D., FULTON, E., TANIGUCHI, Y., SATO, Y., KAWARE, K., OKUDA, H. 1998 *Nature* **394**, 248

BLAIN, A., LONGAIR, M. 1993 *MNRAS*, **264**, 509

BLAIN, A., SMAIL, I., IVISON, R., KNEIB, J.P. 1999a *MNRAS*, **302**, 632

BLAIN, A., KNEIB, J.P., IVISON, R., SMAIL, I. 1999b *ApJ*, **512**, L87

BRINCHMANN, J., ABRAHAM, R., SCHADE, D., TRESSE, L., ELLIS, R., LILLY, S., LE FEVRE, O., GLAZEBROOK, K., HAMMER, F., COLLESS, M., CRAMPTON, D., BROADHURST, T. 1998 *ApJ*, **499**, 112

CALZETTI, D. 1997 *AJ*, **113**, 162

CONNOLLY, A., SZALAY, A., DICKINSON, M., SUBBARAO, M., BRUNNER, R. 1997 *ApJ*, **486**, L11

DICKINSON, M. 1999, in *The Hubble Deep Field* (ed. M. Livio, M. Fall and P. Madau), p219. (Cambridge)

DOWNES, A.J.B., PEACOCK, J.A., SAVAGE, A., CARRIE, D. 1986 *MNRAS*, **218**, 31

DWEK, E., ET AL. 1998 *ApJ*, **508**, 106

EALES, S., LILLY, S., GEAR, W., BOND, J.R., DUNNE, L., HAMMER, F., LE FEVRE, O., CRAMPTON D. 1998 *ApJ*, in press (Paper 1)

FIXSEN, D., DWEK, E., MATHER, J., BENNET, C., SHAFER, R. 1998 *ApJ*, **508**, 123

FOMALONT, E., WINDHORST, R., KRISTIAN, J., KELLERMAN, K. 1991 *AJ*, **102**, 1258

FRAYER, D.T., IVISON, R.J., SCOVILLE, N.Z., YUN, M., EVANS, A.S., SMAIL, I., BLAIN, A.W., KNEIB, J.-P. 1998 *ApJ*, **506**, 7

FRAYER, D.T., IVISON, R.J., SCOVILLE, N.Z., EVANS, A.S., YUN, M., SMAIL, I., BARGER, A., I., BLAIN, A.W., KNEIB, J.-P. 1998 *ApJ*, **514**, L13

GENZEL, R., LUTZ, D., STURM, E., EGAMI, E., KUNZE, D., MOORWOOD, A.F.M., RIGOPOULOU, D., SPOON, H., STERNBERG, A., TACCONNI-GARMAN, L., TACCONI, L.,

THATTE, N. 1998 *ApJ*, **498**, 579

GUZMAN, R., GALLEGO, J., KOO, D., PHILLIPS, A., LOWENTHAL, J., FABER, S.M., ILLING-WORTH, G., VOGT, N. 1998 *ApJ*, **489**, 559

HAEHNELT, M.G., NATARAJAN, P., REES, M.J. 1998 *MNRAS*, **300** 817

HAUSER, M., ARENDT, R., KELSALL, T., DWEK, E., ODEGARD, N., WELLAND, J., FREUN-DENREICH, H., REACH, W., SILVERBERG, R., MODELEY, S., PEI, Y., LUBIN, P., MATHER, J., SHAFER, R., SMOOT, G., WEISS, R., WILKINSON, D., WRIGHT, E. 1998 *ApJ*, **508**, 25

HOLLAND, W.S., ROBSON, E.I., GEAR, W.K., CUNINGHAM, C.R., LIGHTFOOT, J.F., JENNESS, T., IVISON, R., STEVENS, J.A., ADE, P.A.R., GRIFFIN, M.J., DUNCAN, W.D., MURPHY, J.A., NAYLOR, D.A. 1999 *MNRAS*, **303**, 659

HUGHES, D., SERJEANT, S., DUNLOP, J., ROWAN-ROBINSON, M., BLAIN, A., MANN, R., IVISON, R., PEACOCK, J., EFSTATHIOU, A., GEAR, W., OLIVER, S., LAWRENCE, A., LONGAIR, M., GOLDSCHMIDT, P., JENNESS, T. 1998 *Nature*, **394**, 241

IVISON, R., SMAIL, I., LE BORGNE, J-F., BLAIN, A., KNEIB, J-P., KERR, T., BEZECOURT, J., DAVIE, J. 1998 *MNRAS*, **298**, 583

LILLY, S.J., COWIE, L. 1987, in *Infrared Astronomy with Arrays* (ed. Wynn-Williams, G., Becklin, E.), p473. (University of Hawaii)

LILLY, S.J., LE FEVRE, O., CRAMPTON, D., HAMMER, F., TRESSE, L. 1995a *ApJ*, **455**, 50

LILLY, S.J., HAMMER, F., LE FEVRE, O., CRAMPTON, D. 1995b *ApJ*, **455**, 75

LILLY, S.J., LE FEVRE, O., HAMMER, F., CRAMPTON, D. 1996 *ApJ*, **460**, L1

LILLY, S.J., SCHADE, D., ELLIS, R., LE FEVRE, O., BRINCHMANN, J., TRESSE, L., ABRAHAM, R., HAMMER, F., CRAMPTON, D., COLLESS, M., GLAZEBROOK, K., MALLEN-ORNELAS, G., BROADHURST, T. 1998a *ApJ*, **500**, 75

LILLY, S.J., EALES, S.A., GEAR, W., DUNNE, L., BOND, J.R., HAMMER, F., LE FEVRE, O., CRAMPTON, D. 1998b, in *NGST: Scientific and Technical Challenges* (ed. B. Kaldeich-Schrmann), p39. (ESA)

MADAU, P., FERGUSON, H., DICKINSON, M., GIAVALISCO, M., STEIDEL, C., FRUCHTER, A. 1996 *MNRAS*, **283**, 1388

MADAU, P., POZETTI, L., DICKINSON, M. 1998 *ApJ*, **498**, 106

MEURER, G., HECKMAN, T., LEHNERT, M., LEITHERER, C., LOWENTHAL, J. 1997 *AJ*, **114**, 51

PETTINI, M., KELLOGG, M., STEIDEL, C., DICKINSON, M., ADELBERGER, K., GIAVALISCO, M. 1998 *ApJ*, **508**, 539

POZZETTI, L., MADAU, P., ZAMORANI, G., FERGUSON, H.C., BRUZUAL, G. 1998 *MNRAS*, **298**, 1133

PUGET, J-L., ABERGEL, A., BERNARD, J-P., BOULANGER, F., BURTON, W.B., DESERT, F.X., HARTMANN, D. 1996 *A&A*, **308**, L5

RICHARDS, E.A. 1999 *ApJ*, **513**, L9

SANDERS, D., MIRABEL, I. 1996 *ARA&A*, **34**, 749

SAUNDERS, W., ROWAN-ROBINSON, M., LAWRENCE, A., EFSTATHIOU, G., KAISER, N., ELLIS, R.S., FRENK, C. 1990 *MNRAS*, **242**, 318

SAWICKI, M., LIN, H., YEE, H. 1997 *AJ*, **113**, 1

SMAIL, I., IVISON, R., BLAIN, A. 1997 *ApJ*, **490**, L5

SMAIL, I., IVISON, R., BLAIN, A., KNEIB, J-P. 1998 *ApJ*, **507**, L21

SONGAILA, A., COWIE, L.L., LILLY, S.J. 1990 *ApJ*, **348**, 371

STEIDEL, C., GIAVALISCO, M., PETTINI, M., DICKINSON, M., ADELBERGER, K. 1996 *ApJ*, **462**, L17

STEIDEL, C., ADELBERGER, K., GIAVALISCO, M., DICKINSON, M, PETTINI, M. 1998, preprint (astro-ph/9811400)

TRAGER, S., FABER, S., DRESSLER, A., OEMLER, A. 1997 *ApJ*, **485**, 92

TRENTHAM, N., KORMENDY, J., SANDERS, D. 1999, preprint (astro-ph/9901382)

Ages and Metallicities for Stars in the Galactic Bulge

By J A Y A. F R O G E L

Department of Astronomy, The Ohio State University, 174 W. 18th Ave. Columbus OH 43210,
USA

Observations of the stellar content of the Milky Way's bulge helps us to understand the stellar content and evolution of distant galaxies. In this brief overview I will first highlight some recent work directed towards measuring the history of star formation and the chemical composition of the central few parsecs of the Galaxy. High resolution spectroscopic observations by Ramirez *et al.* (1998) of luminous M stars in this region yield a near solar value for [Fe/H] from direct measurements of iron lines. Then I will present some results from an ongoing program by my colleagues and myself which has the objective of delineating the star formation and chemical enrichment histories of the central 100 parsecs of the Galaxy, the 'inner bulge'. We have found a small increase in mean [Fe/H] from Baade's Window to the Galactic Center and deduce a near solar value for stars at the center. For radial distances greater than 1° we fail to find a measurable population of stars that are significantly younger than those in Baade's Window. Within 1° of the Galactic Center we find a number of luminous M giants that most likely are the result of a star formation episode not more than one or two Gyr ago.

1. Introduction

The structure and stellar content of the bulge of the Milky Way are often used as proxies in the study of other galactic bulges and of elliptical galaxies. However, out to a radius of about 2° along the minor axis, and considerably farther along the major axis, the visual extinction is great enough that optical observations are difficult to impossible along most lines of sight. A two degree radius corresponds to 5″ to 10″ for galaxies in Virgo. Thus, if we are to use the stellar content of the inner bulge as a starting point for delineating the global characteristics of the inner regions of nearby spiral bulges and spheroidal galaxies, we must turn to near-infrared observations. This brief review, then, will concentrate on summarizing some of the recent studies in the near-IR of the inner bulge of the Milky Way. Further, it will specifically address two of the most important characteristics of the stars: their ages and metallicities.

2. Within a Few Arc Minutes of the Galactic Center

Krabbe *et al.* (1995) have identified more than 20 luminous blue supergiants and Wolf-Rayet stars in a region not more than a parsec in radius around the center of the Galaxy. The inferred mass of some of these stars approaches 100 $M_\odot$. From this they conclude that between 3 and 7 Myr ago there was a burst of star formation in the central region. They also identified a small population of cool luminous AGB stars from which one can conclude that there was significant star formation activity a few 100 Myr ago as well.

Blum *et al.* (1996a) carried out a K-band survey of the central 2 arc minutes of the Galaxy. They focused on the significant numbers of luminous ($K_0 < 6$) cool stars. These stars are considerably more luminous than would be expected from a typical old stellar population such as is found in Baade's Window, for example. Most of these stars were found by Blum and others to be M stars. With K-band spectra, Blum *et al.* (1996b) were able to distinguish between M supergiants and AGB stars. Such a distinction is of importance because of the implications for the times of star formation. As first demonstrated quantitatively by Baldwin *et al.* (1973), M-type supergiants can be easily distinguished from ordinary giants of the same temperature (or color) via the strengths of the H_2O and CO absorption bands in K-band spectra.

Blum *et al.* (1996b) found only 3 out of 19 stars to be supergiants, one of which is the well known IRS 7. The remainder are AGB stars. From the spectra and the multi-color photometry they concluded that there have been multiple epochs of star formation in the central few parsecs of the Galaxy. The most recent epoch, less than 10 Myr ago, corresponds with that found by Krabbe *et al.* (1995). Other epochs of star formation identified by Blum *et al.* occurred about 30 Myr, between 100 and 200 Myr, and more than about 400 Myr in the past. The majority of stars are associated with the oldest epoch of star formation.

Abundances for the red luminous stars in the region around the Galactic Center are being determined by Ramirez *et al.* (1998) from a full spectral synthesis analysis of high resolution K-band spectra. It will be interesting to compare abundances values for the inner bulge with their results. Based on direct measurements of iron lines in 10 stars they derive a mean [Fe/H] of 0.0 with a dispersion comparable to their uncertainties, about 0.2 dex. This is only a few tenths of a dex greater than the mean [Fe/H] determined for Baade's Window ($\ell = 1°$, $b = -3.9°$) K-giants (Sadler *et al.* 1996; McWilliam & Rich 1994). This small increase in the mean value of [Fe/H] compared with Baade's Window is consistent with the [Fe/H] gradient in the bulge found by Tiede *et al.* (1995) and Frogel *et al.* (1999). The lack of a dispersion in [Fe/H] contrasts with a dispersion of more than an order of magnitude for the K giants in Baade's Window (Sadler *et al.* 1996; McWilliam & Rich 1994). It is, however, consistent with the lack of dispersion found for the M giants in Baade's Window (Frogel & Whitford 1987; Terndrup *et al.* 1991). The fact that [Fe/H] is near solar at the Galactic Center with a star formation rate per unit mass that is considerably in excess of the solar neighborhood value suggests that the rate of chemical enrichment has been quite different at the two locations.

3. The Inner Galactic Bulge

The inner 3° of the Galactic bulge, interior to Baade's Window, will be referred to here as the inner Galactic bulge (corresponding to a projected distance from the Galactic Center within around 400pc). With the 2.5 meter duPont Telescope at Las Campanas Observatory I have obtained JHK images of 11 fields within the inner bulge, three of which are within 1° of the Galactic Center. The two questions to address are: What is the abundance of the stars in this region and is there any evidence for a detectable population of intermediate age or young stars? My collaborators and I are taking two approaches to the abundance question. The first is based on the fact that the giant branch of a metal rich globular cluster in a *K, JK* color magnitude diagram is linear over 5 magnitudes and has a slope proportional to its optically-determined [Fe/H] (Kuchinski *et al.* 1995). Results from work will be summarized here. The second approach, which will give a better answer to the abundance question, is based on the analysis of K-band spectra of about one dozen M stars in each of 11 fields. This is a work in progress.

3.1. *Abundances in the Inner Galactic Bulge*

The best fixed reference point in any measurement of abundances within the inner bulge is the determination by McWilliam & Rich (1994) of a mean abundance of [Fe/H]$= -0.2$ dex for a small sample of K giants in Baade's Window, based on high resolution spectroscopy. Sadler *et al.*'s (1996) spectroscopy of several hundred K giants in Baade's Window yielded a similar result. Both of these analyses measured a spread in [Fe/H] in Baade's Window of between one and two orders of magnitude. The estimate of [Fe/H] for the Baade's Window giants based on the near-IR slope method (Tiede *et al.* 1995) differed from previous near-IR determinations in that they found an [Fe/H] close to the value based on the optical spectra of K giants.

My near-IR survey of inner bulge fields has yielded color-magnitude diagrams that, except for the fields with the highest extinction, reach as faint as the horizontal branch. Thus, with data for the entire red giant branch above the level of the HB we can apply the technique developed by Kuchinski *et al.* (1995) to determine [Fe/H] from the slope of the RGB above the HB. Although the calibration of this technique is based on observations of globular clusters, the applicability of this method to stars in the bulge was demonstrated by Tiede *et al.* (1995) in their analysis of stars in Baade's Window. This method is reddening independent since it depends only on a slope measurement. Based on 7 fields on or close to the minor axis of the bulge at galactic latitudes between $+0.1°$ and $-2.8°$ we derive a dependence of $\langle[Fe/H]\rangle$ on latitude for b between $-0.8°$ and $-2.8°$ of -0.085 ± 0.033 dex/degree. When combined with the data from Tiede *et al.* we find for $-0.8° \leq b \leq -10.3°$ the slope in $\langle[Fe/H]\rangle$ is -0.064 ± 0.012 dex/degree. An extrapolation to the Galactic Center predicts [Fe/H] $= +0.034 \pm 0.053$ dex, in close agreement with Ramirez *et al.* (1998). Also in agreement with Ramirez *et al.*, we find no evidence for a dispersion in [Fe/H] (see Frogel *et al.* 1999 for details).

Analysis of the K-band spectra of the brightest M giants in each of the fields surveyed is nearing completion; the results appear to be consistent with those based on the RGB slope method, namely, an [Fe/H] for Baade's Window M giants close to the McWilliam & Rich value but with little or no gradient as one goes into the central region. Also, the spectroscopic data show little or no dispersion in [Fe/H] within each field.

In summary, several independent lines of evidence point to an [Fe/H] for stars within a few parsecs of the Galactic Center of close to solar. The gradient in [Fe/H] between Baade's Window and the Center is small – not more than a few tenths of a dex. Exterior to Baade's Window there is a further small decline in mean [Fe/H] (e.g. Terndrup *et al.* 1991, Frogel *et al.* 1990; Minniti *et al.* 1995). It remains to be seen whether this gradient arises from a change in the mean [Fe/H] of a single population or a change in the relative mix of two populations, one relatively metal rich and identifiable with the bulge, the other relatively metal poor and more closely associated with the halo. Support for the latter interpretation is found in the survey of TiO band strengths in M giants in outer bulge fields by Terndrup *et al.* (1990), for which they found a bimodal distribution. McWilliam & Rich (1994) proposed an explanation based on selective elemental enhancements as to why earlier abundance estimates of bulge M giants seemed to consistently yield [Fe/H] values in excess of solar. It remains to be understood why no dispersion is observed in measurements of the M giant abundances. It also remains to be determined if the indirect methods used for measuring [Fe/H] are really measuring iron rather than being sensitive to, for example, element enhancements.

3.2. *Stellar Ages in the Inner Galactic Bulge*

If a stellar population has an age significantly younger than 10 Gyr then stars at the top of the AGB will be several magnitudes brighter than they would in an older population. After correction for extinction we found that our fields closer than $1.0°$ to the Galactic Center have significant numbers of bright, red stars implying the presence of a younger component to the stellar population, probably with an age of a few Gyr. This is consistent with Blum *et al.*'s work on the inner few arc minutes of the bulge. Beyond $1.0°$ from the center there is no evidence for such luminous stars (see Frogel *et al.* 1999 for details).

A second test applied to see if there is evidence for a young population in the Galactic bulge was a comparison of the luminosities and periods of bulge long period variables (LPVs) with those found in globular clusters (Frogel & Whitelock 1998). For LPVs of the same age, those with greater [Fe/H] will have longer periods. LPVs with longer periods also have higher mean luminosities. In the past claims have been made for the presence of a significant intermediate-age population of stars in the bulge based on the finding of some LPVs with periods in excess of 500-600 days. It is necessary, however, to have a well defined sample of stars if one is going to draw conclusions based on the rare occurrence of one type of star. The M giants in Baade's Window are just such a well defined sample (e.g. Frogel & Whitford 1987). Frogel & Whitelock (1998) demonstrated that with the exception of a few of the LPVs in Baade's Window with the longest periods, the distribution in bolometric magnitudes of the LPVs from the bulge and from globular clusters overlap completely. Furthermore, because of the dependence of period and luminosity on [Fe/H] and the fact that there has been no reliable survey for LPVs in globulars with [Fe/H] > -0.25, the brightest Baade's Window LPVs could have the same age as the somewhat fainter ones but come from the higher [Fe/H] population.

Finally, observations with the Infrared Astronomical Satellite (IRAS) at 12μm were used to estimate the integrated flux at this wavelength from the Galactic bulge as a function of galactic latitude along the minor axis (Frogel 1998). These fluxes were then compared with predictions for the 12μm bulge surface brightness based on observations of complete samples of optically-identified M giants in minor-axis bulge fields (Frogel & Whitford 1987; Frogel et al. 1990). No evidence was found for any significant component of 12μm emission in the bulge other than that expected from the optically identified M star sample plus normal, lower luminosity stars. Since these stars are themselves fully attributable to an old population, the conclusion from this study was, again, no detectable population of stars younger than those in Baade's Window, i.e. no younger than an age comparable to that of globular clusters.

REFERENCES

BALDWIN, J.R., FROGEL, J.A., PERSSON, S.E. 1973 *ApJ*, **184**, 427

BLUM, R.D., SELLGREN, K., DePOY, D.L. 1996a *ApJ*, **470**, 864

BLUM, R.D., SELLGREN, K., DePOY, D.L. 1996b *AJ*, **112**, 1988

FROGEL, J.A. 1998 *ApJ*, **505**, 659

FROGEL, J.A., TERNDRUP, D., BLANCO, V.M., WHITFORD, A.E. 1990 *ApJ*, **353**, 494

FROGEL, J.A., TIEDE, G.P., KUCHINSKI, L.E. 1999 *AJ*, in press

FROGEL, J.A., WHITELOCK, P.A. 1998 *AJ*, **116**, 754

FROGEL, J.A., WHITFORD, A.E. 1987 *ApJ*, **320**, 199

KRABBE, A., GENZEL, R., ET AL. 1995 *ApJ*, **447**, L95

KUCHINSKI, L., FROGEL, J.A., TERNDRUP, D.M., PERSSON, S.E. 1995 *AJ*, **109**, 1131

McWILLIAM, A., RICH, R.M. 1994 *ApJS*, **91**, 749

MINNITI, D., OLSZEWSKI, E., RIEKE, M. 1995 *AJ*, **110**, 1686

RAMIREZ, S., SELLGREN, K., CARR, J.S., BALACHANDRAN, S., BLUM, R., TERNDRUP, D. 1998, in *The Central Parsecs*, (ed. H. Falcke, A. Cotera, W. Duschl & F. Melia), in press. (ASP)

SADLER, E.M., RICH, R.M., TERNDRUP, D.M. 1996 *AJ*, **112**, 171

TERNDRUP, D.M., FROGEL, J.A., WHITFORD, A.E. 1990 *ApJ*, **357**, 453

TERNDRUP, D.M., FROGEL, J.A., WHITFORD, A.E. 1991 *ApJ*, **378**, 742

TIEDE, G.P., FROGEL, J.A., TERNDRUP, D.M. 1995 *AJ*, **110**, 2788

Integrated Stellar Populations of Bulges: First Results

By SCOTT C. TRAGER[1], J. J. DALCANTON[2],
AND
B. J. WEINER[1]

[1]Carnegie Observatories, 813 Santa Barbara Street, Pasadena CA 91101, USA

[2]Department of Astronomy, University of Washington, Box 351580, Seattle WA 98195-1580, USA

We present first results from an on-going survey of the stellar populations of the bulges and inner disks of spirals at various points along the Hubble sequence. In particular, we are investigating the hypotheses that bulges of early-type spirals are akin to (and may in fact originally have been) intermediate-luminosity ellipticals while bulges of late-type spirals are formed from dynamical instabilities in their disks. Absorption-line spectroscopy of the central regions of Sa–Sd spirals is combined with stellar population models to determine integrated mean ages and metallicities. These ages and metallicities are used to investigate stellar population differences both between the bulges and inner disks of these spirals and between bulges and ellipticals in an attempt to place observational constraints on the formation mechanisms of spiral bulges.

1. Introduction

Current thinking considers two major pathways to the formation of spiral bulges. Simplistically, either the bulge formed before the disk ('bulge-first', e.g. van den Bosch 1998, and these proceedings), or formed *from* the disk ('disk-first', e.g. Combes & Sanders 1981). Previous studies have shown that bulges of big-bulge spirals (like M31) share at least some stellar population properties with mid-sized elliptical galaxies. They fall along the D_n–σ_0 relation (Dressler 1987) and the Fundamental Plane (Bender, Burstein & Faber 1992). Moreover, Jablonka *et al.* (1996) and Idiart *et al.* (1996) find that bulges of spirals (as late as Sc) fall along the Mg–σ_0 relation defined by early-type galaxies, suggesting that bulges share a mass-metallicity relation with elliptical galaxies. Together with kinematical evidence, these observations have led to the idea that mid-sized ellipticals accrete disks from some leftover gas reservoir, forming spiral galaxies (see, e.g. Kauffmann, White & Guiderdoni 1993). Therefore, the stars in spiral galaxy bulges should be similar to the stars in ellipticals (i.e. metal-rich, high [Mg/Fe]; e.g. Worthey, Faber & González 1992; Trager *et al.* 1998b) and unlike the stars in spiral galaxy disks.

Kinematical and surface brightness observations of small-bulge spirals indicate that their bulges share many properties with their disks: Many bulges in these spirals are better represented by a shallower exponential profile than by the steep de Vaucouleurs profile (e.g. de Jong 1996) and many of these bulges exhibit disk-like kinematics (see Kormendy 1993 for an excellent review). Moreover, Peletier & Balcells (1996) and de Jong (1996) find that the color difference between bulge and disk in an individual galaxy is much smaller than the variation in colors across galaxies of a single Hubble type.†
These observations have led these authors and others to propose that bulges (of small-bulge spirals at least) are formed from the stars already present in their underlying disks.

† However, very recent observations by Peletier & Davies (this meeting), using HST WFPC2 and NICMOS imaging of the Peletier & Balcells sample of early-type spirals, have shown that bulges are uniformly red, suggesting that the bulges are uniformly old and metal-rich. Clearly these issues are by no means settled with current observational data.

That is, the stars in spiral galaxy bulges should be similar to the stars in spiral galaxy disks (e.g. ages and metallicities of their inner disks, solar [Mg/Fe]) and unlike stars in ellipticals.

These two scenarios do have testable consequences, and sophisticated techniques now available can provide definitive answers to these questions. We have embarked on a multi-year survey to probe the stellar content of the bulges and inner disks of spirals using absorption-line strengths. These measurements can provide the critical tests needed to understand the basic mechanisms driving bulge formation.

2. The Stellar Populations of Spiral Bulges

2.1. *Sample and Observations*

We have selected a sample of 91 southern face-on spirals, ranging in type from S0/a to Sdm, barred and unbarred, from the ESO (B) survey. These spirals are not interacting, have major axes larger than $3'$, are within 4000 km s^{-1} and are located at $|b| > 20°$. The observations have been made using the long-slit Boller & Chivens spectrograph at Las Campanas Observatory in the blue, giving ≈ 2 Å ($\sigma_{\rm inst} \approx 35$ km s^{-1}) resolution in the 4000–5200 Å region. This spectral range covers portions of both the Lick/IDS (Burstein *et al.* 1984; Worthey *et al.* 1994; Trager *et al.* 1998a) and Rose (1985, 1994) absorption-line strength systems. To date, ten (primarily unbarred) spirals have been observed, most along both major and minor axes, and data from four have been processed. Line-strength profiles for these four galaxies have been calibrated onto the Lick/IDS system using stellar observations in common with Jones (1996), and velocity dispersion profiles and rotation curves have been derived following the Fourier-quotient procedure described by González (1993).

2.2. *The Mg–σ_0 Relation*

As described above, bulge-first formation mechanisms find support in the observation that bulges in galaxies as late as Sc fall along the same Mg$_2$–σ_0 relation as early-type galaxies (Jablonka et al. 1996; Idiart *et al.* 1996), suggesting the existence of a mass-metallicity relation in spirals and a narrow spread in ages at fixed σ_0. We confirm these observational results: our bulges fall on the Mg$_b$–σ_0 relation defined by early-type galaxies (our observations do not cover the red sideband of the broad Mg$_2$ index; see data in Trager *et al.* 1998a). However, the Mg–σ_0 relation is inherently *degenerate to compensating variations in metallicity and age in old stellar populations*—large age spreads can exist if there is a complementary age-metallicity relation (Worthey, Trager & Faber 1996; Trager 1997). This intrinsic age-metallicity degeneracy in Mg$_b$ and Mg$_2$ (and other metal lines and broadband colors) prevents the Mg–σ_0 relation from being an effective stellar population age discriminator. More sophisticated techniques are obviously necessary to distinguish between bulge formation mechanisms.

2.3. *Balmer-Line – Metal-Line Diagrams*

There is a way to break the age-metallicity degeneracy — Balmer lines are more age-sensitive than are metal-line indices (with the possible exception of the G-band; Worthey 1994). In the absence of nebular emission and of large numbers of blue horizontal-branch stars or blue stragglers, Balmer-line strengths reflect the light-weighted mean turnoff temperature of the composite stellar population—i.e., the mean age of the population. Balmer lines do have some intrinsic metal dependence, however, so a clean separation of age and metallicity effects requires diagnostic diagrams combining Balmer line and metal

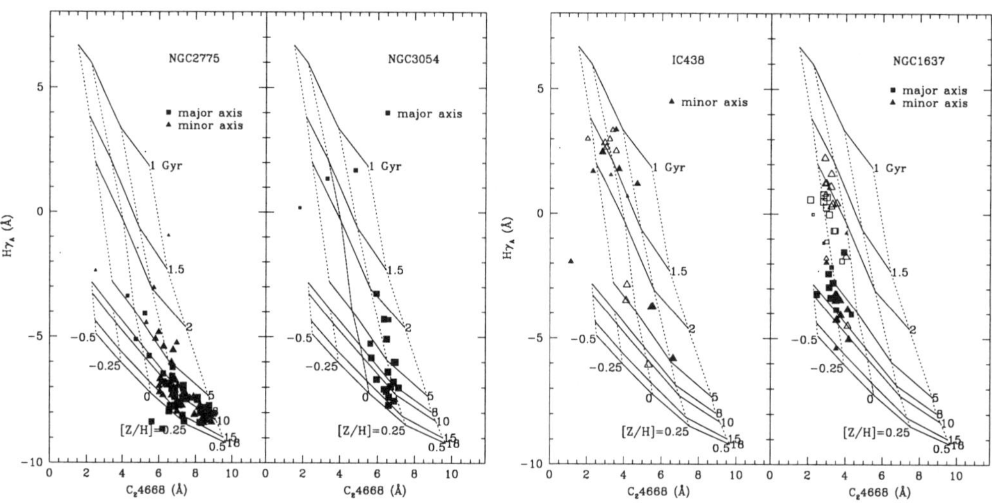

FIGURE 1. Balmer–metal-line diagrams for two Sa–Sb spiral bulges (NGC 2775 and NGC 3054, left panels) and two Sc bulges (IC 438 and NGC 1637, right panels). The Balmer line index $H\gamma_A$ is a sensitive measure of the presence of intermediate-age (1–10 Gyr) stars in an old population. The metal-line index C_24668 is a sensitive measure of the bulk metallicity of an old population. Although both indices are slightly sensitive to both age and metallicity, used together they can effectively measure the mean age and metallicity in an old stellar population (Worthey 1994; Worthey & Ottaviani 1997; Trager *et al.* 1998). Points are coded by the axis along which the slit was placed, and grow smaller going out from the center. Closed symbols are as observed; the open symbols are an attempt to correct for the emission fill-in in the $H\gamma_A$ index using an optimized stellar template (cf. González 1993). Lines represent models of Worthey (1994) and Worthey & Ottaviani (1997): solid lines are contours of constant age and dotted lines are contours of constant metallicity. The Sa/Sb bulges on the left are clearly quite old (10–15 Gyr) and metal-rich (metallicities as high or higher than solar) and are older than their disks by at least a few Gyr. In contrast, the Sc bulges seem younger than the Sa–Sb bulges (< 10 Gyr), and the bulge of NGC 1637 is nearly as young as its disk. Both Sc bulges are also slightly metal-poor (metallicities less than solar), suggesting that a mass-metallicity relation may exist for spiral bulges, as suggested by the Mg–σ_0 relation.

line strengths. Figure 1 presents such diagrams for the first four galaxies analyzed in our sample.

The Balmer/metal-line diagrams for large bulges (Sab–Sbc, $\sigma_0 > 100$ km s^{-1}) suggest that the bulges of these early-type spirals are consistent with having old, metal-rich stellar populations, as seen in our own galaxy ($t \gtrsim 10$ Gyr, [C/H] $\gtrsim 0$; e.g. McWilliam & Rich 1994; Bruzual *et al.* 1997). These bulges are also older than their inner disks by several Gyr, suggesting that these bulges formed early on in the galaxies' histories, and that bulge-first formation is a likely scenario for these galaxies.

On the other hand, small bulges (Sc, $\sigma_0 < 100$ km s^{-1}) seem to have younger and more metal-poor populations ($t \lesssim 10$ Gyr, [C/H] $\lesssim 0$), and at least in some galaxies (NGC 1637), the bulge and inner disk are roughly the same age. Such an observation provides strong support for a common origin of bulge and disk material—that is, for a disk-first formation scenario. However, contamination from emission is difficult to remove, separation of bulge and disk light has not yet been attempted, and small amounts of young or intermediate-age populations can significantly increase Balmer-line strengths (Trager *et al.* 1998b). In these late-type, star-forming, small-bulged galaxies, such effects may obscure truly old bulge populations.

3. Conclusions

The initial indications are that large bulges are genuinely old, metal-rich systems, as expected from our own galactic bulge. The stellar populations of small bulges appear to be younger and more metal-poor than the large bulges, but contamination from emission lines and disk light make these conclusions uncertain for the moment. If these results stand with more data and more extended analysis, we may find that bulges may have more than one formation mechanism—or even more than one mechanism may occur in a single galaxy.

We thank our collaborators Dr. R. O. Marzke and Dr. A. McWilliam for many stimulating conversations during the course of this work. SCT would like to thank the organizers for holding this very interesting workshop and for their financial support.

REFERENCES

BENDER, R., BURSTEIN, D., FABER, S.M. 1992 *ApJ*, **399**, 462

BRUZUAL, G., BARBUY, B., ORTOLANI, S., BICA, E., CUISINIER, F., LEJEUNE, T., SCHIAVON, R. P. 1997 *AJ*, **114**, 1531

BURSTEIN, D., FABER, S.M., GASKELL, C.M., KRUMM, N. 1984 *ApJ*, **287**, 586

COMBES, F., SANDERS, R.H. 1981 *A&A*, **96**, 164

DE JONG, R.S. 1996 *A&A*, **313**, 377

DRESSLER, A. 1987 *ApJ*, **317**, 1

GONZÁLEZ, J.J. 1993, Ph.D. Thesis, University of California, Santa Cruz

IDIART, T.P., DE FREITAS PACHECO, J.A., COSTA, R.D.D. 1996 *AJ*, **112**, 2541

JABLONKA, P., MARTIN, P., ARIMOTO, N. 1996 *AJ*, **112**, 1415

JONES, L.A. 1996, Ph.D. Thesis, Univ. North Carolina

KAUFFMANN, G., WHITE, S.D.M., GUIDERDONI, B. 1993 *MNRAS*, **264**, 201

KORMENDY, J. 1993, in *Galactic Bulges* (ed. H. Dejonghe & H.J. Habing), IAU Symposium **153**, p209. (Kluwer)

McWILLIAM, A., RICH, R.M. 1994 *ApJS*, **91**, 749

PELETIER, R.F., BALCELLS, M. 1996 *AJ*, **111**, 2238

ROSE, J.A. 1985 *AJ*, **90**, 1927

ROSE, J.A. 1994 *AJ*, **107**, 206

TRAGER, S.C. 1997, Ph.D. Thesis, UC Santa Cruz

TRAGER, S.C., WORTHEY, G., FABER, S.M., BURSTEIN, D., GONZALEZ, J.J. 1998a *ApJS*, **116**, 1

TRAGER, S.C., FABER, S.M., GONZÁLEZ, J.J., WORTHEY, G. 1998b, in preparation

VAN DEN BOSCH, F. 1998 *ApJ*, **507**, 601

WORTHEY, G. 1994 *ApJS*, **95**, 107

WORTHEY, G., FABER, S.M., GONZÁLEZ, J.J. 1992 *ApJ*, **398**, 69

WORTHEY, G., FABER, S.M., GONZÁLEZ, J.J., BURSTEIN, D. 1994 *ApJS*, **94**, 687

WORTHEY, G., TRAGER, S.C., FABER, S.M. 1996, in *Fresh Views on Elliptical Galaxies* (ed. A. Buzzoni, A. Renzini & A. Serrano), ASP Conf. Ser. **86**, p203. (ASP)

WORTHEY, G., OTTAVIANI, D.L. 1997 *ApJS*, **111**, 377

HST-NICMOS Observations of Galactic Bulges: Ages and Dust

By **REYNIER F. PELETIER** AND **ROGER L. DAVIES**

Dept. of Physics, University of Durham, South Road, Durham, DH1 3LE, UK

We present a study in B, I and H of a magnitude-limited sample of galactic bulges using WFPC2 and NICMOS. The high spatial resolution of HST allows us to study the dust contents near the center, and stellar populations in dust-free regions. We find extinction in 19/20 galaxies and infer an average central extinction of $A_V = 0.6 - 1.0$ mag. For galactic bulges of types S0 to Sb, the tightness of the $B - I$ vs $I - H$ relation suggests that the age spread among bulges of early type galaxies is small, at most 2 Gyr independent of environment. Comparison with stellar population models shows that the bulges are old. Colors at 1 bulge effective-radius, where we expect extinction to be negligible, suggest that all of these bulges formed around at the same time as bright galaxies in the Coma cluster.

1. Introduction

The formation of the central bulges of spiral galaxies is an unsolved facet of galaxy formation. There are currently two main scenarios for the formation of bulges: the classical picture (e.g. Eggen, Lynden-Bell & Sandage 1962), where bulge formation is described by collapse of a primordial gas cloud into clumps, which then merge together. The disk forms only after the last massive merger via gas infall. In the second scenario, the secular evolution of disks (e.g. Pfenniger & Norman 1990), a bulge is formed through dynamical instabilities in the disk, which first create a bar, and later a bulge. To distinguish between these models one needs to measure the ages of both bulges and disks. In Peletier & Balcells (1996, see also Terndrup *et al.* 1994), it was found that the ages of bulges and inner disks are very similar. Here we use optical and infrared colors to measure the ages of these bulges themselves, which is possible due to the high photometric accuracy of HST data, and the high spatial resolution, which allows us to separate extinction from stellar population effects.

2. Sample, Observations and Data Reduction

Twenty galaxies were observed with HST in Cycle 7 with WFPC2 (F450W and F814W) and NICMOS (F160W). They all are part of the original sample of Balcells & Peletier (1994), which is a complete, B-magnitude-limited sample of early-type spirals (type S0-Sbc) with inclinations larger than $50°$, and for which one side of the minor axis color profile is approximately featureless as seen from the ground. After correcting for the effects of different PSFs in the three bands, calibrated color maps and minor axis color profiles were determined, averaged azimuthally in wedges centered on the H-band nucleus, enabling us to obtain color profiles with an absolute accuracy of 0.02-0.03 mag. Details about the observations and data reduction can be found in Peletier et al. (1999).

3. Dust in the Centers of Bulges

In Figure 1 we show $B - I$ and $I - H$ color maps of 4 of the galaxies, superimposed on H-band contours. The images show black, dusty features near the center. We find that only 3 of the galaxies do not show nuclear dust features, although in some cases a

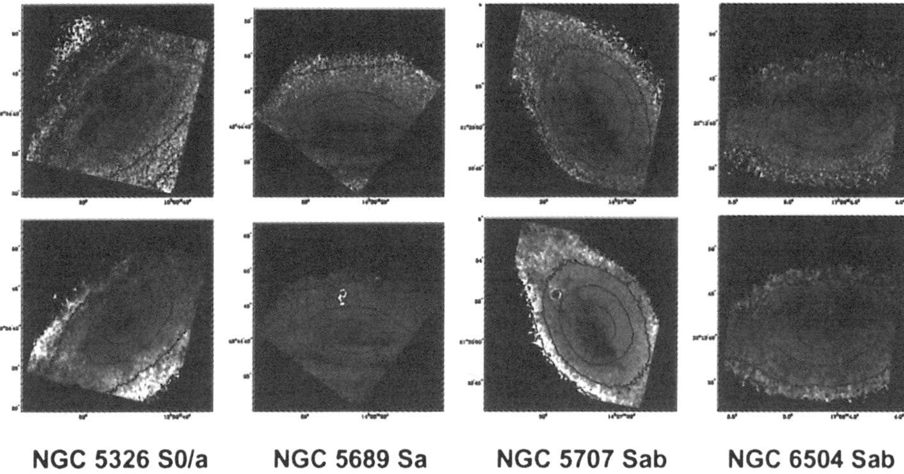

FIGURE 1. Color maps in $B - I$ (upper) and (I-H) of 4 typical galaxies, showing clearly their nuclear dust features. Image sizes are about $25'' \times 25''$.

large foreground dust lane makes nuclear dust difficult to detect. Detailed analysis of the isophotes shows that 2 of these 3 also have extinction. How much reddening is caused by the dust? In Figure 2 we show the colors at the center (filled dots) and at one bulge effective radius (open dots) in a $I - H$ vs. $B - I$ color-color diagram. Figure 2 (b) and (c) show that in all cases the galaxy is redder in the center than at $r_{\rm eff}$, sometimes by very large amounts. Since the vector indicating reddening by dust is almost parallel to the vector indicating changes in metallicity, it is not possible to say exactly how much of the reddening is due to extinction. Lower and upper limits to the internal extinction may be estimated by assuming that the galaxy at 1 $r_{\rm eff}$ is dust-free, and that the dereddened colors would never be redder than NGC 4472, a large nearby giant elliptical. Using the Galactic extinction law (Rieke & Lebofsky 1985) we find an average central extinction of A_V between 0.6 and 1 mag. This is similar to the situation in elliptical galaxies, where van Dokkum & Franx (1995) find evidence for central extinction in 75% of their sample, much more than is found in ground-based studies. The origin of this dust is not clear, but might well be internal, since the amount of dust found can be accounted for well by simple mass-loss from evolved stars in the galaxy (Faber & Gallagher 1976).

4. Stellar Populations and the Formation of Bulges

Having established that extinction at 1 $r_{\rm eff}$ is probably negligible, we will now analyze the colors of the bulges at this position in terms of stellar populations. The first thing to note is the small scatter amongst the open symbols in Figure 2a confirming that extinction is not important at 1 $r_{\rm eff}$. If we exclude the three galaxies with the latest Hubble type (the open crossed symbols in Figure 2c) the stellar populations at $r_{\rm eff}$ form a rather tight sequence in the $B - I$ vs. $I - H$ plane. In Figure 2a we also show a set of Single Stellar Population (SSP) models of Vazdekis *et al.* (1996), with a Salpeter IMF. We see that, independent of the amount of extinction, the tightness of the color-color relation shows that the luminosity-weighted age of the stars is very similar from bulge to bulge. Excluding the three Sbc galaxies the age-spread would be about 1-2 Gyr. We have

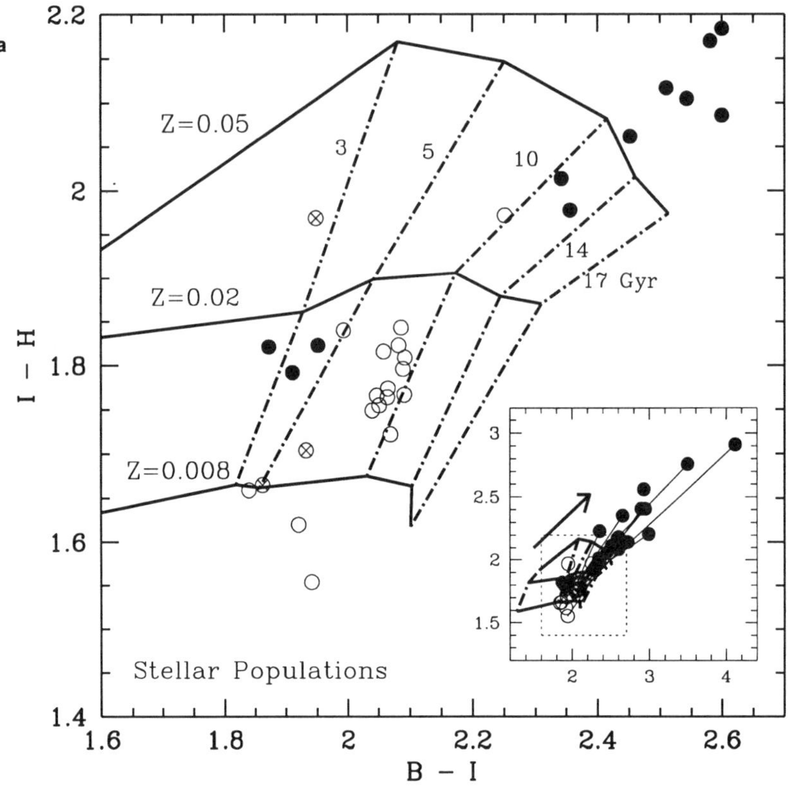

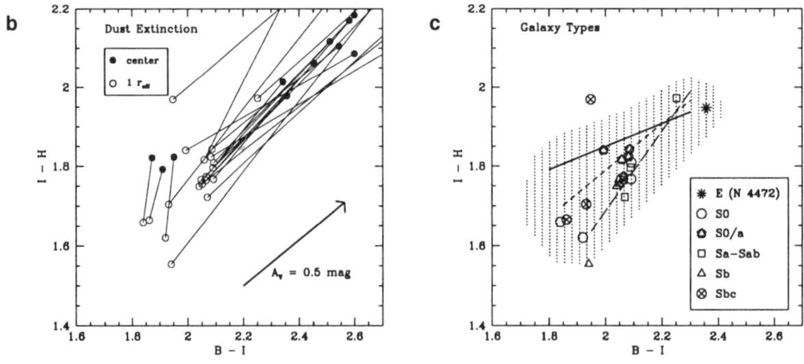

FIGURE 2. Color-color diagrams for the 20 galaxies. In Figure 2 (a) and (b) are displayed
the positions of the center (filled) and 1 bulge effective radius on the minor axis (open circles).
Reddenings vector for a reddening of $A_V = 0.5$ mag (2b) and 1 mag (inset of 2a) are given as
well. Superimposed in Figure 2a are SSP models by Vazdekis *et al.* (1996). Solid lines are
lines of constant metallicity, dashed-dotted lines are loci of constant age. In Figure 2c the same
galaxies at 1 effective radius are plotted, with their symbols coded as a function of morphological
type. Also added is NGC 4472. An attempt has been made to convert the data of Bower et al.
(1992) to $B - I$ and $I - H$, using the models of Worthey (1994, long-dashed), Vazdekis et al
(short-dashed) and an empirical calibration (solid line). Since these calibrations do not agree
very well with each other we can only say that the position of our bulges is consistent with the
colors of early-type galaxies in Coma.

tried to make a comparison with the colors of early-type galaxies in the Coma cluster, by attempting to convert the $V - K$ vs. $U - V$ relation of Bower *et al.* (1992) to our system. Since this is not straightforward (Figure 2c) we can only say that the colors of our bulges are very similar to those of the early-type galaxies in the Coma cluster.

What can we learn from these data about the formation of bulges? The fact that the ages of most of the bulges in this paper are so similar and old makes it very difficult that bulges of early-type spirals (S0-Sb) are formed through secular evolution of disks. In this scenario we expected bulges to undergo major bursts of star formation regularly, to convert gas that has been funneled to the central area through the presence of a bar into stars. This would mean that we would expect to find more young bulges and a large spread in bulge ages. Late type galaxies, starting from type Sbc, might form this way however. The fact that the colors of the majority of our bulges are similar to the early-type galaxies in Coma indicates that our bulges (except maybe the Sbc's) are also old, and formed at redshifts beyond $z = 3$. Since our ages do not depend on galaxy environment, the disks would have to form a number of Gyrs later to avoid conflicts with the star formation history of the universe (Madau *et al.* 1996). Our data are in agreement with the semi-analytic galaxy formation models of Baugh *et al.* (1996), who find that bulges everywhere in the universe are old. The situation in which a large fraction of the bulges is young would not be possible because of the time it takes to form their disks.

This research is based on data from the Hubble Space Telescope. We thank our collaborators Marc Balcells, Paco Prada, Andi Burkert, Alexandre Vazdekis and Yiannis Andredakis for their involvement in this project.

REFERENCES

BALCELLS, M., PELETIER, R.F. 1994 *AJ*, **107**, 135

BAUGH, C.M., COLE, S., FRENK, C.S. 1996 *MNRAS*, **282**, L27

BOWER, R.G., LUCEY, J.R., ELLIS, R.S. 1992 *MNRAS*, **254**, 589

EGGEN, O., LYNDEN-BELL, D., SANDAGE, A.R. 1962 *ApJ*, **136**, 748

FABER, S.M., GALLAGHER, J.S. III 1976 *ApJ*, **204**, 365

MADAU, P., FERGUSON, H.C., DICKINSON, M., ET AL. 1996 *MNRAS*, **283**, 1388

PELETIER, R.F., BALCELLS, M. 1996 *AJ*, **111**, 2238

PELETIER, R.F., BALCELLS, M., DAVIES, R.L., ANDREDAKIS, Y., VAZDEKIS, A., BURKERT, A., PRADA. F. 1999 *MNRAS*, submitted

PFENNIGER, D., NORMAN, C. 1990 *ApJ*, **363**, 391

RIEKE, G., LEBOFSKY, M.J. 1985 *ApJ*, **288**, 618

TERNDRUP, D.M., DAVIES, R.L., FROGEL, J.A., DEPOY, D.L., WELLS, L.A. 1994 *ApJ*, **432**, 518

VAN DOKKUM, P.G., FRANX, M. 1995 *AJ*, **110**, 2027

VAZDEKIS, A., CASUSO, E., PELETIER, R.F., BECKMAN, J.E. 1996 *ApJS*, **106**, 307

WORTHEY, G. 1994 *ApJS*, **95**, 107

Inside-Out Bulge Formation and the Origin of the Hubble Sequence

By FRANK C. van den BOSCH

Department of Astronomy, University of Washington, Box 351580, Seattle WA 98195, USA

Galactic disks are thought to originate from the cooling of baryonic material inside virialized dark halos. In order for these disks to have scalelengths comparable to observed galaxies, the specific angular momentum of the baryons has to be largely conserved. Because of the spread in angular momenta of dark halos, a significant fraction of disks are expected to be too small for them to be stable, even if no angular momentum is lost. Here it is suggested that a self-regulating mechanism is at work, transforming part of the baryonic material into a bulge, such that the remainder of the baryons can settle in a stable disk component. This inside-out bulge formation scenario is coupled to the Fall & Efstathiou theory of disk formation to search for the parameters and physical processes that determine the disk-to-bulge ratio, and therefore explain to a large extent the origin of the Hubble sequence. The Tully-Fisher relation is used to normalize the fraction of baryons that forms the galaxy, and two different scenarios are investigated for how this baryonic material is accumulated in the center of the dark halo. This simple galaxy formation scenario can account for both spirals and S0s, but fails to incorporate more bulge dominated systems.

1. Introduction

Despite considerable progress in our understanding of the formation of galaxies, the origin of the Hubble sequence remains a major unsolved problem. The main morphological parameter that sets the classification of galaxies in the Hubble diagram is the disk-to-bulge ratio (D/B). Understanding the origin of the Hubble sequence is thus intimately related to understanding the parameters and processes that determine the ratio between the masses of disk and bulge. Especially, we need to understand whether this ratio is imprinted in the initial conditions ('nature') or whether it results from environmental processes such as mergers and impulsive collisions ('nurture').

Here a simple inside-out formation scenario for the bulge (a 'nature'-variant) is suggested, and the differences in properties of the proto-galaxies that result in different disk-to-bulge ratios are investigated. A more detailed discussion on the background and ingredients of the models can be found in van den Bosch (1998; hereafter vdB98).

2. The Formation Scenario

In the standard picture of galaxy formation, galaxies form through the hierarchical clustering of dark matter and subsequent cooling of the baryonic matter in the dark halo cores. Coupled with the notion of angular momentum gain by tidal torques induced by nearby proto-galaxies, this theory provides the background for a model for the formation of galactic disks. In this model, the angular momentum of the baryons is assumed to be conserved causing the baryons to settle in a rapidly rotating disk (e.g. Fall & Efstathiou 1980). The turn-around, virialization, and subsequent cooling of the baryonic matter of a proto-galaxy is an inside-out process. First the innermost shells virialize and heat its baryonic material to the virial temperature. The cooling time of this dense, inner material is very short, whereas its specific angular momentum is relatively low. If the cooling time of the gas is shorter than the dynamical time, the gas will condense in clumps that form stars, and this clumpiness is likely to result in a bulge. Even if the low angular momentum material accumulates in a disk, the self-gravity of such a small, compact disk makes it violently unstable, and transforms it into a bar. Bars are efficient in transporting gas inwards, and can cause vertical heating by means of a collective bending instability. Both these processes lead ultimately to the dissolution of the bar; first the bar takes a hotter, triaxial shape, but is later transformed in a spheroidal bulge component. There is thus a natural tendency for the inner, low angular momentum baryonic material to form a bulge component rather than a disk. Because of the ongoing virialization, subsequent shells of material cool and try to settle into a disk structure at a radius determined by their angular momentum. If the resulting disk is unstable, part of the material is transformed into bulge material. This process of disk-bulge formation is self-regulating in that the bulge grows until it is massive enough to sustain the remaining gas in the form of a stable disk. I explore this inside-out bulge formation scenario, by incorporating it into the standard Fall & Efstathiou theory for disk formation.

The *ansatz* for the models are the properties of dark halos, which are assumed to follow the universal density profiles proposed by Navarro, Frenk & White (1997), and whose halo spin parameters, λ, follow a log-normal distribution in concordance with both numerical and analytical studies. I assume that only a certain fraction ϵ_{gf} of the available baryons in a given halo ultimately settles in the disk-bulge system. Two extreme scenarios for this galaxy formation (in)efficiency are considered. In the first scenario, named the 'cooling'-scenario, only the inner fraction ϵ_{gf} of the baryonic mass is able to cool and form the disk-bulge system: the outer parts of the halo, where the density is lowest, but which contain the largest fraction of the total angular momentum, never gets to cool. In the second scenario, referred to hereafter as the 'feedback'-scenario, the processes related to feedback and star formation are assumed to yield equal probabilities ϵ_{gf} for each baryon in the dark halo, independent of its initial radius or specific angular momentum, to ultimately end up in the disk-bulge system. The values of ϵ_{gf} are normalized by fitting the model disks to the zero-point of the observed Tully-Fisher relation. Recent observations of high redshift spirals suggest that the zero-point of the Tully-Fisher relation does not evolve with redshift. This implies that the galaxy formation efficiency ϵ_{gf} was higher at higher redshifts (see vdb98 for details). Disks are modeled as exponentials with a scalelength proportional to λ times the virial radius of the halo (as in the disk-formation scenario of Fall & Efstathiou). The bulge mass is determined by requiring that the disk is stable. Since the amount of self-gravity of the disk is directly related to the amount of angular momentum of the gas, the disk-to-bulge ratio in this scenario is mainly determined by the spin parameter of the dark halo out of which the galaxy forms.

3. Clues to the Formation of Bulge-Disk Systems

Constraints on the formation scenario envisioned above can be obtained from a comparison of these disk-bulge-halo models with real galaxies. A list of ~ 200 disk-bulge systems is adopted, including a wide variety of galaxies: both high and low surface brightness spirals (HSB and LSB respectively), S0, and disky ellipticals (see vdB98 for details). For each galaxy in this sample is calculated, after choosing a cosmology and a formation redshift, z, the spin parameter λ of the dark halo which, for the assumptions underlying the formation scenario proposed here, yields the observed disk properties (scale-length and central surface brightness). The formation scenario is thus used to link the *disk* properties to those of the dark halo. The known statistical properties of dark halos are used to discriminate between different cosmogonies.

The main results are shown in Figure 1, where are plotted the inferred values of λ

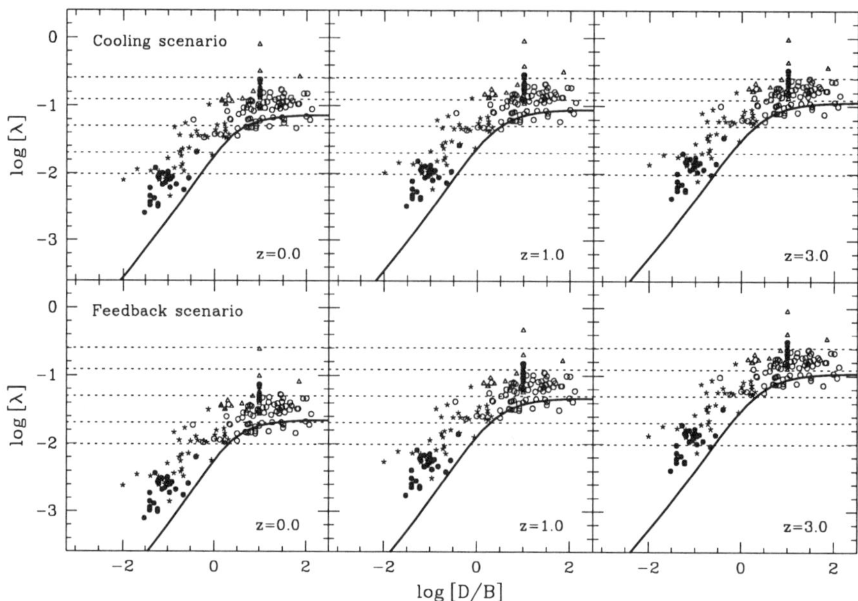

FIGURE 1. Results for a OCDM cosmology with $\Omega_0 = 0.3$. Plotted are the logarithm of the spin parameter versus the logarithm of the disk-to-bulge ratio. Solid circles correspond to disky ellipticals, stars to S0s, open circles to HSB spirals, and triangles to LSB spirals. The thick solid line is the stability margin; halos below this line result in unstable disks. As can be seen, real disks avoid this region, but stay relatively close to the stability margin, in agreement with the self-regulating bulge formation scenario proposed here. The dashed curves correspond to the 1, 10, 50, 90, and 99 percent levels of the cumulative distribution of the spin parameter. Upper panels correspond to the cooling scenario, and lower panels to the feedback scenario. Panels on the left correspond to $z = 0$, middle panels to $z = 1$, and panels on the right to $z = 3$.

versus the observed disk-to-bulge ratio for the galaxies in the sample. The dotted lines outline the distribution function of halo spin parameters of dark halos; it can thus be inferred what the predicted distribution of disk-to-bulge ratios is for galaxies that form at a given formation redshift. Results are presented for an open cold dark matter (OCDM) model with $\Omega_0 = 0.3$ and no cosmological constant ($\Omega_\Lambda = 0$). These results are virtually independent of the value of Ω_Λ, but depend strongly on Ω_0, which sets the baryon mass fraction of the Universe. Throughout, a universal baryon density of $\Omega_b = 0.0125\, h^{-2}$ is assumed, in agreement with nucleosynthesis constraints. The inferred spin parameters are larger for higher values of the assumed formation redshifts. This owes to the fact that halos that virialize at higher redshifts are denser. Since the scalelength of the disk is proportional to λ times the virial radius of the halo, higher formation redshifts imply larger spin parameters in order to yield the observed disk scalelength. In the cooling scenario, the probability that a certain halo yields a system with a large disk-to-bulge ratio (e.g. a spiral) is rather small. This is due to the fact that in this scenario most of the high angular momentum material never gets to cool to become part of the disk. The large observed fraction of spirals in the field renders this scenario improbable. For the feedback cosmogony, however, a more promising scenario unfolds: At high redshifts ($z \gtrsim 1$) the majority of halos yields systems with relatively small disks (e.g. S0s), whereas systems that form more recently are more disk-dominated (e.g. spirals). This difference owes to two effects. First of all, halos at higher redshifts are denser, and secondly, the redshift independence of the Tully-Fisher relation implies that ϵ_{gf} was higher at higher redshifts. Coupled to the notion that proto-galaxies that collapse at high redshifts are preferentially found in overdense regions such as clusters, this scenario thus automatically yields a morphology-density relation, in which S0s are predominantly formed in clusters of galaxies, whereas spirals are more confined to the field.

4. Conclusions

- Inside-out bulge formation is a natural by-product of the Fall & Efstathiou theory for disk formation.

- Disk-bulge systems do not have bulges that are significantly more massive than required by stability of the disk component. This suggests a coupling between the formation of disk and bulge, and is consistent with the self-regulating, inside-out bulge formation scenario proposed here.

- A comparison of the angular momenta of dark halos and spirals suggests that the baryonic material that builds the disk cannot loose a significant fraction of its angular momentum. This argues against the 'cooling scenario' in which most of the angular momentum remains in the baryonic material in the outer parts of the halo that never gets to cool.

- In a low-density Universe ($\Omega_0 \lesssim 0.3$), the only efficient way to make spirals is by assuring that only a relatively small fraction of the available baryons makes it into the galaxy, and that the probability that a certain baryon becomes a constituent of the final galaxy is independent of its specific angular momentum, as in 'feedback scenario'.

- If more extended observations confirm that the zero-point of the Tully-Fisher relation is independent of redshift, it implies that the galaxy formation efficiency ϵ_{gf} was higher at earlier times. Coupled with the notion that density perturbations that collapse early are preferentially found in high density environments such as clusters, the scenario presented here then automatically predicts a morphology-density relation in which S0s are most likely to be found in clusters.

- A reasonable variation in formation redshift and halo angular momentum can yield approximately one order of magnitude variation in disk-to-bulge ratio, and the simple formation scenario proposed here can account for both spirals and S0s. On the other hand, more bulge-dominated systems, i.e., disky ellipticals, have too large bulges and too small disks to be incorporated in this scenario. Apparently, their formation and/or evolution has seen some processes that caused the baryons to loose a significant amount of their angular momentum. Merging and impulsive collisions (e.g. galaxy harassment) are likely to play a major role for these systems.

It thus seems that both 'nature' and 'nurture' are accountable for the formation of spheroids, and that the Hubble sequence has a hybrid origin.

Support for this work was provided by NASA through Hubble Fellowship grant HF-01102.11-97.A awarded by the Space Telescope Science Institute, which is operated by AURA for NASA under contract NAS 5-26555.

REFERENCES

FALL, S.M., EFSTATHIOU, G. 1980 *MNRAS*, **193**, 189

NAVARRO, J.F., FRENK, C.S., WHITE, S.D.M. 1997 *ApJ*, **490**, 493

VAN DEN BOSCH, F.C. 1998 *ApJ*, **507**, 601

Part 3

THE TIMESCALES OF BULGE FORMATION

PART 3: THE TIMESCALES OF BULGE FORMATION

Several observational constraints and plausible scenarios for the formation of bulges are reviewed in this chapter, with particular emphasis given to the timescales involved in making and/or assembling a significant fraction of the stars which ultimately end up in the disk-embedded central spheroids.

Rich opens this excursion into the timescales of bulge formation by summarizing the current knowledge and understanding of this issue as derived from studing the stellar population properties of the bulges of the Milky Way and M31. Carlberg discusses an empirical model in which 'classical' bulges are assembled via gravitational accretion of visible satellite galaxies, and suggests that merger-induced starbursts may generate galactic winds capable of self-regulating this accretion. Quite interestingly, the resulting bulges are found to align along the observed Kormendy relation between bulge density and size. Elmegreen supports a scenario where the bulge formation event is possibly associated with a very powerful (three-dimensional) starburst. However, in contrast with Carlberg's view, he argues against the self-regulation of the bulge-forming starburst event via galactic winds, based on the consideration that the bulge potential may be too deep to allow significant blow-out. Kuijken takes a broader perspective and analyses the different timescales associated with several dynamical processes possibly relevant to bulge formation (including accretion, bar instabilities, stochastic heating, and environmental perturbations). Although several relatively long-lived processes may contribute at some level to the formation of bulges, Kuijken stresses that all the relevant large scale dynamical phenomena will occur on the very short crossing time (which sets the timescale for e.g., the monolithic collapse of a gas cloud into a bulge, or for dynamical instabilities such as bar formation/thickening). It is encouraging that the global picture emerging from the complementary perspectives presented here indicates a growing general consensus for ('classical') bulges having built a significant fraction of their stellar mass on rapid timescales, possibly even shorter than 1 Gyr.

Constraints on the Bulge Formation Timescale from Stellar Populations

By R. MICHAEL RICH

Department of Physics and Astronomy, UCLA, Los Angeles, CA 90095-1562, USA

Within recent years, there has been a confluence of data that favors a large age for the bulges of the Milky Way and M31. A short formation timescale is required by the similarity in ages between the bulge and the old, metal-rich globular clusters. Detailed abundances of bulge giants are consistent with a short enrichment timescale. The bulge of M31 is similarly old and even more metal-rich than the Galactic bulge. There appears to be a strong connection between the M31 bulge and the halo, as metal-rich giants are found in M31 out to great distances. The stellar populations data support a rapid bulge formation timescale, perhaps even less than 1 Gyr.

1. Introduction

"We must conclude, then, that in the central region of the Andromeda Nebula we have a metal-poor Population II, which reaches -3^m for the brightest stars, and that underlying it there is a very much denser sheet of old stars, probably something like those in M67 or NGC 6752. We can be certain that these are enriched stars, because the cyanogen bands are strong, and so the metal/hydrogen ration is very much closer to what we observe in the Sun and in the present interstellar medium than to what is obwserved for Population II. And the process of enrichment probably has taken very little time. After the first generation of stars has formed, we can hardly speak of a 'generation', because the enrichment takes place so soon, and there is probably very little time difference. So the CN giants that contribute most of the light in the nuclear regions of the Nebula must also be called old stars; they are not young."

– Walter Baade, 1963, in *The Evolution of Galaxies and Stellar Populations*, Harvard University Press: Cambridge, p256.

The above paragraph was written nearly 40 years ago, and it quite nicely expresses the conclusions of this paper, and it also encapsulates many of the conclusions found by Renzini at this meeting. One can ask why there has been any doubt that bulges are ancient and were formed early. There are strong observational and theoretical reasons why this view has been challenged over the years, and why there continues to be spirited debate over the age and formation of the bulge. Although the statement is approximately correct, the observational data are not yet good enough to answer the question 'What was the timescale for forming half of the bulge's mass?' that I have been asked to address. I will try to suggest the roadmap to getting answers of the desired precision.

2. Questioning Early Bulge Formation

The central bulge and nucleus is 100 times closer to us than the Andromeda nebula, the nearest other example of such a stellar population. However, our perspective on the bulge suffers from our being embedded in the Milky Way. We must contend with foreground stars scattered at a variety of distances, lying in the foreground disk. Reddening varies from so opaque that obscures the K band (near the nucleus) to being an annoying presence over most of the bulge. On top of this we may add the uncertainty in geometry that comes from the perspective of a single viewing angle. The most secure method of measuring the age of a stellar population is the absolute magnitude of the main sequence turnoff. Photometric crowding and distance will make this a challenging problem in the M31 bulge, perhaps tackled by some future interferometer or maximum-aperture telescope.

Observing the Galactic bulge, we have the difficulty that the foreground veil of the disk main sequence lies in the approximate location of the color-magnitude diagram (CMD) where one expects any young stellar population to be found. Differential reddening and the spatial depth of the bulge broaden the CMD at the turnoff point; finally, there is the wide abundance range of the stars (McWilliam & Rich 1994). If one wants to answer the question of how long it took to form half the mass of the bulge, one must determine the age for the innermost obscured regions. The high surface brightness, metal-rich spheroid population is mostly concentrated within 500 pc of the nucleus, with a vertical scaleheight of approximately 300 pc (Zhao 1996). These are the sightlines most affected by heavy foreground obscuration, and the regions of highest surface brightness where turnoff photometry will be most challenged by crowding.

2.1. *Why an Age Range is Expected: Observations*

2.1.1. *The Central Parsec*

Advances have, of course, occurred since Baade's time. The Galactic Center has been revealed to be a rich stellar population with a wide age range. In particular, most of the luminosity in the central parsecs is due to a huge starburst that is occurring both in the central star cluster, and more widely (Morris & Serabyn 1996). Among the most spectacular indications of this starburst are two massive young ($< 10^7$ yr old) clusters, the Arches and Quintuplet clusters. Presuming that we are not observing this event at some privileged point in time, this activity must have formed substantial mass over a Hubble time. For example, even if the average star formation rate were $0.1 M_\odot \mathrm{yr}^{-1}$ one could produce $10^9 M_\odot$, or a large fraction of the central 500pc. The direct spectroscopic confirmation of Wolf-Rayet stars in the center (cf. Figer *et al.* 1998) and the top-heavy mass function of the Arches cluster (Figure 1 and Figer *et al.* 1999) leaves little doubt that massive star formation is currently in progress. The central black hole and giant molecular clouds are among the means that stars formed in the center could easily be scattered great distances from the nucleus. The logic of how this star formation activity could build the bulge of the Milky Way (and by extension, other spirals) is very hard to ignore.

Because of the obscuration and crowding of images, very little is known about the low end of the initial mass function (IMF) in the Galactic Center. We know that the IMF of the Arches cluster is top-heavy (Figer *et al.* 1999) and that isolated stars such as the Pistol star may exceed 100 $M_\odot$ (Figer *et al.* 1999). There is no observational evidence concerning the low-mass end of the IMF, and therefore it is possible that the star formation activity is not memorialized by long-lived low mass stars. A 4 Myr old cluster (The Quintuplet) barely appears bound and deep imaging does not reveal a population of Solar mass stars clearly associated with the cluster. Rather than invoking a peculiar IMF, the preferred explanation for the apparent demise of Quintuplet is the large tidal field near the Galactic Center, which accelerates dynamical evolution (Kim *et al.* 1999) and would likely strip low mass stars from the cluster preferentially. Perhaps this tidal field also interferes with the formation of low mass stars. The question of whether the observed star formation activity at the center is capable of contributing long lived stars that could build the mass of the bulge is of critical importance. Partially answering that question will probably require deep interferometric imaging of the Arches cluster.

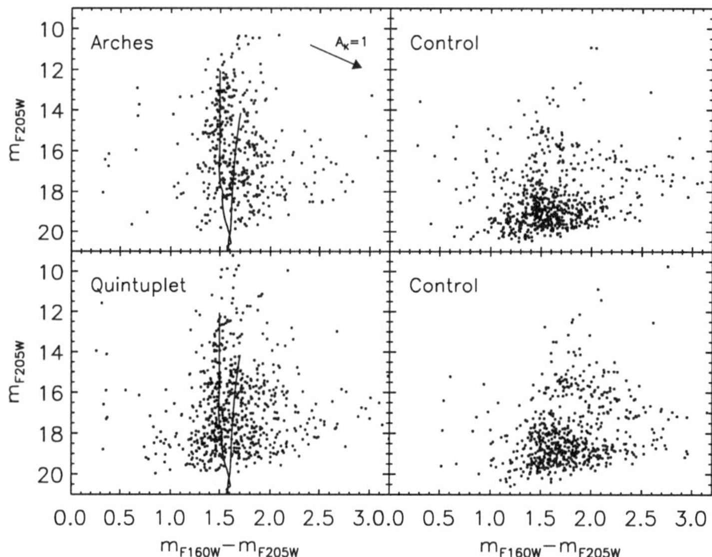

FIGURE 1. Color-magnitude diagram of Galactic Center clusters and adjacent field populations. Left panels illustrate the CMD of the clusters, while the right panels show the field population. The Arches cluster is a compact massive star cluster with age 2 Myr, while the Quintuplet cluster is far less concentrated and may be on the way to tidal disruption. Isochrones (Solar metallicity) are indicated are for the zero-age main sequence and (short vertical line on the righ) a 10 Gyr old population. The fully developed main sequence is clearly evident in the Arches CMD, while the field population CMD's clearly show the red clump. Notice that the clump stars form a locus extended down toward the lower right.

2.1.2. *Spatially Extended Populations of AGB Stars*

Evidence for a widespread intermediate age population in the bulge is based exclusively on identifying luminous asymptotic giant branch stars with intermediate-age turnoff progenitors. The most luminous stars of any kind in a > 1Gyr old stellar population are the dual shell-source giants on second ascent, the AGB stars. Bolometric luminosities can reach $10^4 L_\odot$; discovery of these stars in Magellanic globular clusters confirmed the intermediate-age nature of those populations. For metal-rich populations, the interpretation of these luminous AGB stars is problematic. Metal-rich AGB stars can reach $M_{bol} < -5$, 1.5 mag above the RGB tip, yet still be old stars (Elias & Frogel 1988). In what remains to this day a landmark map of the central few degrees, Catchpole *et al.* (1990) show that the most luminous AGB stars are also the most centrally concentrated, showing a tendency toward some flattening in their spatial distribution as well. Within the central 10 pc, Haller (1992) and Narayanan *et al.* (1996) both argue for a Galactic center AGB extended to $M_{bol} = -6$, a luminosity that inescapably demands intermediate-age progeny.

Perhaps the most persistent problem that requires an intermediate-age population has been the presence of large numbers of luminous Mira variables with periods exceeding 300 days, the approximate upper limit for *bona fide* Population II. Frogel & Whitelock

(1997) argue that the longest period Miras must arise from the metal-rich population. However, in the Solar vicinity such stars are associated (kinematically) with the disk population. As AGB stars and Mira variables will be the only individual stars easily imaged in extragalactic bulges, it would be worthwhile to settle some of these questions. For example, the concepts for a next-generation space telescope include high spatial resolution infrared imaging, ideal for measuring the luminosity function of the AGB in distant galaxies (potentially much beyond 10 Mpc). If we had the ability to interpret the AGB luminosity function, we might be able to constrain age and metallicity in these populations.

Habing and collaborators have measured the kinematics and physical properties of the population of OH/IR stars, finding their kinematics consistent with membership in the bulge. The study of the SiO maser population by Izumiura *et al.* (1995) reaches similar conclusions. The conventional wisdom argues that high metallicity pushes the AGB population toward the oxygen-rich OH/IR stars (as opposed to carbon stars, which are extremely rare) but that these objects are the progeny of an old, metal-rich stellar population.

Discussion of the candidate intermediate-age stars would not be complete without a discussion of the carbon stars. The early surveys by Blanco (1988) showed that these stars are extremely rare, less than 1 in 1000 among the evolved stars. Had Blanco surveyed in the more obscured northern portions of the bulge he might have discovered the thermally pulsing AGB carbon stars associated with the Sagittarius dwarf spheroidal galaxy. Azzopardi *et al.* (1991) discovered 33 carbon stars in various bulge fields; these are early R stars (Tyson & Rich 1991). The origin of the low-luminosity early R carbon stars is an unsolved problem, but these stars are *not* members of the Sgr dwarf nor are they the progeny of a widespread population of intermediate-age stars.

3. Why an Age Range is Expected: Theory

The analysis of early infrared maps of the nuclear region (Blitz & Spergel 1991) showed that observed asymmetries were consistent with the bulge actually being in the shape of a bar whose major axis points nearly along our line of sight. Analysis of the COBE maps and a self-consistent dynamical model (Zhao 1996) also support the existence of a triaxial bulge. The consensus viewpoint is that the bulge is in fact a classical bar.

Bars can be destroyed by the growth of a central point mass, so if the Galactic nucleus has been accreting material as discussed above, it becomes difficult to understand how the bar has survived for a Hubble time. In a rapidly rotating bar (as appears to be the case for the Milky Way) there are orbit families available that can sustain the shape of the bar, yet avoid scattering from a central mass.

One of the most promising scenarios to form the bar would somewhat favor a bulge population that is younger than the oldest stars. The idea is that a massive stellar disk could evolve into a bar through global instabilities (cf. Merritt & Sellwood 1994). Modeling of this process does succeed in producing structures having the vertical scaleheight of the Galactic bulge. Although it is frequently argued that this model would require some age range in the stellar population since the bulge begins as a massive disk, this would only be the case if the mass in the nucleus were large enough to destroy the bar once it formed. This model has frequently been advanced as one of the reasons why the bulge might be intermediate age. The measurement of exponential rather than $r^{1/4}$ surface brightness profiles is not necessarily proof that this mechanism is at work, nor is there any requirement that disks must be younger than spheroids simply because they are disks. The idea that bulges may form in this way is interesting and it would be interesting to see more observational tests.

4. Evidence for an Ancient Bulge

4.1. *Turnoff Photometry*

Given the spatial depth of the bulge, differential reddening, and complicating foreground populations, the best approach to constraining the age of the bulge field population is to compare its luminosity function with that of the old, metal-rich globular clusters. Globular clusters in the central region have the advantage that they are simple stellar populations with the spatial distribution (Barbuy *et al.* 1998), kinematics (Coté *et al.* 1999) and chemistry (Barbuy *et al.* 1999) of the bulge field stars. Ortolani *et al.* (1995) show that the luminosity function of the bulge field population and the metal-rich globular clusters NGC 6553 are identical. In turn, the age-sensitive $\Delta V_{TO}^{HB} = 3.6$mag in the bulge, just as large as is found for the oldest known clusters in the Galaxy. This work is described in more detail in Renzini's contribution.

Although the work of Kiraga *et al.* (1997) suggests that in at least some fields there are bright turnoff stars, we argue that the aforementioned problems make it very difficult to fit isochrones to the CMD, or to work to close the frame limit. The population of apparent bright turnoff stars found by Kiraga *et al.* could be caused by photometric crowding (cf. Renzini 1998) or field blue stragglers, some of which must be present. If such a population were both widespread and contained substantial mass, it would have clearly been detected in the Ortolani *et al.* (1995) study.

Another wide area survey explores this population of bright, turnoff-like stars in another way. Feltzing & Gilmore (1998) find that the number of stars brighter than the turnoff is consistent with their arising in a foreground screen (the disk population), rather than being related to the old turnoff stars. Again, if Kiraga *et al.* were correct and there was a widespread intermediate age population, one would expect it to appear more prominently in this kind of survey.

The presence of the Sgr dwarf population reminds us that there are genuine cases of mixed stellar populations toward the bulge, and it is likely that there is an age range. We simply argue that the vast majority of the stars are very old.

4.2. *Abundances*

Abundances of old stars can constrain the timescale for metals to be enriched. The relative abundances of alpha-elements such as O, Mg, and Ti relative to iron reflect the contribution of massive star supernova (10^6 yr timescale) versus that of the Type I (white dwarf) supernovae (10^8 yr timescale). The original work of McWilliam & Rich (1994) finds that some alpha-elements are enhanced (Mg and Ti) while others (Si and Ca) are not. With much better spectral resolution and S/N from 8-10m telescopes, we will soon be able to specify better the chemistry of bulge stars. In turn, these new data will constrain the timescale of formation for the bulge. Combining the turnoff photometry and constraints from abundance ratios, it may be possible to constrain the formation timescale of the bulge, and perhaps the IMF of the proto-bulge.

4.3. *The Galactic Center Region*

In the very area where a substantial number of intermediate age stars are expected, we obtained deep NICMOS images of the inner 20 pc. Figure 1 shows the Arches cluster and a comparison field. The red clump is very clear, and it is even more evident in other fields that we have imaged. The magnitude difference between the red clump and

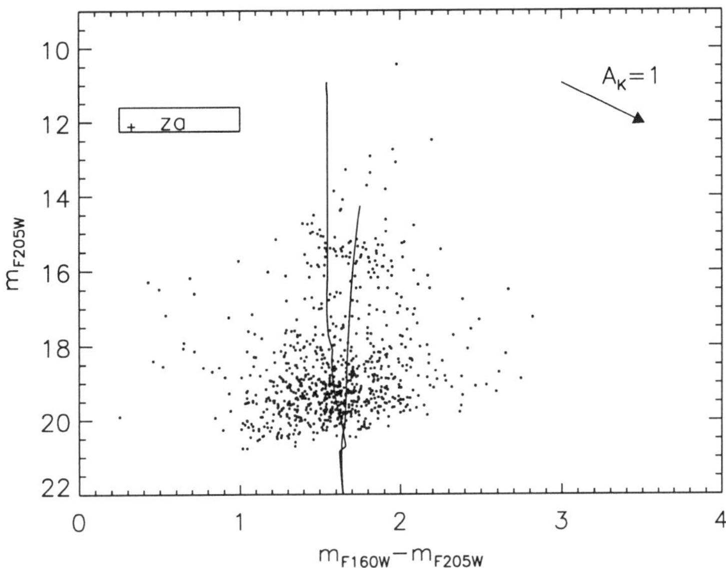

FIGURE 2.　　　Color-magnitude diagram of a field in the Galactic nucleus, at $(l, b) = (0^{\circ}.144, -0^{\circ}.055)$ in the 1.6 micron = m(160) and 2.1 micron = m(205) bands. Notice the fully developed red clump and the turnoff point, which is marked by large numbers of stars. The magnitude difference between the turnoff point and horizontal branch is 3.7 mag, as large as that found in the oldest known globular clusters. The isochrones correspond to a zero-age main sequence and 10 Gyr old population.

the turnoff, ΔV_{TO}^{HB}, is a useful age indicator in old stellar populations. Although the precise calibration is not yet secure, it is well known that the oldest globular clusters have ΔV_{TO}^{HB}=3.5 mag. The most metal-rich globular clusters currently studied, NGC 6553 and 6528, and the field population in the bulge (Baade's Window, at 500 pc from the nucleus) have $\Delta V_{TO}^{HB} = 3.5$ mag. Based on the large amount of ongoing star forming activity in the Galactic center, it is been assumed that the mass has been built by ongoing star formation over a Hubble time. Indeed, the extended luminosity functions of late type stars in the Galactic center region are consistent with at least some intermediate age stars being present there.

Quite surprisingly, we have found that ΔV_{TO}^{HB} is large even in fields less than 50 pc from the Galactic center (Figure 2). An intermediate age population is obvious in only one of our fields, 8 pc from the nucleus (Figure 3). For the first time, we show that the majority of the mass in the nuclear region is comprised by stars > 10 Gyr old.

In the Galactic Center, we find the most luminous star currently known, the Pistol Star, and much other strong evidence of ongoing star forming activity that, with even a modest rate, should have formed easily the total mass of the nuclear region over a Hubble time (§2.1.1). The prominent red clump suggests that while there has been star formation, most of the mass is in an old, underlying stellar population that formed at the same time as the oldest star clusters in the Galaxy. Perhaps the star formation process at the Galactic Center favors formation of the most massive stars, but does not lock up

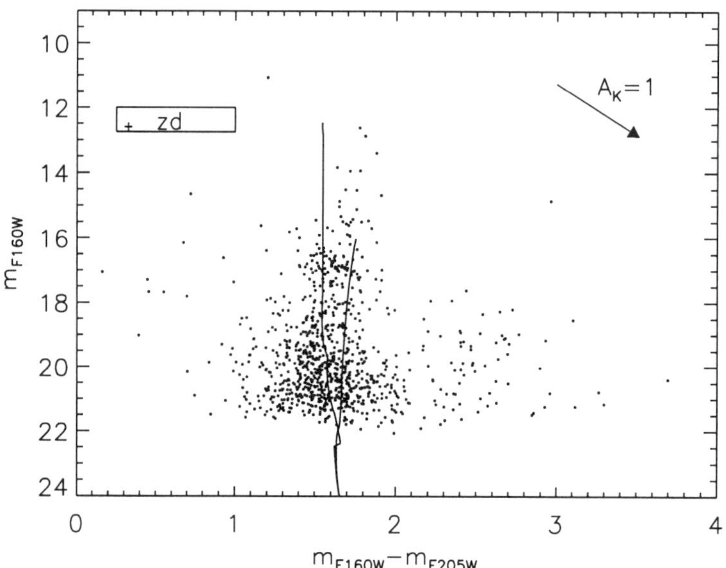

FIGURE 3. Color-magnitude diagram of a field in the Galactic nucleus, at $(l, b) = (0°.044, -0°.046)$ in the 1.6 micron = m(160) and 2.1 micron = m(205) bands. This field is about 8pc from the nucleus. Notice that the gap between the red clump and the turnoff point (which is clear in Figure 2) is filled in with stars which are presumably young and intermediate age main sequence stars. Isochrones as in Figure 2.

mass in the lowest mass stars. Given the tidal forces capable of disrupting star clusters quickly, perhaps before their low mass stars form, the star formation in the nuclear region may produce massive, short-lived stars, but not actually contribute significantly to the mass that forms the nucleus.

5. The Bulge of M31

Because the Milky Way nucleus is 100 times closer than M31, it is possible to reach the main sequence turnoff point, even in the most crowded regions. A 100-m diffraction-limited telescope could succeed in reaching the turnoff in the M31 nucleus, and such an instrument is theoretically possible. For the moment, we must be satisfied with available constraints from photometry of the giants, and integrated spectra. The integrated light of the M31 bulge is as red as that of the most luminous Virgo ellipticals (Sandage, Becklin, & Neugebauer 1969). The M31 bulge is an excellent template for bulge/elliptical populations in general.

From the luminosities of the M31 bulge AGB stars, we may be able to infer whether there is a subpopulation of stars younger than 10 Gyr. There are conflicting reports on this issue: Rich *et al.* (1993) claim that there is a widespread population of luminous AGB stars detected in the IR; some luminous stars are claimed by Rich & Mighell (1995) to have been detected with pre-repair WFPC imaging. These conclusions may be supported by Davidge *et al.* (1997), who use adaptive optics corrected imaging to

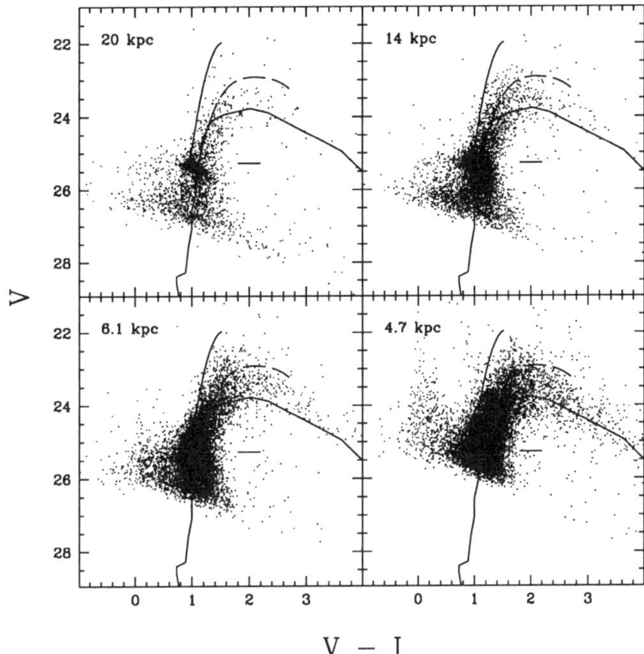

FIGURE 4. Color-magnitude diagrams for field stellar populations in the M31 halo. Giant branch loci have [Fe/H]=−1.6, −0.7, and approximately Solar (in order of descending giant branch tip luminosity). Notice that metal-rich giants are found even in halo fields 20 kpc from the nucleus. The formation of the M31 nuclear/bulge region may have distributed metals over this huge volume.

detect luminous giants in the M31 bulge. However, Renzini (1998) points out that even with high angular resolution, an extremely crowded field such as a bulge population can produce apparently luminous stars which are actually luminosity fluctuations. This effect can produce surprisingly luminous populations of 'intermediate-age AGB stars'. Recently, Jablonka *et al.* (1999) have imaged M31 bulge fields using WFPC2 and find a conventional metal-rich giant branch with a descending slope, similar to that of NGC 6553. They do not image within 100 pc of the nucleus, but they find no evidence for luminous AGB stars. We believe that had a widespread population of intermediate age stars existed, these WFPC2 images would have detected it. As there is little room in either the stellar photometry or integrated light for such a population, we believe that the bulge of M31 is also old and metal-rich, as appears to be the case with the Milky Way bulge. The bulge of M31 formed early and rapidly. However, the ultimate proof will come when and if the turnoff point can be measured in that population.

It is remarkable that the M31 halo has high metallicity (cf. Rich *et al.* 1996). Figure 4 shows that even 20 kpc from the nucleus, stars as metal-rich as ten times Solar abundance are found. How did the metals reach such great distances? One possibility is that the formation of the bulge produced a metal-enriched wind capable of enriching the large volume. Spectroscopy of high redshift galaxies (Steidel *et al.* 1996) shows the P Cygni profiles and classic signatures of the winds (and metal absorption lines) in the stellar continuum. However, it is not entirely clear how to move the metal-enriched material out to these great distances in the halo and get to subsequently form stars. M31 is also not alone in having a metal-rich halo. NGC 3115 (Elson *et al.* 1997) and NGC 5128 (Soria *et al.* 1996; Harris *et al.* 1999) also have significant populations of metal-rich stars in their outer halos. It will be important to increase the sample of halo populations studied, and to explore how spheroid luminosity and halo metallicity are correlated.

6. Conclusions

There is growing evidence that the Galactic bulge is ancient. Based on the age-sensitive magnitude difference between the horizontal branch and the turnoff point, the bulge appears to be as old as the oldest globular clusters in the Galaxy. Most, if not virtually all of the mass of the bulge formed in less than 1 Gyr. It appears as if even the nuclear region of the Milky Way is also largely dominated by an old stellar population. The current mass of the bulge is of order $10^{10} M_\odot$. The implied star formation rates required to form this much mass are of order 10-100 $M_\odot$ yr^{-1}, in good agreement with observed star formation rates in high redshift galaxies. Future space missions may permit more precise constraints from actual turnoff photometry in the nuclear region of the Milky Way and eventually M31. Such measurements might drastically change our views of galaxy formation.

REFERENCES

AZZOPARDI, M., REBEIROT, E., LEQUEUX, J., WESTERLUND, B.E. 1991 *A&AS*, **88**, 265

BARBUY, B., BICA, E., ORTOLANI, S. 1998 *A&A*, **333**, 117

BARBUY, B., RENZINI, A., ORTOLANI, S., BICA, E., GUARNIERI, M.D. 1999 *A&A*, **341**, 539

BLANCO, V.M. 1988 *AJ*, **95**, 1400

BLITZ, L., SPERGEL, D.N. 1991 *ApJ*, **370**, 205

CATCHPOLE, R.M., WHITELOCK, P.A., GLASS, I.S. 1990 *MNRAS*, **247**, 479

COTE, P. 1999 *AJ*, in press

DAVIDGE, T.J., RIGAUT, F., DOYON, R., CRAMPTON, D. 1997 *AJ*, **113**, 2049

ELIAS, J., FROGEL, J. 1988 *ApJ*, **324**, 823

FELTZING, S., GILMORE, G. 1999, in preparation

FIGER, D.F., NAJARRO, F., MORRIS, M., MCLEAN, I.S., GEBALLE, T.R., GHEZ, A.M., LANGER, N. 1998 *ApJ*, **506**, 384

FIGER, D.F., KIM, S., MORRIS, M., SERABYN, E., RICH, R.M., MCLEAN, I.S. 1999 *ApJ*, in press

FROGEL, J.A., WHITELOCK, P.A. 1998 *AJ*, **116**, 754

HALLER, J. 1992, Ph. D. Thesis, University of Arizona

HARRIS, G.L.H., HARRIS, W.E., POOLE, G.B. 1999 *AJ*, **117**, 855

IZUMIURA, H., DEGUCHI, S., HASHIMOTO, O., NAKADA, Y., ONAKA, T., ONO, T. UKITA, N., YAMAMURA, I. 1995 *ApJ*, **453**, 837

JABLONKA, P., ET AL. 1999 *ApJL*, in press

KIM, S., MORRIS, M. 1999, in preparation

KIRAGA, M., PACZYNSKI, B., STANEK, K.Z. 1997 ApJ, **485**, 611

MCWILLIAM, A., RICH, R.M. 1994 *ApJS*, **91**, 749

MERRITT, D., SELLWOOD, J.A. 1994 *ApJ*, **425**, 551

MORRIS, M., SERABYN, E. 1996 *ARA&A*, **34**, 645

NARAYNAN, V.K., GOULD, A., DEPOY, D.L. 1996 *ApJ*, **472**, 183

ORTOLANI, S., RENZINI, A., GILMOZZI, R., MARCONI, G., BARBUY, B., BICA, E., RICH, R.M. 1995 *Nature*, **377**, 701

STEIDEL, C.C., GIAVALISCO, M., PETTINI, M., DICKINSON, M., ADELBERGER, K.L. 1996 *ApJ*, **462**, 17

RENZINI, A. 1998 *AJ*, **115**, 2459

RICH, R.M., MIGHELL, K.J. 1995 *ApJ*, **439**, 145

RICH, R.M., MIGHELL, K.J., NEILL, J.D. 1996, in *Formation of the Galactic Halo, Inside and Out* (ed. H. Morrison & A. Sarajedin), ASP Conf. Ser. **92**, p544. (ASP)

SANDAGE, A.R., BECKLIN, E.E., NEUGEBAUER, G. 1969 *ApJ*, **157**, 55

SORIA, R., ET AL. 1996 *ApJ*, **465**, 79

TYSON, N.D., RICH, R.M. 1991 *ApJ*, **367**, 547

ZHAO, H.S. 1996 *MNRAS*, **282**, 1223

Bulge Building with Mergers and Winds

By RAY G. CARLBERG

Department of Astronomy, University of Toronto, 60 St. George Street, Toronto, Ontario M5S
3H8, Canada

The gravitational clustering hierarchy and dissipative gas processes are both involved in the
formation of bulges. Here I present a simple empirical model in which bulge material is
assembled via gravitational accretion of visible companion galaxies. Assuming that merging
leads to a starburst, I show that the resulting winds can be strong enough to self-regulate
the accretion. A quasi-equilibrium accretion process naturally leads to the Kormendy relation
between bulge density and size. Whether or not the winds are sufficiently strong and long lived
to create the quasi-equilibrium must be tested with observations. To illustrate the model I use
it to predict representative parameter-dependent star formation histories. The bulge building
activity appears to peak around a redshift $z \sim 2$, with tails to both higher and lower redshifts.

1. Introduction

Bulges are stellar dynamical pressure supported systems that generally have much
higher surface brightnesses than galactic disks. They therefore have undergone more
collapse than galactic disks, evidently with the angular momentum barrier removed.
Galaxy merging is an inevitable process that redistributes any pre-merger stars into a
physically dense, but phase density lowered, pressure supported distribution. Stellar
dynamical mergers produce objects with flattenings largely unrelated to their rotation.
In the presence of gas, merging is empirically associated with an often dramatic rise in
star formation. These new stars that are formed in place almost certainly reflect the
chemical history and the dynamical state of the growing bulge. In this paper I calculate
some of the properties of bulges expected on the basis of merging with star formation of
largely gaseous pre-galactic fragments.

The rate of major mergers can be calculated directly from the observed numbers of
close pairs of galaxies. Remarkably, this is now an observational quantity for which there
are measures from low redshift up to the 'U-dropout' population centred around redshift
$z \sim 3$. There are some substantial uncertainties in the various observational quantities
which go into the merger calculation. The details of this calculation will become much
more precise over the next few years as the evolution of the two-point correlation function
becomes better determined.

Mergers are widely observed to induce an intense nuclear starburst. Theoretically this
is at least partially understood (Barnes & Hernquist 1992) on the basis that the strong
dynamical interactions during a merger leads to a loss of angular momentum in a cool
gas, helping to funnel it to build up a dense central gas reservoir from which stars form
at astonishing rates in a starburst. Starbursts in turn develop winds which I here suggest
can lead to the accretion being a self-regulating process, although this is dependent on
the ram-pressure and duration of the wind. Moreover, self-regulating accretion can lead
to quasi-equilibrium star formation in the bulge, which can lead in turn to some of the
observed regularities of bulge properties with mass or size. At this stage the details of this
picture are speculative, but are open to observational refinement, which helps motivate
the calculations presented here.

Here I take the properties of 'classical bulges' to be roughly as follows (e.g.
Wyse, Gilmore & Franx 1997):

"

- Bulges follow the 'Kormendy relation', that is, the characteristic surface brightness correlates strongly with the scale radius (Kormendy 1977, De Jong 1996, Pahre, Djorgovski & De Carvalho 1995).
- The flattening of the figure of bulges is approximately consistent with their rotation (Davies *et al.* 1983).
- Bulge stellar populations are predominantly old, although there are well-documented cases of relatively young bulges.
- Bulges follow a mass-metallicity correlation.

A useful model for bulge formation should be able to provide a physical origin for these properties.

This paper has three main sections. In §2 I discuss the empirically determined rate at which mergers occur as a function of redshift, and in §3 the star burst winds and the effects those winds will have on accreting gas. Section 4 pulls these two together in some specific model calculations.

2. Merger Rate Measures

A host galaxy has a number of near neighbors within radius r and pairwise velocity $|v|$ far above the mean density n_0 (in proper coordinates),

$$N(< r, < v) = 4\pi n_0 (1 + z)^3 \int_0^v \int_0^r \xi(r|z) f(v|z) r^2 v^2 \, dr \, dv, \qquad (2.1)$$

where $\xi(r|z)$ and $f(v|z)$ are the redshift dependent two-point correlation and velocity distribution functions, respectively. I have made the important assumption that the distribution of pairwise velocities is constant over the separations of interest. This is not true in general, but is sufficient for the present application to the relatively small scales, $r \leq 20\,\mathrm{kpc}$, that are of interest for merging. To calculate merger rates one needs estimates of the correlation function on small scales, the pairwise velocity dispersions, and the mean time for a merger to occur within this volume.

2.1. *Close Pairs and the Correlation Function*

The galaxy-galaxy correlation function is accurately modeled as a power-law, $\xi(r) = (r_0/r)^\gamma$. The reliability of this power-law on scales less than about $100h^{-1}\,\mathrm{kpc}$ relevant to galaxy merging is discussed in detail for the SSRS2 (Patton *et al.* 1999a) and CNOC2 samples (Patton *et al.* 1999b). These papers support three important conclusions. First, the power-law extrapolation of the correlation function to $20h^{-1}\,\mathrm{kpc}$ is consistent with the density of pairs measured inside this radius. Second, the R band luminosity function of galaxies in $20h^{-1}\,\mathrm{kpc}$ pairs is consistent with being drawn from the field luminosity function. This property allows the fully general correlation function, $\xi(r, v_{||}, v_\perp, L_1, L_2)$, to be factored into a luminosity function, and a spatial and kinematic correlation function, $\phi(L_1)\phi(L_2)\xi(r)f(v)$ where we drop the distinction between velocities along the line of separation, $v_{||}$, and perpendicular velocities, $v_\perp$ (Peebles 1980). The luminosity factorization glosses over the various lines of evidence (Carlberg *et al.* 1998, Loveday *et al.* 1995) that the theoretically-expected weak increase of correlation with galaxy mass does exist in the correlations. However, this relatively small effect cannot yet be detected in the current samples of close pairs which have not yet broken through the barrier of 100 pairs. The third important result is that there is morphological evidence that $r \leq 20h^{-1}\,\mathrm{kpc}$ pairs are indeed interacting at a level that indicates that these are high probability mergers-to-come. The $20h^{-1}\,\mathrm{kpc}$ scale is

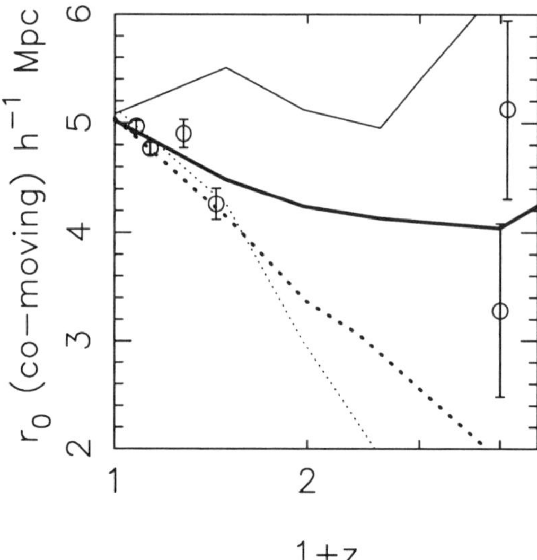

FIGURE 1. Measured co-moving correlation lengths as a function of (one plus) redshift. The points are the LCRS at low redshift, CNOC2 at intermediate redshift, and the Giavalisco *et al.* (1998) close pairs and the 'box counts' of Adelberger *et al.* (1998) at redshift 3. The lines are for the correlation lengths measured in simulations (Colin, Carlberg & Couchman 1997) for galaxy halos (solid lines) and the mass field (dashed), with $\Omega_M = 0.2$, and $\Omega_\Lambda = 0$ (heavy) or $\Omega_\Lambda = 0.8$ (light).

also chosen such that the galaxies are not so strongly interacting that their luminosities, morphologies and colors bear little resemblance to their unperturbed values.

The volume integral of the power-law correlation function, $\xi(r|z) = (r_0(z)/r)^\gamma$, is,

$$4\pi \int_0^r \xi(r|z) r^2 \, dr = \frac{4\pi}{3-\gamma} \left(\frac{r_0(z)}{r} \right)^\gamma r^3. \tag{2.2}$$

The redshift dependence of the average density inside a $r = 20h^{-1}$ kpc neighborhood around a galaxy is estimated using the preliminary CNOC2 correlation $\gamma = 1.8$,

$$r_0(z) = 5.15(1 + z)^{[1-(3+\epsilon)/\gamma]} h^{-1} \, \text{Mpc (co-moving)}, \tag{2.3}$$

where $\epsilon = -0.6 \pm 0.4$ (Carlberg *et al.* 1998). Using this in Equation 2.2 one finds that the integrated density inside $20h^{-1}$ kpc is $1.56[(1 + z)/1.3]^{-\epsilon} n_0 \, \text{Mpc}^3$ (proper units). In Figure 1 is shown the co-moving correlation length as a function of redshift. Also plotted are the correlations of the galaxy mass halos and the particles in simulations (Colin, Carlberg & Couchman 1997). It should be noted that there is a relatively good understanding of why the halo correlation function shows relatively little evolution (Carlberg 1991, Baugh, *et al.* 1998). Observationally it is currently acceptable to take r_0 to be fixed over this redshift interval, or, $\epsilon = -1.2$ for a $\gamma = 1.8$ power-law.

2.2. *Pairwise Velocities*

The CNOC2 velocity distribution function, $f(v)$, at small scales is acceptably modeled as a Gaussian, although an exponential also provides a similar quality fit (Carlberg *et al.* 1998). I will take the velocity to be isotropic, $\sigma_3 = \sqrt{3}\sigma_z$, where σ_z is the velocity dispersion measured along the line of sight. Redshift surveys can be used to measure σ_z. At separations of about $1 \, h^{-1}$ Mpc the pairwise velocity dispersion in the redshift direction is about $300 \, \text{km}\,\text{s}^{-1}$ (e.g. Davis & Peebles 1983,

Marzke, Geller, Da Costa & Huchra 1995, Carlberg *et al.* 1998), constant with redshift over the $z \leq 0.5$ range. If the critical velocity to merge is taken as about $220\sqrt{2}$ km s^{-1} then the fraction of close pairs with sufficiently low velocities to merge is merely 5.1%. This does not accord well with the impression that most close pairs have such large tidal tails that they are almost certainly doomed-to-merge pairs (Toomre & Toomre 1972, Tooomre 1977). In a non-merging, equilibrium distribution the pairwise velocity declines as $\sigma_p^2 \propto r^{2-\gamma}$, where γ is the slope of the power-law correlation. The one dimensional pairwise velocity dispersion at 20 kpc is therefore about 190 km s^{-1}.

The dynamical details of pair mergers in a cosmological setting that includes the tidal fields of surrounding structure have not been studied in detail (but see Carlberg & Couchman 1989), such that one of the best estimates of the critical velocity to merge remains Aarseth and Fall's (1980) value of $v_{mg} = 1.2\sqrt{2}v_p$, where v_p is the 'parabolic' velocity, $v_p = f_p v_c$, at the orbital pericenter, where f_p is at least $\sqrt{2}$ (for a point mass). This 'parabolic' velocity is that for those pairs assured to merge, in the absence of tidal fields (Toomre & Toomre 1972, Aarseth & Fall 1980). For $f_p = \sqrt{2}$ the fraction of the close pairs with velocities low enough to merge is 27%, but this rises quickly with increasing f_p, becoming 41% and 54% at $f_p = \sqrt{3}$ and 2, respectively. I will use an $f_{mg} = 0.5$ (although this number is both empirically and theoretically uncertain).

2.3. *Merger Times of Close Pairs*

As a reference timescale for merging I start from the reference radius of 20 kpc where the time for a circular orbit is 0.62 Gyr at a speed of 220 km s^{-1}. In detail the rate of inflow through the 20 kpc depends on mass and the orbital details, so I use a merger timescale at the reference radius of 20 kpc of 0.3 Gyr (Barnes & Hernquist 1992, Dubinski, Mihos & Hernquist 1999), and the estimate that $f_{mg} = 0.5$ of all 20 kpc pairs have pairwise velocities less than the critical velocity for merging. If the pairwise velocities are substantially higher than the critical velocity for merging, then the merging fraction drops nearly as v^3, which is such a dramatic change that it should be testable via the morphology of galaxies in close pairs.

2.4. *Estimated Merger Rates*

Combining the estimates of clustering, pairwise velocities and the available understanding of the dynamics of merging, it is found that the specific merger rate is $2.4[(1 + z)/1.3]^{-\epsilon} n_0$ Mpc^3 Gyr^{-1}. Taking n_0 as being the CNOC2 galaxy number density to $0.21L_*$ adjusted to $0.1L_*$, $n_0 \simeq 1.1 \times 10^{-2}$ Mpc^{-3} (co-moving), the merging event rate for galaxies above the minimum mass is $1.3 \times 10^{-2}(1+z)^{-\epsilon}$ Gyr^{-1}. The rate of accretion of pre-merger stars onto galaxies as a result of merging is relatively slow. The time scale is $L_*/\mathcal{R}_L \simeq 60$ Gyr, or about 5 Hubble times at $z = 0.3$. This result is based on the directly visible galaxies, $L \gtrsim 0.21L_*$, which contain about 80% of the total stellar luminosity in galaxies. This relatively low rate of accretion of visible galaxies relative to the Hubble time continues on to the $z \simeq 3$ regime (Giavalisco *et al.* 1998, Adelberger *et al.* 1998), where the number densities and the co-moving correlations are similar to those observed for present-day galaxies. That is, for $q_0 = 0.1$ their data indicate a density of $\mathcal{R} < 25.5$ mag Lyman-break galaxies of $n_0 \simeq 2.2 \times 10^{-3}$ Mpc^{-3} and a co-moving correlation length of $5h^{-1}$ Mpc. The self-event rate of this population is 2.0×10^{-2} Gyr^{-1}, only 3.8 times that at $z \simeq 0.3$. Since the cosmic time at $z \simeq 3$ is about 20% of that at $z = 0.3$ the relative impact on the hosts of these self-mergers is small adding perhaps 5-10% more mass over the entire $z = 0$ to 3 range. For the lower-luminosity galaxies inferred to be in this redshift range from the HDF (i.e., about 2 magnitudes fainter than those with spectroscopic redshifts), the volume-density is about a factor 20 higher, but the implied cross-correlation length is about a factor $\sqrt{3}$ smaller (Steidel *et al.* 1999), where it is assumed that the cross-correlation depends on the product of the relative biases. This implies that the high/low-luminosity merger rate is about $0.5\,\mathrm{Gyr}^{-1}$, which is large enough to build a galaxy over a $z \simeq 1 - 4$ interval. In the intermediate redshift range the $M \gtrsim 0.2M_*$ galaxies cannot self-merge to significantly alter the mass function. At higher redshift the lower-luminosity Lyman-break galaxies rise very steeply in number, $\alpha \simeq -1.8$ (Steidel *et al.* 1999). These large numbers completely change the situation, allowing their mergers to substantially alter galaxy masses.

3. Starburst Winds

There is a remarkably simple physical description of what happens when star formation is rapid in a relatively small volume. The inevitable outcome is a very strong wind. Chevalier & Clegg (1985, hereafter CC) simplify the situation to the equilibrium solution of mass injection at a rate $\dot{M}$, with accompanying energy injection, $\dot{E}$, in a sphere of radius R. Since it turns out that the wind velocities exceed $1000\,\mathrm{km}\,s^{-1}$, gravity can be ignored in a first approximation. CC provide a full solution at all radii. Here we are mainly interested in the asymptotic solution at large radii, where one can cast the CC solution in terms of the terminal wind velocity, u and the mass injection rate, $\dot{M}_w$,

$$\rho_w = \frac{\dot{M}_w}{4\pi u r^2}. \tag{3.1}$$

This wind produces a ram pressure of,

$$P_w \equiv \rho_w u^2 = \frac{u\dot{M}_w}{4\pi r^2}, \tag{3.2}$$

where R is the size of the region into which the mass and energy are injected.

As representative numbers I will take $R = 10^{21}$ cm, about 1/3 kpc, and $\dot{M}_w = 10^{27}\,\mathrm{g}\,\mathrm{s}^{-1}$, about $15\,\mathrm{M}_\odot\,\mathrm{yr}^{-1}$, approximately the mass injection rate expected during a burst of star formation of $1500\,\mathrm{M}_\odot\,\mathrm{yr}^{-1}$. Following CC I use $u = 2000\,\mathrm{km}\,s^{-1}$. In the central region,

$$\rho_w \simeq 0.296\frac{\dot{M}_w}{uR^2}, \tag{3.3}$$

or 1.5×10^{-24} g, or a number density of about $1\,\mathrm{cm}^{-3}$.

The cooling time $t_{cool} = 3kT/(n\Lambda(T))$ at $T = 10^8$K (where the cooling rate is about $\Lambda \simeq 3 \times 10^{-23}\,\mathrm{cm}^3\,\mathrm{s}^{-1}$) is about 4×10^7 yr. The flow time across the region is only 1.6×10^5 yr, so the hot wind does not have time to cool. Denser regions in pressure equilibrium will cool more quickly so that the ISM is unstable and bound to consist of the hot wind phase and one or more cool phases. Many of the aspects of this situation are discussed in Ikeuchi & Norman (1991).

3.1. *Ram Pressure Stripping*

A major objective of this paper is to estimate the ram pressure stripping of the hot wind on infalling objects. The effects of ram pressure stripping are calculated, but note that transport processes, such as turbulence and heating, could help to increase the rate of gas removal (Nulsen 1982). The calculation proceeds in a series of steps. First I derive for this specific case the fairly standard result that the wind will have a very high momentum flux. The wind has a sufficiently low density that it will move out before cooling. The infalling objects are taken to be angular momentum conserving, but maximally dissipated disks in dark halos, approximated as truncated Mestel disks. The fractional stripping can be easily estimated for these objects. The strength of the starburst wind is calculated under the assumption that star formation is occurring on timescale comparable to the crossing time of the bulge. Bringing these elements together gives an expression for the fractional mass of an infalling object which succeeds in joining the bulge, Equation 3.13.

The ridge line of Kormendy's relation is $\mu_B = 3.02 \log r_0 + 19.74$ B mag arcsec^{-2} (Kormendy 1977, De Jong 1996), where μ_B is the B band surface brightness and r_0 is the bulge scale radius in kpc. For a constant mass-to-light ratio this translates to $\rho \propto r_0^{-2.2}$. This density-radius relation would be slightly weakened if we allowed for a decrease in mass-to-light for lower luminosity systems. The implied densities are high enough that bulges are self-gravitating. If the rate of infall is, on the average, regulated by the starburst winds, then the fact that the infall rate is very insensitive to the mass of the host implies that the characteristic radius and density of the bulge will scale roughly as $\rho \propto R^{-2}$, as in Equation 3.11. This relation is roughly the ridge line of the Kormendy relation. This is physically easy to understand. The total bulge star formation rises as the bulge gas density, ρ_b, to the 3/2 power. For a given accretion rate a rise in bulge density will increase the SFR, and hence the outgoing wind, which temporarily reduces the accretion, allowing the gas density to be reduced.

The ram pressure rises as r^{-2} with decreasing radius. At R, the outer radius of the star forming volume, the surface density below which stripping occurs rises to its maximum,

$$\Sigma \frac{V_c^2}{R} = \frac{u \dot{M}_w}{4\pi R^2},$$ (3.4)

where V_c is the circular velocity of the incoming gas in its dark halo. For the starburst numbers above, $\Sigma = 0.16 \, \mathrm{g \, cm^{-2}}$ is the maximum surface density that can be blown away via ram pressure alone. For comparison, the central surface density of a disk galaxy is about $1 \, \mathrm{g \, cm^{-3}}$. Thus, the effects of this wind would be significant even on a disk like that of the Milky Way if it were completely gaseous.

If the gas collapses inside an isothermal halo with a velocity dispersion σ_s and an angular momentum parameter $\lambda \simeq 0.05$ to centrifugal equilibrium, its surface density increases by about a factor of 10^2 over that of the projected isothermal halo,

$$\Sigma(d) \simeq 10^2 \Omega_b \frac{\sigma_s^2}{2dG},$$ (3.5)

where d is the radial coordinate in the disk. For the typical case discussed here, $\Omega_b \simeq 0.01$, $H = 100 \, \mathrm{km \, s^{-1} \, Mpc^{-1}}$, this Mestel disk has a surface density profile

$$\Sigma(d) \simeq 0.7 \left(\frac{\sigma_s}{100 \, \mathrm{km \, s^{-1}}} \right)^2 \left(\frac{10^{21} \, \mathrm{cm}}{d} \right) \mathrm{g \, cm^{-2}}.$$ (3.6)

The total mass in a Mestel disk of the above form diverges. If we cut the disk off at the radius of the last orbit that can have come from the outermost virialized part of the halo, we can estimate a total mass. The halo is virialized inside approximately r_{200}, inside of which the mean density is $200\rho_c(z)$, where $\rho_c(z) = 3H(z)^2/(8\pi G)$. For an isothermal sphere, $M(r) = 800/3\pi\rho_c(z)r_{200}^2 r$. The isotropic velocity dispersion that maintains this sphere in equilibrium is $\sigma_1^2 = 400/3\pi G \rho_c(z) r_{200}^2$. Using the definition of $\rho_c(z)$, it follows that, for the isothermal sphere,

$$r_{200} = \frac{\sqrt{2}}{10} \frac{\sigma_s}{H(z)}.$$ (3.7)

This is a physical (proper) distance. In this potential the circular velocity is $V_c = \sqrt{2}\sigma$.

The total gas mass in the disk (assuming that there are few stars) is $200\rho_c \Omega_b \frac{4}{3}\pi r_{200}^3$, or,

$$M_m = \Omega_b \frac{2\sqrt{2}}{10GH(z)} \sigma_s^3, \tag{3.8}$$

which is needed to find σ_s (required in the evaluation of d/r for stripping, Equation 3.13).

3.2. *Starburst Rates*

Both empirical evidence and theoretical considerations suggest that the timescale for star formation should be proportional to the local dynamical timescale (Lehnert & Heckman 1996, Kennicutt 1998). The available data indicate that $t_{sfr} \simeq R/v_b$, where v_b is the local circular velocity, which in the case of a stellar bulge may be due to self-gravity, not the dark matter background. Therefore $\dot{M}_{SFR} = Mv_b/R$, and I will take the wind as $\dot{M}_w = \epsilon_w \dot{M}_{SFR}$. The mass can be expressed as $M = 4\pi/3\rho R^3$, where ρ refers to the gas density.

The wind that results from this starburst has a ram pressure,

$$P_w = \epsilon_w v_b u\, \rho \frac{4\pi}{3} \frac{R^2}{r^2}. \tag{3.9}$$

3.3. *Self-Regulating Starbursts*

The starburst wind will blow away infalling surface densities smaller than,

$$\Sigma \frac{V_c^2}{r} < \epsilon_w v_b u\, \rho \frac{4\pi}{3} \frac{R^2}{r^2}. \tag{3.10}$$

In equilibrium this leads to a balance of the bulge which creates the wind and the infall,

$$\rho R^2 \simeq \Sigma \frac{3}{4\pi\epsilon_w u} \frac{V_c^2 r_{200}}{v_b}. \tag{3.11}$$

If bulges are to be self-gravitating, then $v_b \gtrsim V_c$, and generally they are found to have circular velocities comparable to those of the disk. The important thing to note is that the quantity ρR^2 is completely determined by the starburst, whereas the RHS is completely determined by the physics of infall.

Mestel disks derived from satellite halos of σ_s, and $\sigma_b^2 = 2\pi G\rho R^2$, are stripped at radii beyond

$$\frac{d}{r} > \frac{\sigma_s^2}{\sigma_h^2} \frac{3}{4\epsilon_w} \frac{v_b}{u} 10^2 \Omega_b. \tag{3.12}$$

For typical numbers I take $\sigma_s = \sigma_h/2$ and $\epsilon_w = 10^{-2}$, and find that stripping occurs at $d/r > 1.3$ (which is 99.5% of the disk mass for $r = R$ and $R \simeq 1/3$ kpc).

The equation for stripping of Mestel disks can be used to predict the rate of successful accretion in the presence of a starburst. I simply multiply M_m with the ratio of d/r for stripping. The accreted mass is simply

$$M_{\mathrm{accrete}} = \frac{d}{r} \frac{R}{10 r_{200}} M_m. \tag{3.13}$$

4. Realizations of Bulge Formation Histories

Combining merging rates and stripped fractions to build bulges with redshift is now straightforward. I present a simple Monte Carlo simulation to illustrate the model. The simulation starts when the universe is about 0.5 Myr old, prior to significant galaxy creation or merging. An $H_0 = 65$, $\Omega_M = 0.2$, $\Lambda = 0$ cosmology is adopted, although the results are not very sensitive to the precise choice of cosmology. The mergers occur

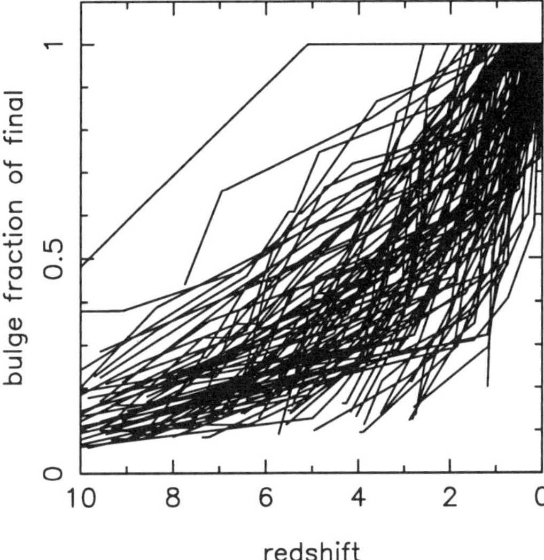

FIGURE 2. Mass redshift history for 100 realizations of bulge building. The calculation assumes that suitable pre-galactic objects, disks in dark halos, are available at the beginning of the calculation.

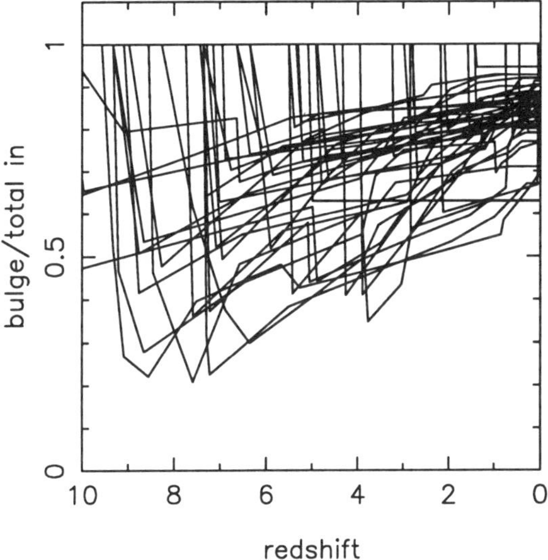

FIGURE 3. Fraction of infalling gas mass successfully accreted versus redshift for the 100 realizations of bulge building.

at a rate $\mathcal{R}_n(1 + z)^m$, independent of mass. Time is divided into 0.5 Myr intervals and the probability of an accretion event is the rate per unit time multiplied with the time interval. An accretion event occurs if a [0,1] random number generator produces a number less than this probability. The accreted objects begin with a mass drawn from $\phi(M) \propto M^\alpha e^{-M}$, where the normalizing constants are unity. This therefore assumes that the Mestel disks in their dark halos are largely present when the calculation is turned on. Clearly this is not accurate at large redshift, but is arguably a useful assumption over the redshift zero to about four range (Steidel *et al.* 1999). There is no presumption that the

TABLE 1. Merger-Wind Simulation Parameters

Parameter	Symbol	Default Value
Minimum satellite mass	M_{min}	0.1
Mass function slope	α_M	-1.8
Current merger rate	$\mathcal{R}_n$	0.01
Merger-redshift index	m	1.2
Gas fraction	Ω_{gas}	0.01
Bulge size	R_b	10^{21} cm
Wind duration	T_w	10^8 yr
Mass loss efficiency	ϵ_w	0.01

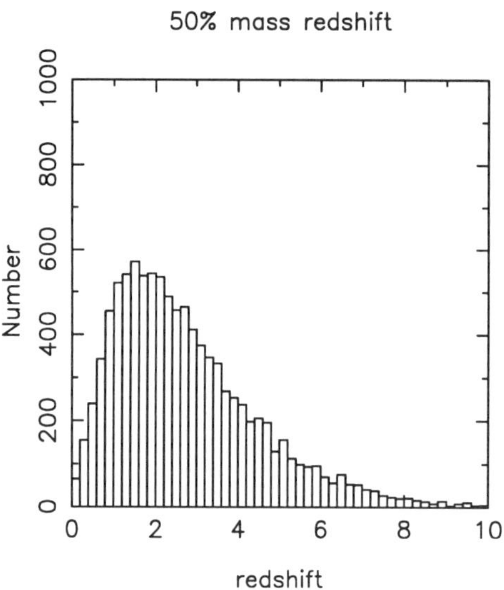

FIGURE 4. Redshift of 50% mass assembly for 10000 realizations of the standard model.

characteristic mass is as large as that of a full galaxy today, however the characteristic mass does need to be comparable to a bulge mass, since we find that the final masses are distributed around unity, the characteristic mass of the infalling objects. The wind is assumed to blow uniformly for a duration of T_w from the previous merger. If a new accretion event is generated during this interval, then the mass of the incoming satellite is reduced according to the stripping equation. The parameters and their default values are outlined in the following table.

In detail there are many (mostly minor) complications that are swept under the rug here. The model is quite naïve in that it assumes that there is a ready supply of gas containing companion galaxies with roughly a galaxy mass distribution. Observation seems to support this as being true from redshift zero to about four, which covers most of the activity here. Mergers wouldn't do much at all if $\alpha \gtrsim -1.2$, as is observed at low redshift. Here a steeper α has been chosen to both mimic the increase in gas content with decreasing mass and to take into account that α does become steep in the redshift 3–4 range (Steidel *et al.* 1999). The gas content of galaxies should likely vary with redshift, whereas we have simply taken them to be all gas. Likely the pairwise velocity dispersion decreases somewhat with increasing redshift, which will diminish the fraction of low redshift pairs that merge. Overall all these effects would tend to decrease

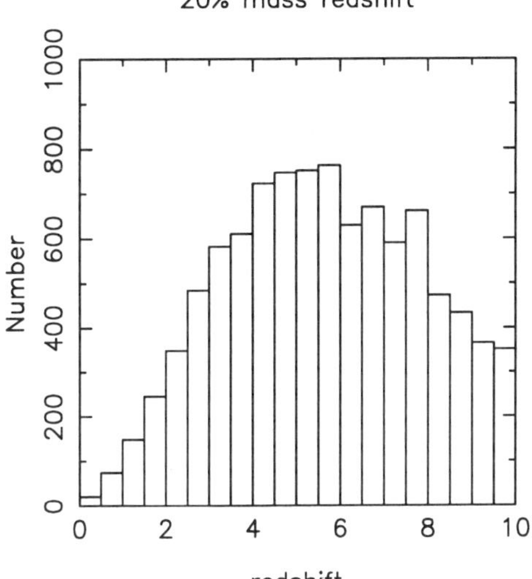

FIGURE 5. Redshift of 20% mass assembly for 10000 realizations of the standard model. Comparison with Figure 4 emphasizes the very large spread in redshifts of formation.

the importance of merging below redshift one. The natural tendency of the model is to have little activity at low redshift anyway, so the basic character of the results should not depend on these simplifications. In any case, the main purpose of this 'toy' star formation history is to examine the basic viability of the model, not to fine tune the parameters.

Overall, several interesting results are found. First, the mass buildup predicted by this simple model seems to be very roughly in accord with the requirements of 'classical' bulges. Figure 4 shows that about 10% are more than 50% formed at redshift 5. However, about 3% are only half formed at redshift 0.5. Although the median time of half assembly is a reassuring redshift of about two, there is a tremendous spread of formation times. The redshifts of 20% assembly are shown in Figure 5.

The standard wind has a lifetime of 10^8 years, which in many cases limits infall to about 2/3 of what it would normally be (Figure 3). The limitation of infall would help drive bulges towards the Kormendy relation. Winds are even more effective if the smallest mass to be merged into a bulge is reduced to 3% of M_*, rather than the assumed standard $0.1 M_*$. This is shown in Figure 6. The basic merger rate is the same, so the redshifts of assembly are not greatly altered. However, because the bulge is being built of more, smaller, units, the buildup has less dispersion in time.

5. Conclusions

The bulge formation history is predicted here using the observed density of nearby (gas-rich) galaxies with masses comparable to the eventual bulges. Further, it has been argued that starburst winds will have a significant effect on the accreting gas. A straightforward assessment of these results is that the merger history appears to be roughly in accord with what is known about star formation histories and bulge ages. An additional step, not taken here, is to use these assembly histories to predict the color distribution of bulges as a function of redshift. These then become a simple but powerful test of the model. The attraction of the merger model is that it is based on observations, which now

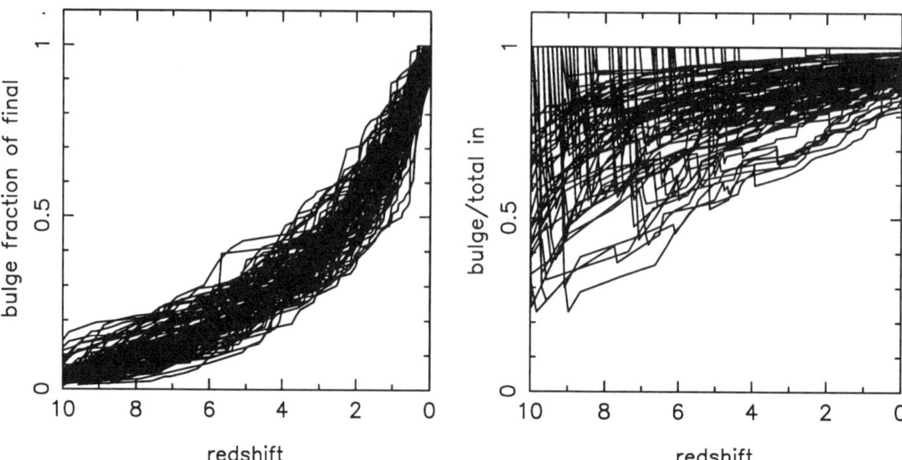

FIGURE 6. The assembly of a bulge built from pre-galactic objects with a minimum mass of $0.03M_*$, rather than the standard $0.1M_*$ and wind lifetime of 10^7 yr. The left panel is the analog of Figure 2 and the right panel is the analog of Figure 3. The assembly here is smoother and more affected by the starburst winds. Although there is less dispersion in formation times, the median 20% and 50% time are not significantly different from the standard model.

provide a very basic measure over the redshift zero to four range, although the masses of the incoming objects are not well quantified at the highest redshifts. Beyond redshift four, this model is likely becoming less reliable, since the assumption that the pregalactic fragments are in place has no basis in observation and even in a low density Universe, the Press-Schechter approach would indicate that the halos in which these reside are becoming less numerous. Furthermore the gas content probably evolves with redshift. An attraction of this approach is that the need for additional physical parameters can be driven by observation.

The relevance for bulges of the strong starburst winds that we have advocated is far less clear at this stage. The attraction of the idea is that it promotes the development of the Kormendy relation in a self-gravitating system. The tests of this model will require fairly detailed observations of bulges at high redshifts, particularly concentrating on the heavily reddened objects in which starbursts generally occur. The presence of starburst winds and their effects can potentially be detected in optical emission lines, X-ray observations of hot gas (e.g. Heckman *et al.* 1999) and ultimately in *resolved* observations of the molecular gas (e.g. Frayer *et al.* 1999) which is known to exist.

This research was supported by NSERC of Canada. I thank the Carnegie Observatories, Pasadena, for their hospitality during the time when this work was initiated.

REFERENCES

AARSETH, S.J., FALL, S.M. 1980 *ApJ*, **236**, 43

ADELBERGER, K.L., STEIDEL, C.C., GIAVALISCO, M., DICKINSON, M., PETTINI, M., KELLOGG, M. 1998 *ApJ*, **505**, 18

BARNES, J.E., HERNQUIST, L. 1992 *ARAA*, **30**, 705

BAUGH, C.M, BENSON, A.J., COLE, S., FRENK, C.S., LACEY, C.G. 1998 *MNRAS*, submitted (astro-ph/9811222)

CARLBERG, R.G., COUCHMAN, H.M.P. 1989 *ApJ*, **340**, 47

CARLBERG, R.G. 1991 *ApJ*, **367**, 385

CARLBERG, R.G., YEE, H.K.C., MORRIS, S.L., LIN, H., SAWICKI, M., WIRTH, G., PATTON, D., SHEPHERD, C.W., ELLINGSON, E., SCHADE, D., PRITCHET, C.J., HARTWICK, F.D.A. 1998 *Phil. Trans. Roy. Soc.*, in press

CHEVALIER, R.A., CLEGG, A.W. 1985 *Nature*, **317**, 44

COLIN, P., CARLBERG, R.G., COUCHMAN, H.M.P. 1997 *ApJ*, **390**, 1

DAVIES, R.L., EFSTATHIOU, G., FALL, S.M., ILLINGWORTH, G., SCHECHTER, P.L. 1983 *ApJ*, **266**, 41

DAVIS, M., PEEBLES, P.J.E. 1983 *ApJ*, **267**, 465

DE JONG, R.S. 1996 *A&Ap*, **313**, 45

DUBINSKI, J., MIHOS, J.C., HERNQUIST, L. 1999 *ApJ*, submitted (astro-ph/9902217)

FRAYER, D.T., ET AL. 1999 *ApJ*, **514**, L13

GIAVALISCO, M., STEIDEL, C.C., ADELBERGER, K.L., DICKINSON, M.E., PETTINI, M., KELLOGG, M. 1998 *ApJ*, **503**, 543

HECKMAN, T.M., ARMUS, L., WEAVER, K.A., WANG, J. 1999 *ApJ*, in press (astro-ph/9812317)

IKEUCHI, S., NORMAN, C.A. 1991 *ApJ*, **375**, 479

KENNICUTT, R.C. JR. 1998 *ApJ*, **498**, 541

KORMENDY, J. 1977 *ApJ*, **218**, 333

LEHNERT, M.D., HECKMAN, T.M. 1996 *ApJ*, **472**, 546

LOVEDAY, J., MADDOX, S.J., EFSTATHIOU, G., PETERSON, B.A. 1995 *ApJ*, **442**, 457

MARZKE, R.O., GELLER, M.J., DA COSTA, L.N., HUCHRA, J.P. 1995 *AJ*, **110**, 477

NULSEN, P.E.J. 1982 *MNRAS*, **198**, 1007

PAHRE, M.A., DJORGOVSKI, S.G., DE CARVALHO, R.R. 1995 *ApJ*, **453**, L17

PATTON, D.R., MARZKE, R.O., CARLBERG, R.G., PRITCHET, C.J., DA COSTA, L.N. 1999, in preparation

PATTON, D.R., CARLBERG, R.G., MARZKE, R.O., YEE, H.K.C., LIN, H., MORRIS, S.L., SAWICKI, M., WIRTH, G.D., SHEPHERD, C.W., ELLINGSON, E., SCHADE, D., PRITCHET, C.J. 1999, in preparation

PEEBLES, P.J.E. 1980 *The Large Scale Structure of the Universe*. (Princeton)

STEIDEL, C.C., ADELBERGER, K.L., GIAVALISCO, M., DICKINSON, M., PETTINI, M. 1999 *ApJ*, in press (astro-ph/9811399)

TOOMRE, A., TOOMRE, J. 1972 *ApJ*, **178**, 623

TOOMRE, A. 1977, in *Evolution of Galaxies and Stellar Populations* (ed. B.M. Tinsley & R.B. Larson), p401. (Yale)

WYSE, R.F.G., GILMORE, G., FRANX, M. 1997 *ARAA*, **35**, 637

Role of Winds, Starbursts, and Activity in Bulge Formation

By B R U C E   G.   E L M E G R E E N

IBM Research Division, T.J. Watson Research Center, P.O. Box 218, Yorktown Hts NY 10598, USA

The starburst phase of nuclear disk evolution may not be directly related to bulge formation, but the bulge formation event itself may have been a starburst, acting at the maximum possible rate allowed by the total virial density for a few internal crossing times. Starbursts in bulges differ from those in disks because the bulge potential is too deep to allow significant self-regulation by blow-out. The total luminosity of a bulge-forming starburst is comparable to that observed in distant galaxies, when the bulges are supposed to have formed.

1. Starburst Models and the Formation of Bulges

If the duration of bulge formation is as short as some recent data suggest (e.g., Renzini 1999, these proceedings), then star formation in the bulge must have occurred very rapidly, perhaps in only a few internal crossing times. This implies a star formation rate of several hundred $M_\odot$ yr^{-1} for less than 10^8 years. Such an event would be called a starburst if viewed in a primordial galaxy, so it is natural to wonder if any of the starburst regions that are observed today could be undergoing processes similar to what happened in bulges in the early Universe.

Wada, Habe & Sofue (1995) suggested that a starburst in the nuclear disk of a galaxy could generate an expanding shell of gas because of pressures from supernovae and winds. They proposed such a shell would turn into stars and mix, forming a bulge in only a few orbits. Thus in their model, a disk starburst can lead directly to the formation of a three-dimensional bulge.

Hasan, Pfenniger, & Norman (1993) and Friedli & Benz (1993) modeled bulge formation with a dissolving bar. They suggested that gas accretion along the bar produces a starburst in the disk, and that continued accretion produces a critical mass in the nucleus when the bar orbits no longer reinforce the bar. Then the bar dissolves and the stars spread out to form a bulge. Combes *et al.* (1990), Pfenniger & Friedli (1991), and Friedli & Benz (1995) considered bulge formation from a bar independent of a starburst, noting that bar stars can get injected into the third dimension following vertical resonances.

These bulge formation mechanisms have only weak connections to starbursts, since the star formation in the burst is confined to the disk, as directly observed in normal starburst galaxies, while the bulge has to be made by some activity that operates perpendicular to the disk.

We consider here another point of view, that the formation of the bulge is itself a starburst, operating with a prodigious rate of star formation in three-dimensions, perhaps in the manner suggested by Eggen, Lynden-Bell, & Sandage (1962). That is, if the bulges of some galaxies formed during a three-dimensional accretion or collapse phase when the galaxy formed, and the resulting star formation was very rapid, then the galaxy, viewed at a very early time, would appear to be a starburst, although different in some respects from today's two-dimensional starbursts.

To understand the properties of such three-dimensional starbursts, we developed a theory of star formation in three-dimensions, with a threshold for the onset of star formation, a predicted star formation rate, and a condition for self-regulation or termination of star formation (Elmegreen 1999). At the present time, most theories of star formation on a galactic scale are for two-dimensions, since they employ a 2D disk instability condition, and self-regulation by disk thickening.

The main points of the 3D theory are that there should be a threshold for star formation in three-dimensions as there is in two-dimensions, based on gas density rather than column density, and that the overall star formation rate depends on the density too. The density at the threshold gives the volume emissivity of energy from star formation. This is analogous to the energy per unit area produced by a 2D burst, but the three dimensional nature of this energy input is significantly different than for 2D because of the way hydrostatic equilibrium is maintained. Pressure equilibrium in a self-gravitating system depends on the surface density of matter, rather than the density. Thus the gas pressure is approximately $G\sigma^2$ for mass column density σ. If star formation is a three-dimensional process but pressure balance is two-dimensional, then the condition for pressure equilibrium gives a critical thickness or a critical velocity dispersion. This critical dispersion is such that galaxies or parts of galaxies with potential wells deeper than the square of the critical velocity dispersion cannot generate enough pressure from star formation at the threshold rate to blow out the gas and halt further star formation. Potential wells shallower than this critical value can have regulated star formation at some rate less than critical.

For conventional star formation pressures, the critical velocity dispersion is higher than the typical dispersion in a galaxy disk, but lower than the dispersion in a galaxy bulge. For this reason, bulge formation should proceed at a maximum rate, giving an intense burst of star formation for a relatively short time, whereas normal disk formation can regulate itself, giving a lower rate of star formation for a longer time.

2. A Critical Density for Star Formation in Three Dimensions

A gas spheroid inside a dark matter spheroid settles to the center of the combined potential well as the gas cools. There the gaseous self-gravity gets stronger and stronger as the gas becomes denser, and eventually the internal gas pressure cannot support the spheroid against self-gravity. At this point star formation should begin at a very fast rate.

The condition for equilibrium of a gas sphere embedded in a background dark matter potential was investigated by Spitzer (1942) in a different context, with stars replacing dark matter in his calculation. However, this distinction is unimportant to the equilibrium condition, which he wrote as a maximum gas mass able to support itself in a background potential for a given gaseous velocity dispersion. The existence of a maximum gas mass applies even for an isothermal gas sphere, unlike the case for a pure isothermal sphere in the absence of background stars or dark matter. (Recall that a pure isothermal sphere has infinite mass, because it extends out to infinite radius with a density distribution that varies as r^{-2}.) The background potential in the star-gas case provides an additional inward force, much like the external pressure in the case of a pressure-bounded isothermal sphere. External pressure truncates an isothermal gas sphere at finite mass, and the gas inside that radius cannot tell if the additional inward force is from overlying matter or from the boundary pressure. But if the external pressure is too great, then the self-gravitating core cannot support itself, and the sphere must collapse. The same is true when there is a background potential from dark matter, even where there is no pressure boundary or gas truncation radius. The additional inward force from dark matter offsets the gas stability if the velocity dispersion is low enough (or the gas mass high enough), and the sphere must collapse.

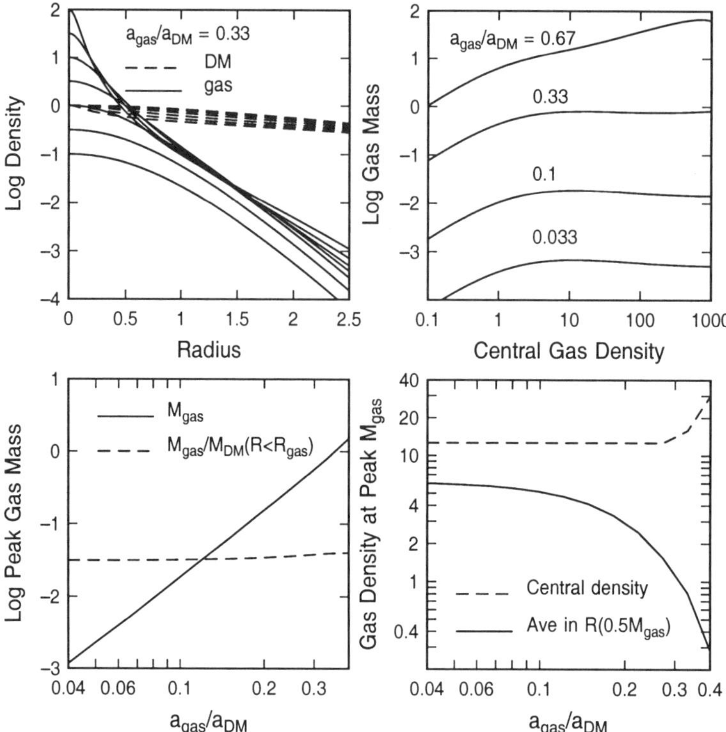

FIGURE 1. (top left) Density profiles for gas and dark matter in a spherical potential, considering different central gas densities. (top right) Gas mass versus central density for four values of the ratio of the gas to the dark matter velocity dispersion. (lower left) Maximum gas mass, and ratio of the maximum gas mass to the dark matter mass in the gas sphere, versus velocity dispersion ratio. (lower right) Central gas density at the maximum stable gas mass (solid line) and average gas density inside the radius containing half the total gas mass (dashed line), versus velocity dispersion ratio.

We show in Figure 1 various solutions for the density and mass in a two-component sphere in which dark matter and gas separately satisfy the conditions of pressure equilibrium, $\nabla P_{gas} = \rho_{gas} g$ and $\nabla P_{DM} = \rho_{DM} g$, for gravitational acceleration g, pressure P, and density ρ in the form of gas and dark matter (DM). The two components are coupled by their mutual contribution to the gravity in Poisson's equation: $\nabla \bullet \mathbf{g} = -4\pi G \left(\rho_{gas} + \rho_{DM}\right)$. We use normalized quantities in the Figure, with the basic scales provided by the central dark matter density, $\rho_{DM,0}$, the dark matter velocity dispersion, $a_{DM} = \left(P_{DM}/\rho_{DM}\right)^{1/2}$, and the gravitational length and mass scales given by these quantities, namely, $a_{DM}/\left(4\pi G \rho_{DM,0}\right)^{1/2}$ and $a_{DM}^3/\left(3\left[4\pi\right]^{1/2} G^{3/2} \rho_{DM,0}^{1/2}\right)$, respectively.

The top left diagram plots various density profiles for dark matter (dashed lines) and gas (solid lines), considering different central gas densities. The top right diagram shows the total gas mass as a function of the central gas density for various ratios of the gaseous velocity dispersion to the dark matter dispersion. When the gaseous velocity dispersion is small, the total gas mass peaks at an intermediate value of the central gas density, but as the gaseous dispersion approaches the stellar dispersion, the total mass of the gas sphere approaches infinity, as it does for an isothermal sphere.

The lower left diagram shows the peak total gas mass versus the velocity dispersion ratio as a solid line, and the ratio of the total gas mass to the dark matter mass inside the gas sphere as a dashed line. The gas mass ratio is about 0.03 at low dispersion ratio. The final result is the gas density ratio, plotted in the lower right. When the central gas density exceeds about 12.6 times the central dark mass density, or the average gas density inside the half-mass gas radius exceeds about 6 times the central dark matter density, there are no stable solutions which permit the addition of more mass to a cool gas component of a gas+dark matter galaxy. Any additional gas mass leads to catastrophic collapse of the gas sphere.

If write the total gas+dark matter density as the virial density, then this result gives a critical average gas density $\rho_{crit} \sim 6\rho_{DM}$, which implies

$$\rho_{gas} \sim \rho_{vir}. \tag{2.1}$$

This condition implies that once the centralized gas density dominates the total density in all forms, including dark matter, the gaseous component in a growing galaxy goes catastrophically unstable and forms stars.

The same condition applies to a flat disk if we rewrite the usual stability condition, involving the Toomre Q parameter, in terms of density rather than column density, and if we write the epicyclic frequency that appears in Q in terms of the background galaxy density ($\kappa \sim 2.9\,[G\rho_{vir}]^{1/2}$ for a flat rotation curve and galaxy virial density ρ_{vir}).

The association of a critical gas density with the total virial density makes sense because the self-gravity of lower density material is easily resisted by tidal forces, shear, and virialized turbulent motions. Once the gas density comes to within a factor of ~ 2 of the total virial density, gaseous self-gravity begins to dominate all these other processes.

3. A Threshold Star Formation Rate and Luminosity

The maximum star formation rate per unit volume in a region is the efficiency of star formation on a local level, ϵ, times the gravitational rate, $(G\rho)^{1/2}$, multiplied by the density. If the density is comparable to the virial density, then this gives

$$SFR \approx \frac{\epsilon \rho_{vir}}{t_{dyn}} \sim \epsilon G^{1/2} \rho_{vir}^{3/2} \sim \frac{\epsilon}{G t_{dyn}^3}, \tag{3.1}$$

where $\epsilon \sim 0.1$ is the fraction of the gas that is turned into stars in each major star formation event. The dynamical time is $t_{dyn} \sim R/V$ for galactocentric radius R and rotation or virial speed V. For $t_{dyn} = 10^7 t_{dyn,7}$ years, this rate becomes

$$SFR \approx \frac{220\epsilon}{t_{dyn,7}^3} \ \text{M}_\odot \ \text{kpc}^{-3} \ \text{yr}^{-1}. \tag{3.2}$$

The total rate at which stellar mass is added to the bulge is this SFR multiplied by the bulge volume, $4\pi R^3/3$. However, R^3 appears in the denominator of the SFR inside t_{dyn}^3, so it cancels out of the equation, leaving only V^3 in the numerator. Thus we can write for the total mass rate:

$$\frac{dM_{bulge}}{dt} = SFR \times \frac{4\pi R^3}{3} = 10^3 \epsilon V_{100}^3 \ \text{M}_\odot \ \text{yr}^{-1}, \tag{3.3}$$

where $V_{100} = V/100$ km s^{-1}. This remarkably simple expression says that it takes $\sim M_{bulge}/\left(10^3 \epsilon V_{100}^3\right) \sim 0.01 M_{bulge}/V_{100}^3$ years to make a bulge of mass M_{bulge} and velocity dispersion V_{100}, regardless of the size of the bulge, where M_{bulge} is in M$_\odot$ and $\epsilon \sim 0.1$. In the Milky Way, $V_{100} \sim 2$ and $M_{bulge} \sim 10^{10}$ M$_\odot$ (Gilmore, King, & van der Kruit 1989), so it took $\sim 10^7$ years to make most of the bulge (i.e., within 1 kpc) at the threshold rate given by the virial density. This also equals t_{dyn}/ϵ for these parameters.

Evidently, the formation of a typical galactic bulge could be mostly over within $10-100$ million years once it starts, i.e., within $\sim t_{dyn}/\epsilon$ for a complete conversion of gas into stars. This result is consistent with recent observations of rapid bulge formation in the Milky Way (Renzini 1999, these proceedings).

The mass formation rate corresponds to a certain luminosity during bulge formation that can be determined using the conversion from star formation rate to volume emissivity given by Meurer *et al.* (1997). The result is

$$L \sim 4 \times 10^{12} \epsilon \left(\frac{V}{100 \text{ km s}^{-1}}\right)^3 L_\odot. \tag{3.4}$$

The starburst luminosity during bulge formation depends only on the depth of the potential well, measured as a virial velocity dispersion. For merging galaxies, the dynamical speed can be $V \sim 200$ km s^{-1} or more, leading to star-formation luminosities of $L \sim 10^{12}$ L$_\odot$ for times of $t_{dyn} \sim 10^8$ years ($t_{dyn}/\epsilon \sim 10^9$ years if the residual gas from inefficient star formation continues to make stars). For the Milky Way bulge, with $V_{100} \sim 2$, $L \sim 3 \times 10^{12}$ L$_\odot$ during its formation with $\epsilon \sim 0.1$. For small bulges in which $V \sim 100$ km s^{-1}, $L \sim 10^{11}$ L$_\odot$. The total formation time is about $t_{dyn}/\epsilon \sim 10R/V \sim 10^7$ years. These luminosity values are consistent with observations of the Lyman α emitting galaxy at $z = 3.4$ found by Giavalisco *et al.* (1995), who suggest they are witnessing the formation of a galactic bulge or elliptical galaxy (see also Lilly 1999, these proceedings).

4. Self-Regulation of Star Formation

Star formation may not be able to proceed at the maximum rate given above if the pressures from young stars disrupt the gas and lower the density to below the critical value. Such self-regulation has been applied to galactic disks (Goldreich & Lynden-Bell 1965; Franco & Shore 1984) and dwarf galaxies (Dekel & Silk 1986). Does it also apply to bulges?

Meurer *et al.* (1997) estimated the pressure from a star-forming region to be

$$P_{SF} \sim 7.2 \times 10^{-11} \text{dyne cm}^{-2} \times \left(\frac{SFR_{2D}}{\text{M}_\odot \text{ kpc}^{-2}\text{yr}^{-1}}\right) \tag{4.1}$$

$$= 1.1 \times 10^7 \text{dyne cm}^{-2} \times \left(\frac{SFR_{2D}}{\text{gm cm}^{-2}\text{s}^{-1}}\right). \tag{4.2}$$

This pressure is proportional to the two-dimensional star formation rate, SFR_{2D}, which is the galactic thickness $2H$ times the 3-dimensional rate from equation 3.1. Thus, at threshold (in cgs units),

$$P_{SF} \sim 1.1 \times 10^7 \epsilon \left(G\rho_{vir}\right)^{1/2} \rho_{vir} 2H. \tag{4.3}$$

This disruptive pressure should be compared to the self-gravitational pressure that holds the region together, which is proportional to the square of the mass column density.

For a disk, $P_{grav,disk} \sim (\pi/2) G\sigma_{gas}\sigma_T$ for total column density in the gas layer, σ_T. This equation follows from the scale height $H^2 = a^2_{gas}/(2\pi G\rho_T)$ and from the relation $\sigma = 2H\rho$; then $P_{grav} = \rho_{gas}a^2_{gas}$, as expected. If we take $\rho_{gas} \sim \rho_{vir}$ at threshold, and $\rho_T \sim 3\rho_{vir}$ (which is the case in the Solar neighborhood), then $P_{grav,disk} \sim 6\pi G\rho^2_{vir}H^2$.

For a sphere, $P_{grav,sph} \sim (\pi/2) G\rho_T\rho_{gas}4H^2$ again if we define $H = a_{gas}/(2\pi G\rho_T)^{1/2}$ in the same way. But now $\rho_T \sim \rho_{vir}$ for the bulge problem, and $\rho_{gas} \sim \rho_{vir}$ inside the half-mass radius. Thus $P_{grav,sph} \sim 2\pi G\rho^2_{vir}H^2$.

To cover both cases, we write this pressure as

$$P_{grav} = \alpha\pi G\rho^2_{vir}H^2, \tag{4.4}$$

where $\alpha = 6$ and $\rho_T = 3\rho_{vir}$ for a disk, and $\alpha = 2$ and $\rho_T = \rho_{vir}$ for a sphere.

After combining these results, we see that the star formation pressure exceeds the self-gravitational binding pressure when

$$1.1 \times 10^7 \epsilon > (\pi/2)\,\alpha\,(G\rho_{vir})^{1/2}\,H \sim 1.7 a_{gas}, \tag{4.5}$$

where 1.7 is the average between $(1.5\pi)^{1/2} = 2.2$ for a disk and $(0.5\pi)^{1/2} = 1.2$ for a sphere. The coefficient on the left is in cgs units, namely, cm s^{-1}. Thus star formation pressure can blow apart a region at the threshold density if the gaseous velocity dispersion, in km s^{-1}, is less than a certain limit,

$$a_{gas} < 65\epsilon \ \ \mathrm{km}^{-1}\ \mathrm{s}^{-1}. \tag{4.6}$$

For typical $\epsilon \sim 0.1$, star formation can regulate or stop itself in a disk or dwarf galaxy where the gaseous velocity dispersion is low, $a_{gas} \sim 7$ km s^{-1}, but it cannot regulate or stop itself in a young galactic bulge or elliptical galaxy, where the velocity dispersion from the potential well is large, $a_{gas} \geq 100$ km s^{-1}. This result implies that *bulge formation should occur at close to the maximum rate with little regulation by star formation pressures.*

5. Conclusions

We have applied some standard concepts about star formation in normal galaxy disks to the formation of galactic bulges in the early Universe. These should be appropriate for the model in which the bulge forms in a spheroidal cloud following the accretion and compression of gas by self-gravity and/or protogalaxy interactions. The discussion does not apply to alternative bulge formation models in which the stars are added one at a time from a vertical disk instability, or where the stars come from the disk in another way, e.g., by first forming a thick bar that later dissolves.

We have suggested that there is a critical density for star formation to begin in a three-dimensional system, and this density is equal to several times the virial density from dark matter. This result leads to rather simple expressions for the star formation rate and luminosity, which turn out to depend only on the depth of the potential well. We also suggested that pressures from young stars in rapidly forming bulges are not enough to halt or slow the star formation process by winds or supernovae at the threshold star formation rate. This is unlike the situation for galaxy disks or dwarf galaxies, which, because of their lower velocity dispersions, are more easily disturbed by star formation pressures. Thus the bulges of galaxies could have formed very quickly, in only a few times 10^7 to perhaps 10^8 years.

REFERENCES

COMBES, F., DEBBASCH, F., FRIEDLI, D., PFENNIGER, D. 1990 *A&A*, **233**, 82

DEKEL, A., SILK, J. 1986 *ApJ*, **303**, 39

EGGEN, O.J., LYNDEN-BELL, D., SANDAGE, A. 1962 *ApJ*, **136**, 748

ELMEGREEN, B.G. 1999 *ApJ*, **517**, in press

FRANCO, J., SHORE, S.N. 1984 *ApJ*, **285**, 813

FRIEDLI, D., BENZ, W. 1993 *A&A*, **268**, 65

FRIEDLI, D., BENZ, W. 1995 *A&A*, **301**, 649

GIAVALISCO, M., MACCHETTO, F.D., MADAU, P., SPARKS, W.B. 1995 *ApJ*, **441**, L13

GILMORE, G., KING, I., VAN DER KRUIT, P. 1989, in *The Milky Way as a Galaxy*, p222. (Geneva Observatory)

GOLDREICH, P., LYNDEN-BELL, D. 1965 *MNRAS*, **130**, 97

HASAN, H., PFENNIGER, D., NORMAN, C. 1993 *ApJ*, **409**, 91

MEURER, G.T., HECKMAN, T.M., LEHNERT, M.D., LEITHERER, C., LOWENTHAL, J. 1997 *AJ*, **114**, 54

PFENNIGER, D., FRIEDLI, D. 1991 *A&A*, **252**, 75

SPITZER, L. JR. 1942 *ApJ*, **95**, 329

WADA, K., HABE, A., SOFUE, Y. 1995 *PASJ*, **47**, 1

Dynamical Timescales of Bulge Formation

By **KONRAD KUIJKEN** [1,2]

[1] Kapteyn Institute, PO Box 800, 9700 AV Groningen, The Netherlands

[2] Visiting Scientist, Dept. Theoretical Physics, University of the Basque Country

The relevant dynamical processes for bulge formation are reviewed: collapse, accretion, bar formation, stochastic heating and external forcing. All of these processes take place at some level, but it appears hard to escape the conclusion that bulges formed quickly and early.

1. Introduction

There are many beautiful examples of galaxies with prominent bulges to be found in any atlas of normal galaxies. Superficially bulges appear to share many properties with elliptical galaxies: there are many similarities in shape, stellar population and stellar dynamics between ellipticals and bulges (see Wyse, Gilmore and Franx 1997 for a review). To the eye, bulges appear to be a quite distinct component of disk galaxies. They appear as high surface brightness, concentrated central objects in the, often considerably larger and fainter, disk. Often also, they have quite a distinct colour, and, as the name suggests, they are considerably fatter than the disk.

Tempting as it is to consider bulges as a completely separate subsystem of galaxies, it would be wrong to ignore the relation between a bulge and the surrounding galaxy. Gravity links all components of a galaxy, and the central position of the bulge means that it will certainly be influenced by the other parts. Also, the bulge is the natural place for dissipated material to end up, and this is likely to have a profound effect on the star formation history of the bulge. It is also where almost all stellar populations will have their greatest density, making it hard to disentangle the central regions into clean well-separated components. For these reasons it is important to work with clear definitions of what really constitutes the bulge—or at least to recognize that one is not working within a strict definition at all! Since I know of no satisfactory definition, I shall not adopt one here; instead I will focus on those processes likely to affect the central kpc or so of a galaxy like the Milky Way.

Many aspects of bulges have been discussed at this meeting. The present review is intended as a summary of the dynamical processes that are likely to be relevant for bulge formation, with particular emphasis on the timescales over which they operate. I will discuss cooling, accretion, dynamical instabilities, phase mixing, external forcing and internal heating mechanisms that affect the central regions of disk galaxies.

2. A Crude Mass Model of the Galactic Bulge

As a reference a simple model is adopted which describes the Galactic Bulge reasonably well, though by no means perfectly. In particular no attempt is made to describe the innermost part of the bulge, which are dominated by a massive object (e.g., Eckart & Genzel 1997), or its flattening. It is a spherical exponential model: exponential profiles have been shown to give good descriptions of the central regions of many disk galaxies (Andredakis & Sanders 1994). Thus, it has a density profile

$$\rho = 10e^{-r/400\,\mathrm{pc}}\,\mathrm{M_\odot pc^{-3}} \tag{2.1}$$

which corresponds to a total mass of $1\text{--}2\times10^{10}\,\mathrm{M_\odot}$ (Kent 1992). More sophisticated analyses result in similar density profiles and masses (Blum 1995, Kent 1992, Dwek *et al.* 1995, Binney *et al.* 1991). The dynamical timescale at a radius of 1kpc (the time required to cross the bulge at the orbital speed of $(GM/r)^{1/2} \sim 175\,\mathrm{km\,s^{-1}}$) is around 10^7 yr. Apart from factors of order unity (which in the case of accidental resonances can be very small, though), this is also the timescale for orbital oscillations in the radial and vertical directions. Therefore any instability which operates as a consequence of collective responses of stellar orbits will try to act on timescales of this order.

Note that in the central nuclear region of the Galaxy the dynamical time is only a few 1000 yr.

3. Cooling

A natural way of forming dense objects such as bulges out of an initially nearly homogeneous universe is to invoke some form of dissipation. The timescale for dissipation of gas in the context of galaxy formation was considered by Rees and Ostriker (1977), who compared the cooling time of a gas cloud to the free-fall time $t_{\rm ff}$, which is the time in which it would collapse under its own gravity in the absence of pressure. For a spherical cloud of mass M and radius $R(t)$ starting at rest we have:

$$\frac{d^2 R}{dt^2} = -\frac{GM}{R^2} \quad \Rightarrow \quad t_{\rm ff} = \left(\frac{2}{3\pi G\rho}\right)^{1/2}, \tag{3.1}$$

whereas the cooling time is a more complicated function of density (see Figure 1). Wherever the cooling time is shorter than the collapse time, the former may be ignored and the gas cloud is able to cool sufficiently rapidly to collapse under its own gravity. It is evident that on galactic, and especially on galactic bulge scales, cooling is sufficiently efficient for it not to impede collapse of cold gas clouds. Therefore any time a bulge mass of gas is assembled, it will collapse on a dynamical time unless a source of heating exists which is able to support the cloud against collapse. The most plausible such source is vigorous star formation in the collapsing gas: the issue is considered further by Carlberg (this volume).

4. Accretion

Accretion of satellites or globular clusters takes place because of dynamical friction, primarily against the dark halo surrounding the galaxy. In the Galaxy, the dynamical friction on the Magellanic Clouds will lead to their absorption in about 10^{10} yr. In general, a satellite of mass $M_{\rm sat}$ orbiting at radius R in the halo, on a near-circular orbit, will experience a deceleration

$$\dot{v} \sim -5\frac{GM_{\rm sat}}{R^2} \tag{4.1}$$

(e.g., Binney and Tremaine 1987), leading to an angular momentum loss rate of

$$\dot{L} = -R\dot{v} \sim -5\frac{GM_{\rm sat}}{R} \tag{4.2}$$

and hence a sinking timescale of

$$\frac{R}{\dot{R}} \sim 10^{10}\,\mathrm{yr} \left(\frac{R}{10\,\mathrm{kpc}}\right)^2 \left(\frac{v_{\rm circ}}{200\,\mathrm{km\,s^{-1}}}\right) \left(\frac{M_{\rm sat}}{10^8\,\mathrm{M_\odot}}\right)^{-1}. \tag{4.3}$$

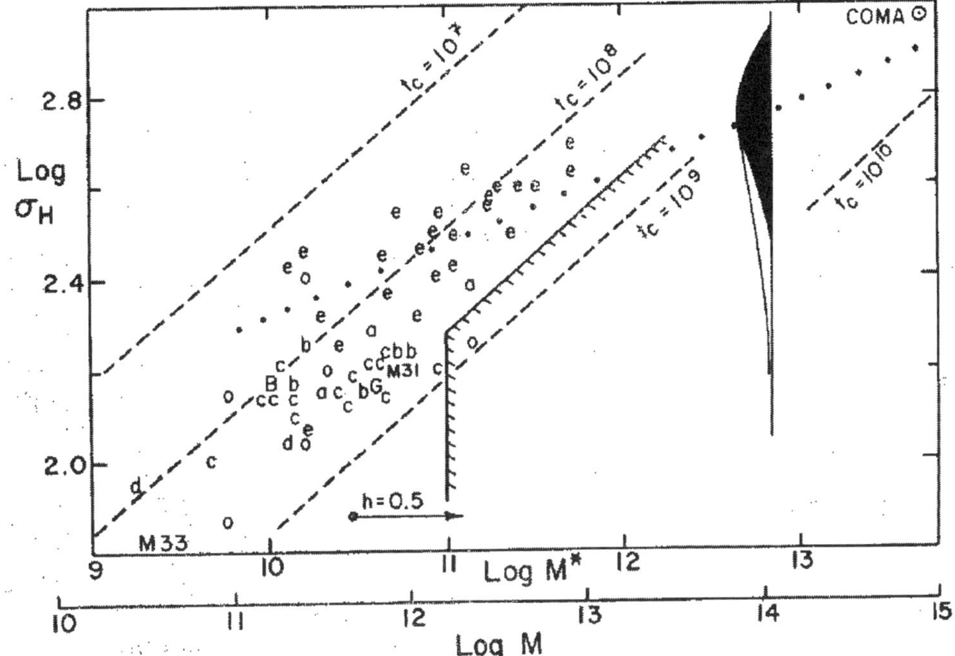

FIGURE 1. The cooling diagram for self-gravitating gas clouds, from Gunn (1982). Mass of baryons M^* is plotted versus temperature (velocity dispersion σ_H of the halo, in $\mathrm{km\,s^{-1}}$). Diagonal dashed lines indicate lines of constant free-fall time t_c, in years. The hashed region is bounded by the curve on which cooling and free-fall timescales are equal: these objects can cool sufficiently fast that they can undergo free-fall collapse. The symbols indicate different types of galaxies for which data were available at the time. The Galactic bulge is indicated by a B.

Accretion of satellites will thus have only a limited impact on present-day galaxies; however at early times when separations were smaller, such accretion/merging may have taken place on a timescale of Gyr.

In reality these estimates need to be modified because the satellite is tidally stripped as it sinks. Galactic tidal forces on a satellite orbiting at radius R strip stars outside a radius r in the satellite where

$$\bar{\rho}_{\mathrm{bulge}}(< R) \sim \bar{\rho}_{\mathrm{sat}}(< r) \tag{4.4}$$

and thus as the satellite sinks and the tidal fields get stronger, the satellite is progressively stripped of its envelope. This reduces its mass and the dynamical friction on the satellite, and as a consequence accretion onto the central bulge region is slowed and reduced. Hence accretion of stellar material by the Galaxy is unlikely to produce a significant bulge at the present time: satellites take too long to make it into the center, and they lose much mass before they get there. In a nice paper Syer and White (1998) have shown how tidal stripping naturally leads to a central cusp with density $\sim R^{-1}$ when a galaxy halo is built from a succession of accreted satellites, confirming that most of the mass does not end up in the center.

A further constraint on accretion is provided by the disk: a massive satellite accretion event will significantly heat the disk, whereas observed stellar disks are typically rather cold. Tóth and Ostriker (1992) have shown that the total amount of mass accreted by the Milky Way in the last 5Gyr (since the stellar disk formed) at radii inside the solar circle can be limited by the present-day kinematic energy in random motions of the disk to about 4% of the disk mass. A caveat on this result was given by Huang & Carlberg (1997) who showed that much of the energy could be absorbed by a large-scale disturbance of the disk such as a bodily tilt, rather than only in random motions; nevertheless an accretion origin of a bulge as important in mass as that of the Galaxy requires this accretion to take place before most of the stellar disk formed.

In fact, some accretion does take place in disk galaxies, as witnessed by the prevalence of counter-rotating nuclear gas disks in S0 galaxies (Bertola *et al.* 1992). It is difficult to see how this gas could not have an external origin, or how it does not imply that at least an equal number of galaxies have suffered prograde accretion events of a similar magnitude. Also some examples exist of retrograde stellar disks, indicating substantial infall of retrograde material over long time periods (Rubin *et al.* 1992, Merrifield and Kuijken 1994, Braun *et al.* 1994, Bertola *et al.* 1996). Since in a retrograde accretion of a gas-rich system substantial dissipation and cancellation of angular momentum between the pre-existing ISM and the accreted gas is expected, such events would have been prime candidates for dumping large amounts of gas into the middle of the galaxy. Nevertheless, an unbiased survey of some 30 S0 galaxies (Figure 2) revealed no new cases of counter-rotating galaxies, allowing strict limits to be set on the importance of retrograde accretion for the formation of stellar disks (Kuijken *et al.* 1996).

5. Instabilities and the Relation between Bulges and Bars

We now turn to an issue which is particularly relevant for the Galactic Bulge, but may have wider implications: that of the relation between bulges and bars.

There is now considerable evidence that the Galaxy is barred. This evidence comes from many directions, starting with de Vaucouleurs' (1964, 1970) mainly morphological arguments, but in the last decade augmented with a variety of photometric, kinematic and star-counting evidence as well as arguments based on the microlensing surveys towards the bulge. Much of this evidence has been reviewed before (e.g., Kuijken 1996), and will not be repeated here: suffice it to say that several very different diagnostics all place the major axis of the bar at least in the same Galactic quadrant (quadrant I, where $0 < l < 90°$), though there are still disagreements over the details of the bar shape and its pattern speed. The true situation may well be that there are several overlapping distortions of the central kpc of the Galaxy, including lop-sidedness and one or even two bars.

What is relevant here is the prediction from galactic dynamics that bars are quite generically unstable to buckling, distortion which forms in the plane of a disk as a result of instability will quickly thereafter thicken out of the plane into a structure with rather angular, 'boxy' isophotes (Combes and Sanders 1981, Combes *et al.* 1990, Raha *et al.* 1991). The buckling process, and an explanation of why it is so generic, is decribed in nice physical terms by Merrifield (1996). Numerical simulations show that the buckling occurs on a timescale of at most 10 orbital times.

The link between boxy isophotes and bars was argued to be observationally plausible because both phenomena are seen with comparable frequency in galaxy catalogues (Sellwood and Wilkinson 1993). Nevertheless it has proved hard to establish directly because the boxy isophotes are only visible in edge-on systems and bars require a more face-on view: only in one object, NGC 4442 (Bettoni & Galetta 1994) could both features be detected simultaneously. However, by searching edge-on galaxies for the kinematic signatures of the resonances associated with the pattern rotation of bars, it has now been shown (Kuijken and Merrifield 1995, Bureau 1998, Merrifield and Kuijken 1999) that

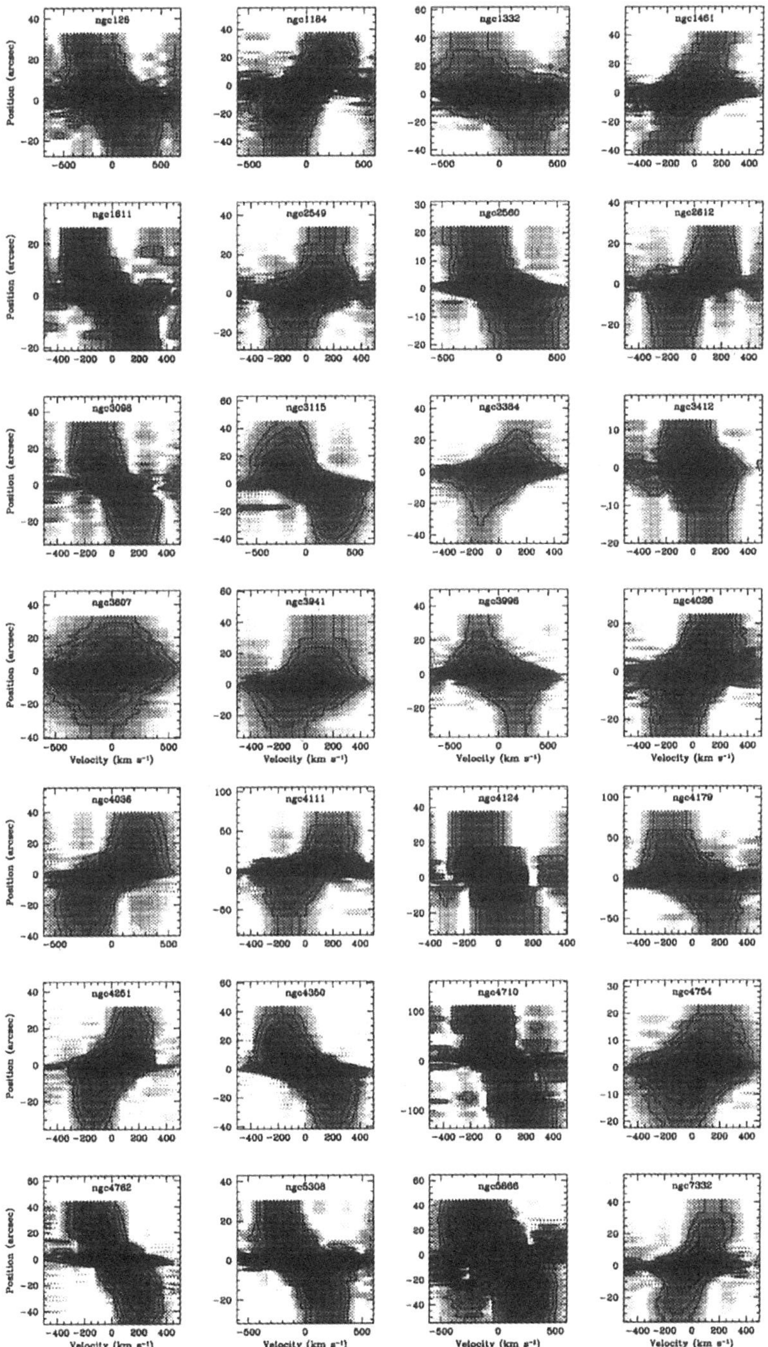

FIGURE 2. A subset of the S0-Sa galaxies surveyed by Kuijken *et al.* (1996) for evidence of counterrotating stellar disks. In these panels, the density of stars as a function of projected radius and line-of-sight velocity is shown for each galaxy. No cases of counter-rotating disks were found in this randomly selected sample of edge-on galaxies from the RC3 catalogue. Note that also the bulges generically rotate in the same sense as the disks.

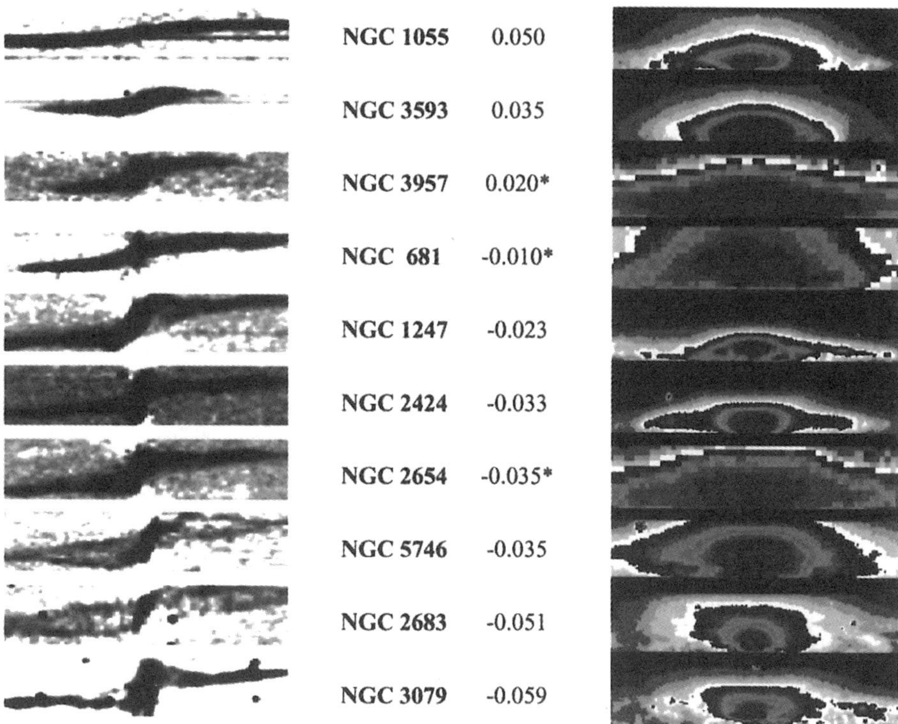

FIGURE 3. Line-of-sight Hα kinematics (left) of a sample of edge-on galaxies with different bulge isophote shapes (right), from Merrifield and Kuijken (1999). Bars produce gaps and multiple components in the line profiles, since the resonances associated with their pattern rotation carve up the orbital structure. Such gaps are clearly seen in the galaxies with the more rectangular and peanut-shaped bulge isophotes, shown on the right and not in those with round or elliptical ones. These data thus confirm the predicted relation between bars and box- or peanut-shaped bulges.

edge-on galaxies with boxy isophotes do indeed exhibit barred kinematics, whereas those with rounder isophotes do not (see Figure 3).

Could bars, and in particular their vertical instability, be responsible for all bulges? The evidence from stellar populations is that bulges tend to be significantly older than the surrounding disks (see papers by Rich, Renzini, and Peletier & Davies in this volume), and therefore it is unlikely that bars are forming bulges at the present day. Nevertheless this mechanism offers some intriguing possibilities. Bar formation occurs on a dynamical timescale, but may occur late: studies of the HDF indicate that bars are more prevalent after a redshift of 0.5 than at earlier times (Abraham *et al.* 1999). A comparable study of bulge-to-disk ratio, though more challenging, would be an interesting complement to such a study.

Bar formation is also a secular process which alters the structure of the surrounding disk. Bars will generate radial flow patterns capable of funneling the ISM towards the center of the galaxy. The added mass in the center may in turn affect the orbit structure of the bar itself: as show by Hasan *et al.* (1993), bars may be destroyed by a small concentrated nuclear mass through a process of scattering of orbits away from those orbit families required to support the bar figure. Thus, a feedback loop might form: a disk forms a bar, which generates radial inflow, which forms a central mass, which destroys the bar. The process would leave the disk hotter than it was originally, but if it is able to cool through accretion of fresh ISM the cycle might restart.

Rix (this meeting, unpublished) has suggested that this cycle may not, in fact, recur; it may not even run completely once. Nevertheless, it is clear that

(a) galaxies form bars rather readily, but bars are found preferentially after redshift 0.5;

(b) bars are related to box-shaped bulges;

(c) the stellar populations of bulges appear to be older than those of the surrounding disks.

Hence at least some of the stars now seen in bulges high above the disk must have been put there through the action of bar instability and its buckling.

6. External Forcing

The Galaxy is orbited by at least three massive satellites: the two Magellanic Clouds and the Sgr dwarf. The latter is clearly being disrupted tidally by the Galaxy, and may be on its final orbit before disintegrating into the halo. However, the Magellanic clouds will survive for $\sim 10^{10}$ yr more.

Weinberg (1995, 1998) has been studying the effect of these massive orbiting satellites on the Galaxy halo and disk. His findings show that the effects may be much stronger than naively expected, through a combination of accidental resonances between the LMC orbit frequencies and natural frequencies of the Galactic disk, and the amplifying effect of the self-gravity of the halo on its response to the LMC's disturbance. Weinberg shows that when these effects are taken into account, the tidal effect of the LMC on the disk may be sufficiently amplified for it to cause the observed warp amplitude of the Milky Way disk. He also analysed the in-plane effect of the disturbance on the disk, and found a significant disturbance both in the $m = 1$ lop-sided terms and in the $m = 2$ bar terms. However, the difference between these disturbances and the bars discussed above is that the driving frequency is now set by the LMC orbit, resulting in timescales several orders of magnitude longer than those of intrinsic bar instability. In particular, these bars will have a very slow pattern rotation.

Measurements of pattern speeds of bars tend to favour the shorter pattern speeds expected of bars formed through disk instability: for a review see Elmegreen (1996); since then the pattern speed of the S0 galaxy NGC 4596 has also been shown to be high (Gerssen *et al.* 1999).

In summary, external disturbance of disks may provide convincing explanations of some phenomena such as warps or lopsidedness. However, tides are not very effective at small radii, and their timescales are rather long for them to effectively couple to the central dynamics of a galaxy.

7. Phase Space Density Constraints

Some constraints on the dynamical history of a galaxy come from Liouville's theorem, which states that along stellar orbits the phase space density (the density in coordinate and velocity space combined) is conserved. However, in some circumstances the phase flow can be so convoluted that empty parts of phase space become mixed in with populated regions. The net result is therefore that phase space density may decrease as a result of phase mixing, but it cannot increase (the result is analogous to stirring some milk into a cup of coffee: the milk will mix and its density will decrease irrevocably).

Liouville's theorem holds only in the absence of dissipation, i.e., in collisionless dynamics.

Now, bulges are characterized by higher central densities than disks, whereas their velocity dispersions are broadly similar to the inner parts of the surrounding disks. Hence the phase space densities of bulges are generally higher than those of the inward extrapolations of disks, and we can conclude that bulges as we see them are *not* the result of dissipationless evolution of the disk (Carlberg 1986; Wyse 1998). Or can we?

In order to make the above argument, we need to know what the initial density is of a disk. To illustrate how uncertain this value is, let's return to an old scheme for disk formation due to Mestel 1963, and revisited by Gunn (1982) and van der Kruit (1987). In this picture a disk is formed as a result of detailed angular momentum conservation in the collapse of a uniform gas sphere in solid-body rotation. We assume the disk carries little mass compared to a dark halo gravitational potential, which we take to be of the simple 'Binney' form (asymptotically flat rotation curve of amplitude v_0, inner core of radius a)

$$\Psi(r) = v_0^2 \ln(a^2 + r^2)^{1/2}. \tag{7.1}$$

The gas sphere has a radius chosen so that the angular momentum at its edge corresponds to that of a particle on a circular orbit in the dark halo potential at radius $r = 1$. All disk material settles into a plane perpendicular to the angular momentum vector of the sphere, with each gas parcel's radius defined by its angular momentum. The result is shown in Figure 4 for various choices of the core radius a of the dark matter potential. Note that current predictions are that dark halos are cuspy (Dubinski and Carlberg 1991; Navarro etal 1997).

While for $R > 0.2$ most solutions are close to the observed exponential disks with cutoff radius near 4.5 exponential scale lengths, the solutions with small halo core radius have much higher central densities than the inward extrapolations. As the core radius approaches zero, in fact the central disk density increases without limit.

The point of this exercise is not to claim that these are realistic ways of forming galaxy disks, but rather to show that there may be a lot of freedom in the central densities of proto-disks as they collapse into dark matter halos. Hence it may be that the formation of a gas disk during a galaxy's collapse is able to raise the central phase space density to levels at least as high as those observed in galaxy bulges at the present. Arguments that phase space constraints rule out formation of bulges out of disks thus have to be treated with care, especially since the sort of evolution that may turn a disk into a bulge will be most likely to occur in the highest-density disks.

8. Heating

Another possible origin of the high random energies of the bulge stars is gravitational scattering. It is well-known that the two-body relaxation time for a stellar system consisting of N equal-mass stars is

$$\frac{0.1N}{\ln N} t_{\rm cross}, \tag{8.1}$$

which for the bulge ($N \sim 10^{10}$, $t_{\rm cross} \sim 10^7$ yr) is many Hubble times. However, some sort of scattering does take place in stellar disks, as witnessed by the steady increase of the velocity dispersion of stars in the solar neighbourhood with their age (e.g., Wielen 1977). The probable cause of this heating is not two-body relaxation between stars, but

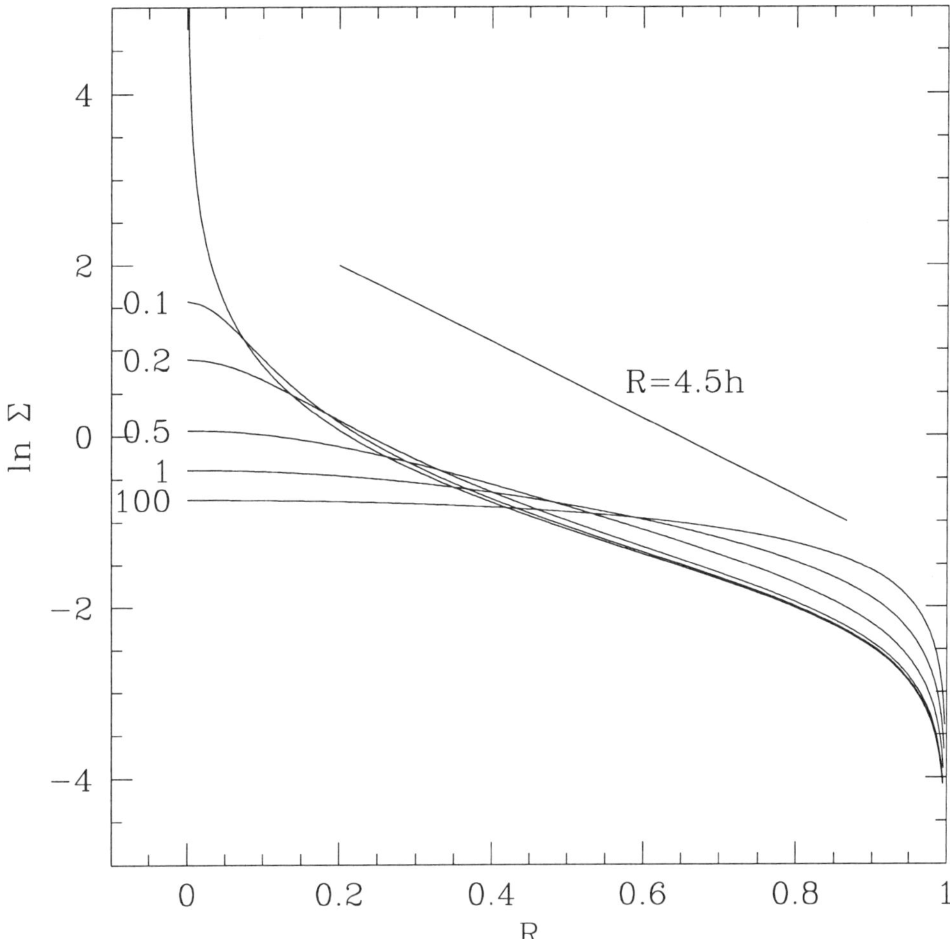

FIGURE 4. The surface density profile of disks made from detailed angular momentum conservation during the collapse of a homogeneous gas sphere in solid body rotation into the gravitational potential of a dark halo of core radius a. Note that while the density profile is roughly exponential at large radii, with a cutoff near 4.5 scale lengths, the central density is a strong function of halo core radius.

rather a combination of scattering off spiral structure and off molecular clouds. The combination of these effects was studied by Jenkins and Binney (1990), who showed that after a time

$$5 \left(\frac{h_{cl}}{60 \, \mathrm{pc}} \right)^4 \left(\frac{\nu}{90 \, \mathrm{km \, s^{-1} kpc^{-1}}} \right)^3 \left(\frac{\Sigma_{cl} \overline{M_{cl}^2} / \overline{M_{cl}}}{10^6 \, \mathrm{M_\odot^2 pc^{-2}}} \right)^{-1} \mathrm{Gyr} \qquad (8.2)$$

the velocity dispersion of disk stars doubles as a result of scattering by molecular clouds. Again, this is a rather slow effect, but the parameters are adopted to conditions in the solar neighbourhood. In the central kpc of the Galaxy we find a smaller scale height h_{cl} of clouds, a higher vertical oscillation frequency ν, and a lower surface density of molecular material. Since the timescale depends rather strongly on some of these factors it is hard to estimate them accurately, but it seems safe to say that it will be measured in Gyr.

Molecular clouds will affect the central dynamics in another way. These massive objects experience dynamical friction against the surrounding stars. As a total cloud mass of $M_{\rm cl}$ sinks from radius R_i to R_f, it loses potential energy $\Delta E \sim M_{\rm cl} v_{\rm circ}^2 \ln(R_i/R_f)$. This energy is converted to kinetic energy of the stars, who therefore experience an increase in their velocity dispersion $\Delta\sigma_*$ of

$$\frac{\Delta\sigma_*}{\sigma_*} = \frac{1}{2}\frac{\Delta E}{KE_*} \sim \frac{v_{\rm circ}^2 M_{\rm cl} \ln 10}{3 M_* \sigma_*^2} \sim 3\frac{M_{\rm cl}}{M_*} \tag{8.3}$$

(taking $\sigma_* \simeq 100\,{\rm km\,s}^{-1}$, $v_{\rm circ} \simeq 200\,{\rm km\,s}^{-1}$, and $R_i \simeq 1\,{\rm kpc}$, $R_f \simeq 100\,{\rm pc}$).

This result tells us that any inflow of clouds into the bulge will also increase the dispersion of those stars already there. Even a moderate cloud mass on the order of 10% of the present bulge mass will substantially increase the velocity dispersion in the central disk, and therefore puff up the disk.

9. Conclusions

The fundamental dynamical timescale for a bulge is the crossing time. A typical value is $t_{\rm cr} \sim 1\,{\rm kpc}/100\,{\rm km\,s}^{-1} = 10^7$ yr. All large scale dynamical phenomena in a bulge will occur on this, very short, timescale. It sets the timescale for monolithic collapse of a gas cloud into a bulge (we have seen that cooling efficiency is not a problem), and for dynamical instabilities such as bar formation and the subsequent thickening of the bar.

Below I list the processes which may affect bulge structure which occur over longer timescales. All of these occur at some level, but there is no clear evidence that any of them are a dominant factor in the formation of bulges. Furthermore, stellar population evidence also points in the direction of early and rapid formation of bulges.

(*a*) *Accretion.* This is unlikely to be significantly altering bulge structure at present. Galaxies are surrounded by large dark halos which provide dynamical friction that will bring satellites into the center of a galaxy on 1–10 Gyr timescales, but tidal stripping will mean that much of the satellite mass will be lost before it makes it into the middle. Only very centrally concentrated objects would stand a chance of adding significantly to the bulge, but timescales are very long.

(*b*) *Bar instability.* We see many disks with bars, and theory predicts that these bars will quickly buckle and inject stars into the space above the central disk, resulting in a boxy isophote profile. There is now evidence that this predicted relation between box-shaped bulges and bars does indeed hold in nature. Given the ubiquity of bars this process would seem to have the potential of being an important factor in the formation of bulges, but in fact stellar population comparisons of boxy bulges and the surrounding disk show that the disk is almost invariably younger than the bulge; hence if the bars did create the bulge, they must have done so a long time ago and the disk must have formed many new stars since then. At present I would say that the conflicting evidence from dynamics (boxy bulges are clearly related to bars) and from stellar populations (the bulge and disk appear unrelated) means that the importance of this dynamical process, which must surely take place in barred galaxies, is still not clear.

(*c*) *Perturbation by satellites.* Massive satellites of a galaxy may perturb the inner regions through tidal effects, and sometimes accidental resonances with the central regions can lead to surprisingly strong couplings. Nevertheless, the strongest effects are seen in the outer regions of the disk, where dynamical timescales are closer to the orbital frequencies of the distant satellites. Because these frequencies are low, satellites do offer the possibility of forcing very long timescale evolution of their host galaxy, but effects in the central kpc are likely to be small.

(*d*) *Stochastic heating.* Molecular clouds sinking into the central regions of the galaxy will deposit much of their potential energy into the stellar populations. If, for example as a result of bar-like streamlines, a significant amount of mass in clouds flows into the central regions, this will lead to a substantial increase in the velocity dispersion of the bulge. In a thin disk, the result would be a central thickening of the disk into a bulge-like component.

In summary, bulges offer a rich variety of dynamical processes, even if it is not clear how many of these processes are important for the overall structure of the bulges we observe. A systematic study of bulge properties as a function of redshift would be an important pointer towards the importance, or otherwise, of bar-related evolution to bulge formation.

REFERENCES

ABRAHAM R.G., MERRIFIELD M.R., ELLIS R.S., TANVIR N., BRINCHMANN J. 1999 *MNRAS*, submitted

ANDREDAKIS, Y.C., SANDERS, R.H. 1994 *MNRAS*, **267**, 296

BERTOLA, F., BUSON, L.M., ZEILINGER, W.W. 1992 *ApJ*, **401**, L81

BERTOLA, F., CINZANO, P., CORSINI, E.M., PIZZELLA, A., PERSIC, M., SALUCCI, P. 1996 *ApJ*, **458**, L67

BETTONI, D., GALLETTA, G. 1994 *A&A*, **281**, 1

BINNEY, J., GERHARD, O.E., STARK, A.A., BALLY, J., UCHIDA, K.I. 1991 *MNRAS*, **252**, 218

BINNEY, J., TREMAINE, S. 1987 *Galactic Dynamics.* (Princeton)

BLUM, R.D. 1995 *ApJ*, **444**, L91

BRAUN, R., WALTERBOS, R.A.M., KENNICUTT, R.C.J., TACCONI, L.J. 1994 *ApJ*, **420**, 558

BUREAU, M. 1998, Ph. D. Thesis, Australian National University

CARLBERG, R.G. 1986 *ApJ*, **310**, 593

COMBES, F., DEBBASCH, F., FRIEDLI, D., PFENNIGER, D. 1990 *A&A*, **233**, 82

COMBES, F., SANDERS, R.H. 1981 *A&A*, **96**, 164

DE VAUCOULEURS, G. 1964, in *IAU* **20** (ed. F.J. Kerr & A.W. Rodgers), p88

DE VAUCOULEURS, G. 1970, in *IAU* **38** (ed. W. Becker & G. Contopoulos), p18

DUBINSKI, J., CARLBERG, R.G. 1991 *ApJ*, **378**, 496

DWEK, E., ET AL. 1995 *ApJ*, **445**, 730

ECKART, A., GENZEL, R. 1997 *MNRAS*, **284**, 598

ELMEGREEN, B. 1996, in *Barred Galaxies* (ed. R. Buta, D.A. Crocker & B.G. Elmegreen), ASP Conf. Ser. **91**, p197. (ASP)

GERSSEN, J., KUIJKEN, K., MERRIFIELD, M.R. 1999 *MNRAS*, in press

HASAN, H. , PFENNIGER, D., NORMAN, C. 1993 *ApJ*, **409**, 91

HUANG, S., CARLBERG, R.G. 1997 *ApJ*, **480**, 503

GUNN, J.E. 1982, in *Astrophysical Cosmology Proceedings*, p259. (Pontificia Academia Scientiarum)

JENKINS, A., BINNEY, J. 1990 *MNRAS*, **245**, 305

KENT, S.M. 1992 *ApJ*, **387**, 188

KUIJKEN, K. 1996 in *Barred Galaxies* (ed. R. Buta, D.A. Crocker & B.G. Elmegreen), ASP Conf. Ser. **91**, p504. (ASP)

KUIJKEN, K., FISHER, D., MERRIFIELD, M.R. 1996 *MNRAS*, **283**, 543

KUIJKEN, K., MERRIFIELD, M.R. 1995 *ApJ*, **443**, L13

MERRIFIELD, M.R. 1996, in *Barred Galaxies* (ed. R. Buta, D.A. Crocker & B.G. Elmegreen), ASP Conf. Ser. **91**, p179. (ASP)

MERRIFIELD, M.R., KUIJKEN, K. 1994 *ApJ*, **432**, 575

MERRIFIELD, M.R., KUIJKEN, K. 1999, in preparation

MESTEL, L. 1963 *MNRAS*, **126**, 553

NAVARRO, J.F., FRENK, C.S., WHITE, S.D.M. 1997 *ApJ*, **490**, 493

RAHA, N., SELLWOOD, J.A., JAMES, R.A., KAHN, F.D. 1991 *Nature*, **352**, 411

REES, M.J., OSTRIKER, J.P. 1977 *MNRAS*, **179**, 559

RUBIN, V.C., GRAHAM, J.A., KENNEY, J.D.P. 1992 *ApJ*, **394**, L9

SELLWOOD, J.A., WILKINSON, A. 1993 *Rep Prog Phys*, **56**, 173

SYER, D., WHITE, S.D.M. 1998 *MNRAS*, **293**, 337

TOTH, G., OSTRIKER, J.P. 1992 *ApJ*, **389**, 26

VAN DER KRUIT, P.C. 1987, in *The Galaxy*, p27 (Dordrecht: Reidel Publishing Co.)

WIELEN, R. 1977 *A&A*, **60**, 263

WEINBERG, M.D. 1995 *ApJ*, **455**, L31

WEINBERG, M.D. 1998 *MNRAS*, **299**, 499

WYSE, R.F.G. 1998 *MNRAS*, **293**, 429

WYSE, R.F.G., GILMORE, G., FRANX, M. 1997 *ARA&A*, **35**, 675

Part 4

PHYSICAL PROCESSES IN BULGE FORMATION

This section focuses on physical processes involved in forming bulges, with particular emphasis on the creation/destruction of bars, star-formation thresholds and supernova-driven winds, gas dynamics and the transition to a multi-phase ISM.

Pfenniger leads off with a general introduction to galactic bars and their possible role in forming bulges. Bars provide a natural mechanism for moving stars out of the plane of the disk and gas into the center of a galaxy; they are likely to come and go. In principle, a galaxy could go through multiple episodes of creation and destruction of bars over a Hubble time.

Bureau *et al.* and Lütticke & Dettmar present evidence for roughly 50% of bulges being boxy/peanut-shaped and embedded in hosts with possibly more companions; Bureau *et al.* also argue that boxy/peanut-shaped bulges are mostly thick bars viewed edge on. Hasan discusses the role of a secondary bar in creating vertical resonances which enable stars to leave the galactic plane, thereby forming a bulge. Support to theories of bar-driven transport and constraints to the timescale of the possible bar dissolution come from the findings of Sakamoto *et al.*, who show that barred spirals have higher central molecular gas than non-barred galaxies. Other dynamical processes possibly relevant in the evolution of bulges are discussed by Valluri, who shows that stochastic orbits in a triaxial bulge will result in the evolution of its shape from triaxial to axisymmetric, and Ciotti, who suggests that the observed centrally steep density profiles of bulges may be a consequence of the shape of the underlying dark matter distribution.

While bars and secular evolution appear likely to contribute to bulge formation, the stellar populations anticipated from repeated cycles of bar creation and destruction are unlikely to look like the uniform, α-element enhanced old populations in 'classical' bulges discussed by Renzini and Rich earlier in this book. Spaans stresses that an efficient formation of stars depends strongly on the presence of metallic atoms and molecules, and investigates the evolution with cosmic time of star formation in a multi-phase ISM. Wada & Norman examine the nature of the ISM in the inner region of the galactic disk using two-dimensional hydrodynamical simulations that include star-formation feedback. They find that the ISM forms a quasi-stable filamentary structure. The resulting system cannot be described as a simple two- or three-phase medium with pressure equilibrium between the phases. The implications of the resulting picture for bulge formation need to be evaluated.

The Role of Bars for Secular Bulge Formation

By DANIEL PFENNIGER

Geneva Observatory, University of Geneva, CH-1290 Sauverny, Switzerland

The purpose of this paper is to describe the dynamics in bars as predicted by classical mechanics, and its relationship with bulge structure and evolution. From physical ground the tight dynamical relationship between small bulges (spheroidal looking structures about as large or smaller than the disk scale-length) and bars leads to doubt that they are decoupled and dynamically independent structures, as often assumed in bulge-disk decompositions or bulge modeling. For big bulges (spheroidal structures much larger than the disk scale-length) the link with bars is looser or indirect, such as a common origin through a merger. Simulations show that big bulges may result from mergers any time after disk formation. *Therefore, dynamics indicates that the age of bulges is not a generally well defined concept for bulges, because the formation of a large fraction of bulges can occur much later than the formation of their present material content, the stars.* More likely, if bulges are made for a substantial part of a mixture of disk material older than 1 Gyr (otherwise they would not be called bulges) the age spread of the stars is comparable to the age of the oldest stars, thus the ill-defined definition of bulge ages. In turn, the better understanding of all the possible origins of bulges contributes to a revision of the interpretation of the Hubble sequence, which appears as a broad aging sequence of individual galaxies, from late to early types. Aging is not tightly related to time, but to the number of 'events', such as mergers, bar formation and destruction, and the fraction of gas transformed into stars.

---

1. Introduction

1.1. *Bulges and Epistemology*

Astronomers are certainly used to taking an extra-terrestrial point of view and considering human activities as if they would not belong to them. This is not far from the Kuhnian attitude which takes science as object of study. Now, contemplating how astronomers work and interact may be quite instructive and may be used to correct our perception of celestial objects gained by proper astronomical observations and theoretical works. In some way, the 'personal aberration' that visual observers long ago used to take into account can, for the benefit of scientific progress, sometimes be evaluated on a whole group of scientists including ourselves. The topic of bulges provides good examples of possible group aberrations that I will discuss below.

A repeating sociological pattern, well known in epistemological circles, leading to a possible group aberration occurs as follow. One or a few pioneer scientists, already or becoming influential, begin to explore a subject and by necessity adopt tentatively simple working hypotheses. Over the years the subject is worked out by many followers in more details, but the initial working hypotheses have survived. After several years, in the community minds the hypotheses belong to the pool of common wisdom, hardly distinguishable from the well verified facts. The longer the working hypotheses remain undisputed, the harder it becomes to challenge them and to adopt alternative hypotheses, and the less the next generation of scientists perceives the fragility of the initial still unverified hypotheses. Fortunately, sooner or later new advances lead to a revision of the hypotheses prematurely adopted as truth.

1.2. *Origin of Bulges as Subject of Study*

In the case of bulges, the initial starting point of their study is very simple. As we know, the mere visual perception of a frequent central luminous concentration in the inner parts of a disk galaxy, e.g., by inspections of photographic plates, triggers automatically interrogations on their nature. Edge-on views of disk galaxies, such as NGC 891, suggest particularly that these light concentrations are thicker than their surrounding disks, therefore the term 'bulge'. The name describes well the observed structure, therefore it has found a rapid adoption.

A cautionary remark is appropriate here: the scale-height inferred in edge-on disks from projecting mass has little relationship with the local scale-height as viewed from the plane, even when ignoring extinction effects. Indeed, the local density scale-height

$$h_z(x,y) = \frac{\int_0^\infty dz\, z\, \rho(x,y,z)}{\int_0^\infty dz\, \rho(x,y,z)} \qquad (1.1)$$

in many models and N-body models is constant or even *decreases* toward the galaxy center, while in the disk edge-on view the scale-height inferred from the projection along one more coordinate (say y) corresponding to the bulging perception:

$$H_z(x) = \frac{\int_0^\infty dz\, z \int_{-\infty}^\infty dy\, \rho(x,y,z)}{\int_0^\infty dz \int_{-\infty}^\infty dy\, \rho(x,y,z)} \qquad (1.2)$$

usually increases toward the center. Thus a bulge may correspond in 3D in fact to a mass depression toward the galaxy plane.

For a long time a precise definition of bulge has not been felt necessary, which led subsequently to numerous discussions about the differences with overlapping terms such as 'nucleus', 'spheroid', or 'halo'. Here I will argue that a 'bar' can also be mistaken as a bulge. The most famous case is the Milky Way bulge which is now recognized to be as well a bar with a boxy shape.

1.3. *Models vs. Observations*

To rationalize such light concentrations, bulge models had to be set up. To begin with, obviously simple hypotheses were adopted such as a spherical symmetry and time-independence (i.e., a stable equilibrium), discarding the possible significance of frequent slight departures to strict symmetry seen in projection (meaning that in 3D the distortion may be potentially much more pronounced). Because bulges are by definition surrounded by a spiral galaxy, the simplest way to model the combination of both differently *named* structures is to suppose their independence, or weak dynamical coupling. The different morphologies between a bulge and disk suggest distinct initial conditions. The scenario appearing the most plausible to explain bulges was for a long time to suppose that they formed before their surrounding disks in a rapid collapse. In this classical view the meaning of bulge ages makes sense, because the formation time is a small fraction of the time up to now.

In fact, as usual in macroscopic objects, attentive inspection of observational data reveals rapidly that few bulges are perfectly symmetric, although in comparison with disks the symmetry may appear 'fairly good'. Some bulges may be boxy, some others contain double cores, some may be shifted from the disk geometric centre, etc. Observing with further wavelengths often reveal still more irregular features, such that a minority of good cases remain in which the original assumption of sphericity is valid, say, at the percent level. The question is therefore to quantify the observed effects: is a departure from sphericity at a given level significant with respect to the *theory* (i.e., stellar dynamics) adopted for explaining the equilibrium of the perfectly symmetric models?

Beside morphology irregularities, colors can also disturb the apparent order of the bulge class. Some central regions of spirals do contain concentrations of stars, but they are too blue to fit in the adopted view that bulge must be old, so are rather ranged in the category of starbursts. These anomalous blue bulges cannot be 'fiducial' bulges since they contradict the definition, but obviously semantics does not help in understanding how they can fit in the broader class of 'central concentrations of stars surrounded by a disk'.

A disturbing fact is that a large fraction of the late type spirals do not possess any bulge. This proves at least that spirals can form first, without a preceding bulge. As the fraction of detected low-surface brightness galaxies can only increase with better instruments, the bulges appear less and less as 'normal', but rather as associated with the bright galaxy class, i.e., with galaxies that have succeeded in transforming faster a sizable part of their gas into stars.

In the following I will present the different known facts about the dynamics of the regions occupied by bulges which suggest strongly that bulges and galaxies in general are less symmetric and steady than envisioned by our predecessors. A major 'feature' of most face-on disk galaxies is a central bar or oval, that appears today to exist in a large majority of spirals at a dynamically significant level of non-axisymmetry, particularly over time-scales of order of 10 Gyr.

2. Basic Tools: Stellar Dynamics

2.1. *Hypotheses*

When assuming a theoretical framework, consistency requires us to consider all its predictions, not only the convenient or easy ones (say, the virial theorem). Stellar dynamics has been used for a long time in astronomy; its results permeate, sometimes implicitly, the astronomical knowledge such that it is often not immediately obvious where basic but fragile hypotheses have been made.

In comparison, classical mechanics has been applied from its beginning to celestial mechanics. Up to the invention of special relativity, classical mechanics was believed to be absolutely true†, so people were used to drawing the predictions of mechanics as far as possible, which is a sound practice even if the theory is imperfect.

In contrast, galactic astronomers did not draw dynamics as far as celestial mechanicians did. Among the many reasons why predictions in galactic dynamics could not be taken as seriously as in celestial mechanics, is that severe simplifying hypotheses had to be made in order to be able to make any prediction at all. Indeed, galactic dynamics and stellar dynamics are necessarily many-body problems, which are not solvable by classical pen and paper calculations. In principle, the physics included in stellar dynamics is as clean as in celestial mechanics; the difficulty of calculating without simplifying assumptions is just much larger.

In addition, although not perceived before the 30's as a problem, the dark matter, dominant only in the outer parts of luminous galaxies, leaves much freedom in the models. In the recent years it has become clear that in bulges and the inner luminous regions of galaxies the uncertainties associated with the stellar populations and with diverted light by dust are sufficiently large to account for the modest fraction of dark matter.

---

† Newton's 'laws' were often taken on the same level of truth that religious truth.

2.2. *Simplifications*

Typically, in galactic or stellar dynamics the system complexity is reduced by assuming a steady and stable state, and a high degree of symmetry. These assumptions have little theoretical ground. Indeed, all galactic disks depart from strict axisymmetry and the visible irregularities suggest time dependence often to some not so negligible degree, particularly over multi-Gyr time-scales.

Beside global symmetries, local symmetries are also assumed for convenience. An example of fundamental hypothesis made without discussion in stellar dynamics is that the distribution function $f(\vec{x}, \vec{v}, t)$ is smooth, i.e., that it is a function of phase space that can be differentiated at least once (in order to be able to use Boltzmann's equation). A priori f is a distribution in the mathematical sense and can include singular or non-differentiable functions (fractals or else). Namely, the N-body description

$$f(\vec{x}, \vec{v}, t) = \sum_{i=1}^{N} m_i \, \delta\big(\vec{x} - \vec{x}_i(t)\big) \, \delta\big(\vec{v} - \vec{v}_i(t)\big), \qquad (2.1)$$

where the $\{x_i(t), v_i(t)\}$ are the body trajectories, is just the simplest example of distribution function, which is also a distribution in the mathematical sense, quite accurate in fact despite being singular, but difficult to work with the classical tools of analysis. In fluid dynamics the assumption of differentiability is justified because the relaxation time (of order of the collision time of molecules) is much shorter than the physical time-scale of interest, and the mean-free path much shorter than the considered scales, so collective correlations may be assumed to be locally destroyed, leading to local homogeneity in space and Maxwellian distribution in velocity space. But in galactic dynamics the relaxation time-scale is precisely assumed to be much longer than the time-scale of interest, and the mean-free path much larger than the system, so the building of correlations and inhomogeneities is not necessarily erased everywhere. In short, the assumption of differentiability of f must be seen as a working hypothesis adopted for convenience and by lack of alternative and better justified assumptions. At the only place where a local measure of a galactic distribution function is possible, the solar neighborhood, no clear indications of a smooth distribution function are found: instead the vertex deviation, stellar streams, etc., suggest on the contrary that the real situation is more complex than assumed in static and symmetric models.

Yet whole branches of galactic astronomy are based on these assumptions more or less explicitly. Stellar populations models often assume isolation and steady state over several Gyr (e.g., Matteucci & Brocato 1990) which is granted only if the complexity of dynamics can be absorbed by a stable equilibrium in an axisymmetric galaxy over this time interval. Similarly, dynamical mass models of the Milky Way nearly always assume perfect axisymmetry and steady states (e.g., Bahcall & Soneira 1980; Bienaym et al. 1987; Dehnen & Binney 1998).

2.3. *The Role of Computers*

The situation has markedly changed with the increasing performances of electronic computers, that over the last 50 years multiplied our ability to calculate by several billion, a drastic technological revolution indeed. We can now investigate models that are much less dependent on simplifying assumptions, and singular distribution functions close to the above one are then adopted (the $m\delta(\vec{x})$ distribution is often smoothed by a Plummer mass model). Let us remind that in N-body simulations of galaxies, a particle does not represent a star but a sample of the mass in the Monte-Carlo sense.

Over the years, the various techniques invented to perform N-body simulations (tree-

or grid-based potential solvers, brute force direct summations by hardware, etc.) have shown that consistent, reproducible results at the global level can be obtained. However, since the 60's it was clear that at the individual particle trajectory level, the N-body problem is too sensitive to perturbations, too chaotic to yield robust *detailed* predictions (Miller 1964). Only statistical results can be trusted. The important point is that one can now extend the predictions of classical mechanics to systems that were in the past too complex to be handled. The results from N-body simulations, when well checked from the various numerical pitfalls, must be taken as seriously as analytical results. Each approach has domains for which it is the most effective.

An illumination brought mostly by N-body experiments is about the formation of galaxies and other objects formed by gravitational instability. Nothing such as the synchronous collapse of myriad of rapidly decoupled pre-galactic clouds of typical galactic mass, as imagined up to the computer experiments, does occur. Instead, multi-scale pancakes and filamentary structures surrounding expanding voids form in time such that rapidly the speed of evolution is very asynchronous: some regions are nearly frozen while some other regions are much ahead in the process of structure formation. The general picture of structure formation by gravitational collapse has been completely modified by the N-body experiments.

2.4. *Time-Dependent Galactic Dynamics*

The assumption allowing us to sweep 'dynamics under the rug' and thus allowing us to work in the reassuring world of statics, is the assumption of stability of stellar systems. In contrast with the solar system which lived for billions of crossing times, galaxies only existed for a few tens of crossing times. Therefore the assumption of quasi-stationarity is necessarily much coarser, because secular effects (i.e., effects occurring over timescales of order of hundred of crossing times) must still be active. Except for time-scales relatively short (say $< 1\,\mathrm{Gyr}$, a couple of galactic rotations) galactic dynamics must be seen as a time-dependent dynamics, a point largely ignored still today.

As noted above, the best way to grasp the complex effects involved by collective interactions in stellar systems with a minimum of simplifying assumptions is the N-body approach. N-body simulations have taught us that the main assumptions made in analytical works, stability and axisymmetry, are very often too coarse. Every new instability is a factor invalidating the assumption of time-independence. Now that the N-body techniques have been well checked, it is largely time to adopt the point of view that these techniques are more powerful than analytical or semi-heuristic pen and paper approaches, so the latter type models should always be checked with the former before taking firm conclusions (for example galaxy formation scenarios still assuming spherical symmetry or axisymmetry should be viewed as very uncertain unless confirmed by simulations potentially able to break the assumed symmetries).

2.5. *Chaos and its Topology*

Since stellar systems are never strictly symmetric, one must understand what happens when strictly symmetric models are perturbed. Generically, chaotic regions grow in phase space around *resonances*, which occur when frequencies associated with integrable motion are commensurate. The lower the resonance order and the larger the perturbation are, the wider the chaotic region. The chaotic region in phase space takes the shape of a web surrounding the islands of regular motion. This is a generic property of Hamiltonian systems, fundamental to understand since completely general. Chaos permeates phase space connecting widely separated islands of stability, therefore chaos is the very seed of evolution, the origin of time dependent, unstable, or 'turbulent' phenomena. Such facts

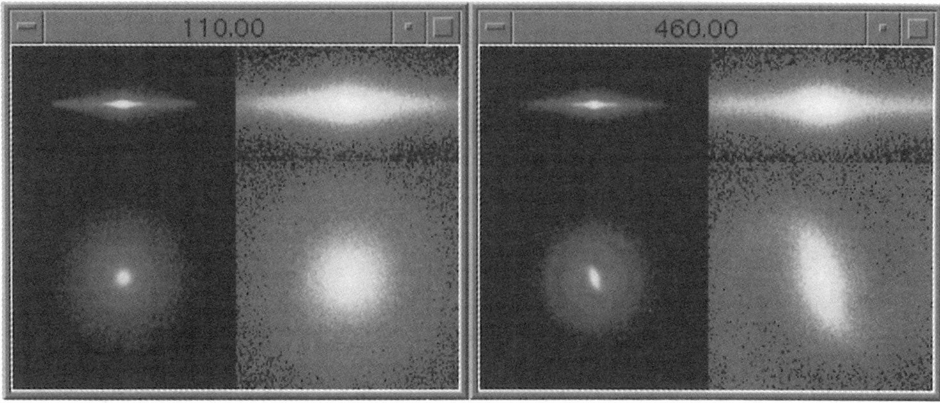

FIGURE 1. The now classical spontaneous formation of a bar, seen regularly in N-body simulations of self-gravitating disks. The time in Myr is indicated at the top. *Left:* The initial axisymmetric disk of collisionless particles (200'000) is shown face-on and end-on with a zoom on the central region. The amount of velocity dispersion is just critical to let the disk gravitational instabilities to develop. *Right:* After 350 Myr a bar is well developed while spiral arms tend to fade away. Such non-dissipative bars are robust and tend to survive at least for a time equivalent to the Hubble-time. They occur in every method used to calculate the gravitational forces.

were not widely known up to the 80's. Before, very often physicists believed that motion had to be always regular, quasi-periodic.

3. Elements of Bar Dynamics

3.1. *Bar Discovery by Computer*

Another unexpected discovery in the 60's due to computer simulations was that bars embedded in stellar disk form spontaneously and easily (perhaps the first to contribute to this discovery was P.O. Lindblad 1960b). Since then a lot has been learned about these non-linear structures. Before the possibility to calculate the consequences predicted by stellar dynamics free of the usual assumptions of stability and symmetry, these structures were largely mysterious, except for von Weizsäcker (1951) who lucidly expressed in a single paragraph the essential physics required to understand stellar bars. It turned out that no further physical ingredients that classical mechanics are required. In particular central explosions or magnetic fields were sometimes invoked to explain bars but are inessential factors.

The easy formation of bars in computer simulations was first often considered by many as a disease, because it disturbed the implicit dogma that normal galaxies must be stable and symmetric. Of course computer simulations were not free of shortcomings, but further program and method verifications, increased numerical accuracy and resolution steadily confirmed the result that bars form easily. For example to test a new computer, we recently increased the number of particles by a factor 40 (to $8 \cdot 10^6$) to check that earlier calculations of bars (Pfenniger & Friedli 1991) were indeed reproducible.

In general instabilities find the source for breaking symmetries from small scale 'noise', which is exponentially amplified to macroscopic sizes by the instability. Some parameters of the end-results appear therefore as random, some others appear reproducible. In the case of bars the phase of the bar is random, while the amplitude is largely deterministic.

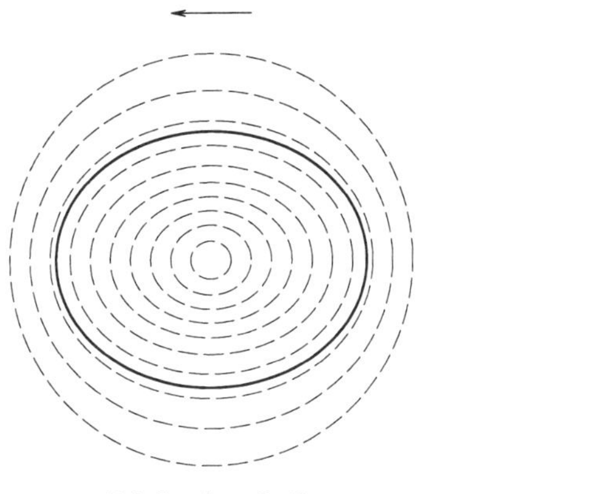

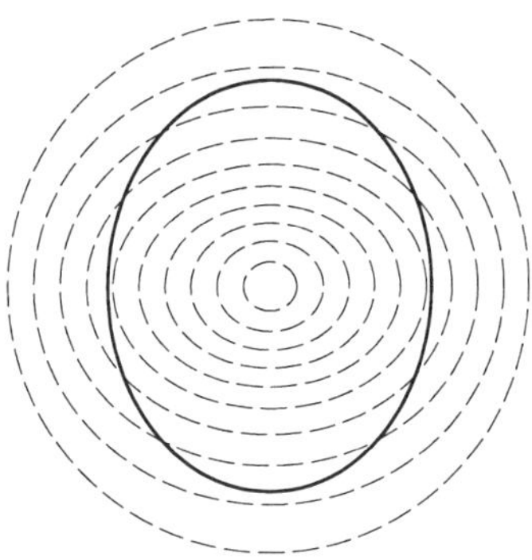

FIGURE 2. The fundamental reason for rapidly rotating bars to exist is that the loop orbits inside the bar tend to be elongated as the mass, they are consistent with the density distribution (left). On the contrary, in non-rotating bars, the loop orbits are elongated perpendicularly to the density (right).

The bar length depends mostly on the main scale initially in the system, the disk scalelength.

So bars do form easily in galactic disks over the linearly rising part of the rotation curve. This result was 'counter-intuitive' for people used to believe that more symmetric systems are more 'natural', or more stable. Bars just disprove this belief, and the reason is now well understood: a gravitational system is unstable if it is kinematically too cold. Another famous very symmetric but always unstable model is the infinite homogeneous isothermal medium (a classical model of the Universe) which is always Jeans unstable at finite temperature.

The next attempt to dismiss bars was to invoke massive dark halos of hot particles supposed to suppress the bars. This possibility turned out also not to be sufficient to suppress bars. Recent work along this line by Debattista & Sellwood (1998) suggest instead that the mere existence of numerous bars make difficult to assume that dark halos can be dynamically dominant inside stellar disks†.

In short, the spontaneous or triggered formation of bars was a first case showing that Hubble types can change well after the formation of a disk galaxy (change from SA to SB). This simple fact and its immediate consequences, that galaxies are not so steady, took 2–3 decades to be widely appreciated.

3.2. *The Self-Consistency Requirement: A Strong Constraint*

The non-trivial aspect of bars, when rotating at a suitable speed, is that they belong to the rare structures that are self-consistent: the potential generated by the mass distribution constrains trajectories to move on orbits reproducing the mass distribution (see Figure 2). In addition the bar shapes are fairly stable, which is also far from trivial. Indeed, if one takes an arbitrary mass distribution (say, a cube), the probability to be self-consistent is very small. Even a sphere which for symmetry reason must have orbits

† A similar conclusion is reached by considering the shape of the rotation curves (Salucci & Persic 1999). In the optical regions of galaxies the amount of dark matter is sufficiently small to be explained by the uncertainties about stellar populations and dust extinction.

compatible with the spherical shape is not necessarily feasible because in addition to the orbit shapes, the *weight* of these orbits must be everywhere positive. Added to these constraints is the stability of the systems which further restrict the domain of long-lived astrophysical systems. As example of impossible system, a hollow 2D stellar dynamical ring (without central body) is always unstable because in the interior part the gravitational force is directed outward, circular orbits do not exist! Actually, the only robust systems found in computer simulations resemble a lot to real galaxies.

What is also found is that real rotating bars are not isolated structures, they need to be embedded in a larger disk which serves as a reservoir of angular momentum. Angular momentum needs to be exchanged when a bar forms, and this occurs at best with spiral waves in the disk. This process is much faster that any other process often invoked in accretion disks to export angular momentum (magnetic fields, non-gravitational turbulence).

3.3. *Phase Space Structure*

Generally, to understand the dynamics of stellar systems, it is important to grasp the structure of their 6D phase space $\{\vec{x}, \vec{v}\}$. This is a daunting requirement for most minds. Fortunately, the main periodic orbit families help greatly, they can be imagined as the fundamental backbone of all the other orbits, so their knowledge helps greatly to understand the general organization of the 6D phase space. Comprehensive reviews about orbits and dynamics in bars are the ones of Contopoulos & Grosbøl (1989), and Sellwood & Wilkinson (1993).

Indeed, and this is a general property of Hamiltonian systems, in a given class of potentials phase space is foliated in sub-regions with similar topological properties. The smallest of these regions are *orbits*, which are the ensemble of phase-space points visited by a star starting at a given point (going both in the future and in the past). Orbits are the smallest elementary building blocks making a stellar system; understanding orbit morphologies is also understanding the elementary bricks by which stellar systems are constructed. The dimensionality of orbits (curves, surfaces, volumes, multi-fractals) depends on the number of *integrals of motion*, i.e., of constraints implicitly contained in the potential symmetries and in Newton's equations of motion. For example, the independence of the potential with respect to time constrains the energy (a function of $\{\vec{x}, \vec{v}\}$), to be conserved along any orbit. Or, a symmetry of revolution about an axis implies the conservation of the angular momentum component parallel to the axis. For such a reason the motion of a particle in any spherical potential is constrained to remain in a plane perpendicular to its angular momentum vector.

Looked in this way, bar dynamics has little to do with sphere dynamics (possessing at least 4 global integrals), or disk dynamics (2 global integrals plus 1 third quasi-integral) for two main reasons: first, bars do not possess any continuous spatial symmetry, and furthermore, they are thought to rotate rapidly (almost everyone agrees that bars must rotate rapidly), then the potential depends on time, energy is no longer a constant of motion. But for a constant rotational speed one global integral subsists, the Jacobi integral, which takes a simple form in the rotating frame of reference. Assuming the rotation axis to be along the z-axis, it reads,

$$H(\vec{x}, \dot{\vec{x}}) = \tfrac{1}{2}\dot{\vec{x}}^2 + \Phi(\vec{x}) - \tfrac{1}{2}\Omega_p^2\left(x^2 + y^2\right). \tag{3.1}$$

where Φ is the total potential, independent of time in the rotating frame, and Ω_p is the bar pattern frequency. H tends toward the energy integral when Ω_p tends toward zero, but the Ω_p^2 term, when non-zero, changes completely the phase space structure. A fixed value of H determines a hard to visualize 5-dimensional surface in the 6-dimensional phase

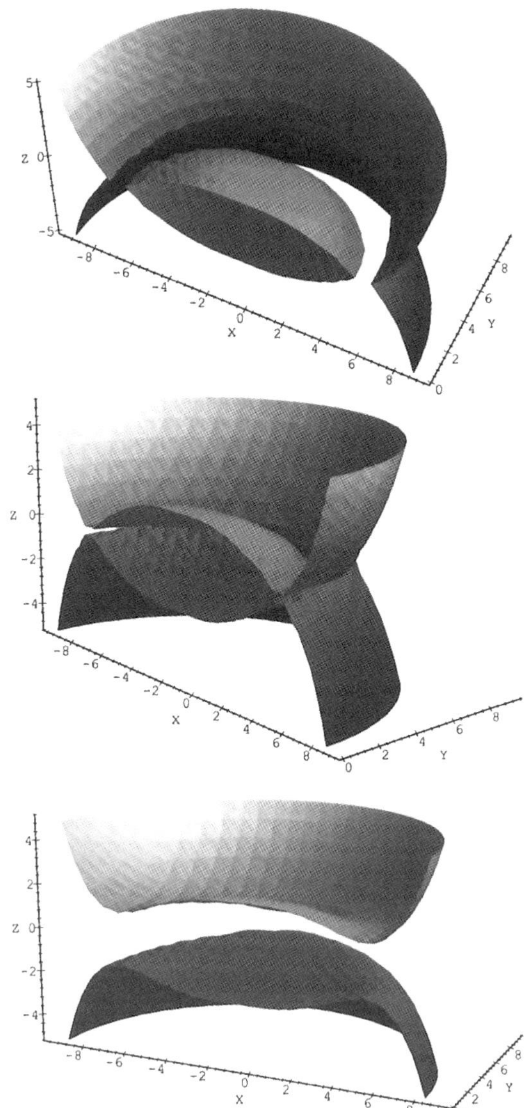

FIGURE 3. The aim of this figure is to show that bar dynamics is radically different from the spheroid or sphere dynamics as commonly thought for bulges. In a z-rotating bar (aligned with the x-axis), the structure of the zero-velocity surface in space due to the Jacobi integral confines motion differently from the energy zero-velocity surfaces in spheroids (concentric spheroids). From top to bottom, the Jacobi integral H has increasing values. *Top:* At low H, space is divided in two distinct allowed regions, the bar and the outer halo-disk beyond the corotation radius located between the two surfaces. *Middle:* At a critical value of H the two pieces are in contact at the $L_{1,2}$ Lagrangian points at the end of 5the bar 8und along the x-axis. Stars with higher H values can eventually escape to arbitrary large distances through the holes. This possibility concerns particularly chaotic orbits. *Bottom:* At still higher values of H the zero-velocity surface breaks down in two pieces at the Lagrangian points $L_{4,5}$ along the bar short axis (along the y-axis in the galaxy plane). At still higher values of H only the regions above the bar poles remain forbidden.

space. Yet a frontier in space can be drawn between allowed and forbidden regions when the velocity vanishes for a fixed value of H. Figure 3 shows how the zero-velocity surface in configuration space constrains motion in a rotating bar when the value of H changes. In comparison, the shapes of the zero-velocity surface in spherical or spheroidal stellar systems following from the energy integral are always concentric spheres or spheroids, i.e., convex closed surfaces. Therefore, in such systems stellar motion is bounded (below the escape energy). In contrast, above a given value of H, motion in bars is unbounded even if the energy is negative. Stars able to cross the Lagrangian points near the tips of the bar are also susceptible to reach distances much larger than the bar, not only in the galactic plane, but also out of the plane (see below and Ollé & Pfenniger 1998).

3.4. *Main Resonances in Bars Forces 3D Dynamics*

In bars the most important resonances associated with chaotic motion are the corotation resonance, located slightly beyond the bar end, and the inner Lindblad resonance, located well inside the bar. These resonances are wide because bars are major perturbations of axisymmetric systems. The most notable effect of resonances in bars is to exist both for motion in the galaxy plane as well as for motion transverse to it (Pfenniger 1984). Therefore bars offer a natural and inescapable way (recall that we discuss predictions of classical mechanics, not from a less well understood branch of physics) to increase the motion of stars originally in the plane to several kpc heights. Anyone using easy tools of classical mechanics, such as the virial theorem, should take the above more subtle considerations on the same level of seriousness.

In summary, inside bars it is impossible to maintain a 2D dynamics because the integral keeping motion close to the plane in disks, the third integral, is destroyed by major resonances. Therefore the dynamics of bars is fully three-dimensional, precisely in regions where bulges are usually existing. The question is no longer to ask whether bulges may be influenced by bars, but rather whether the distinction of bulge and bar makes a physical sense in view of their unavoidably closely coupled dynamics.

3.5. *Remark about Dissecting Galaxies in Components*

The above considerations lead to the following remark. Often galaxy observers decompose additively spirals in overlapping components such as disk, bulge and eventually bar, implicitly taking as granted that the components are physically distinct and their dynamical properties independent from each others. In fact this is far from being a good assumption in view of the relatively similar gravitational forces that each component induces on the other ones. Only one potential is felt by the whole system, and the phase-space structure is directly affected by the potential integrals or quasi-integrals. Stars in a bulge surrounded by a bar have a dynamics lacking from the angular momentum conservation constraint that exists in an isolated bulge. Or, since circular orbits inside a bar do not exist, there is no reason to assume that a galactic disk extends down to the center through the bar-bulge region if the orbits supporting a disk do not exist.

4. Secular Effects

4.1. *Exponential Disks Produced by Bars*

An early result of N-body simulations that has not been properly appreciated is the finding by Hohl (1971), confirmed later by more detailed calculations (Pfenniger 1990; Pfenniger & Friedli 1991), that the tendency of stellar disks to adopt an exponential profile can be entirely due to the action of a bar. Contrary to alternative models involving dissipative ingredients (gas friction, star formation), the stellar dynamical result requires no fine tuning of parameters and is based on well known physics: pure stellar disks surrounding a bar relax toward an exponential profile. The longer the disk-bar interaction the farther out the exponential profiles extends.

Of course, other processes mildly relaxing disks can eventually also contribute to shape optical disks toward an exponential profile. The exponential profile, like the $R^{1/4}$ profile for elliptical galaxies, appears to be an attractor among all the possible shapes. A deeper theoretical understanding is presently lacking, the exponential disk attractor is an empirical result well observed in simulations. Among all the possible processes leading to an exponential disk, bars look as one of the most unavoidable, since they are so ubiquitous.

4.2. *Stellar Diffusion in Barred Galaxies*

To understand at the particle level what happen to stars moving in a barred potential and belonging to the chaotic web, we have performed calculations about the horizontal and vertical diffusion of orbits in barred potentials (Pfenniger 1985; Martinet & Pfenniger 1987; Oll & Pfenniger 1998). In strong bars one finds a fast $(0.1 - 1$ Gyr$)$ radial and vertical *diffusion* of material initially inside the bar near the Lindblad resonance or at the periphery of bars around the corotation resonance. Figure 4 shows a case of diffusion for particles starting near the corotation region. As a consequence, since the Milky Way does have a sizable bar, any stars near the Sun with kinematical properties leading them below 4–5 kpc from the Galactic center are with large probability on such diffusing and chaotic orbits that can also reach large distances beyond the solar radius $(> 20$ kpc with exponentially decreasing density$)$.

In short, in barred galaxies stellar motion can be markedly different from circular motion, and the frequent assumption of closed box over several Gyr for material near the Sun is dubious since the Milky Way is barred.

4.3. *The Bar Bending Instability*

Another robust result from stellar dynamical N-body models is that stellar bars are frequently subject to a transverse bending instability which evolves toward a peanut-shaped bar (Combes & Sanders 1981; Combes et al. 1990; Pfenniger & Friedli 1991; Raha et al. 1991) (Figure 5). The rationale at the origin of the bending instability and the evolution toward a peanut shape lies in the orbit shapes and the vertical inner Lindblad resonance. Depending on the viewing angle the bar may appear as a round, boxy or peanut-shaped bulge, resembling much the real ones, as illustrated by Bureau at this conference (see also Bureau & Freeman 1999). As noted by Raha et al., the initial amount of vertical velocity dispersion in the bar exceeds the fire-hose instability criterion (Araki 1985), so the bar bending instability is of different nature. Actually, apart from being a transverse instability, the bar bending instability has few common points with a fire-hose instability as pushed forward by, e.g., Merritt (1999), or Raha et al. (the initial phase-space structure in the bar, split by strong vertical orbital resonances, is very different from the one of a fire-hose unstable plan-parallel, homogeneous and non-rotating sheet).

4.4. *Bar Destruction by a Weak Dissipative Component*

Finally, real galaxies are never purely dissipationless. At the minimum due to stellar winds a few percent of the mass is in a gaseous form. At the percent level a dissipative component can have dramatic effects such as in a bar. The scenario that we thought about with Colin Norman and Hashima Hasan while working at STScI a few years ago was to consider the effect of a weak dissipation in bars (Hasan & Norman 1990; Pfenniger

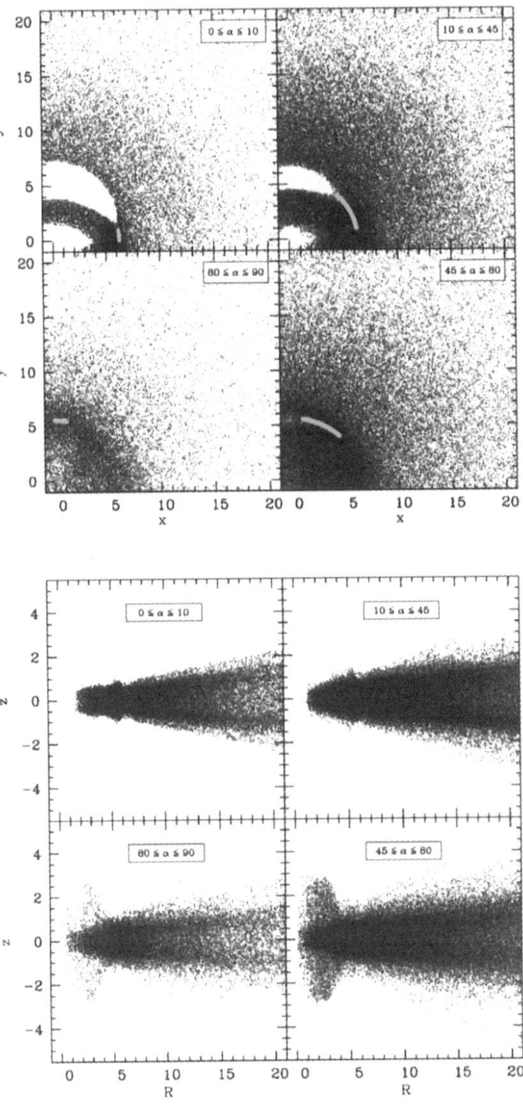

FIGURE 4. Diffusion in a barred potential for stars starting in the plane near corotation (Ollé & Pfenniger 1998). The bar is aligned with the x-axis with semi-axes $\{a, b, c\} = \{6, 3, 2.5\}$ kpc. The initial conditions are located in the small grey patches at $z = 0$ near the corotation ellipse in the (x, y) frames. The initial velocity dispersion is close to a realistic values in a real stellar disk. *Top:* The positions of stars are sampled at regular time interval and projected on the galaxy plane. The star trajectories are integrated until they reach $R = 50$ kpc, or, more rarely, $t = 10$ Gyr. *Bottom:* The same in the meridional plane (R, z).

These plots show that motion for such initial conditions is markedly unbounded in the radial direction, but also to some sizable extent in the vertical direction in the bulge region. Stars initially near the galaxy plane reach heights above the galactic plane of order of $2 - 3$ kpc in a couple of crossing times. So disk material in a barred galaxy must populate both the bulge and a thick disk.

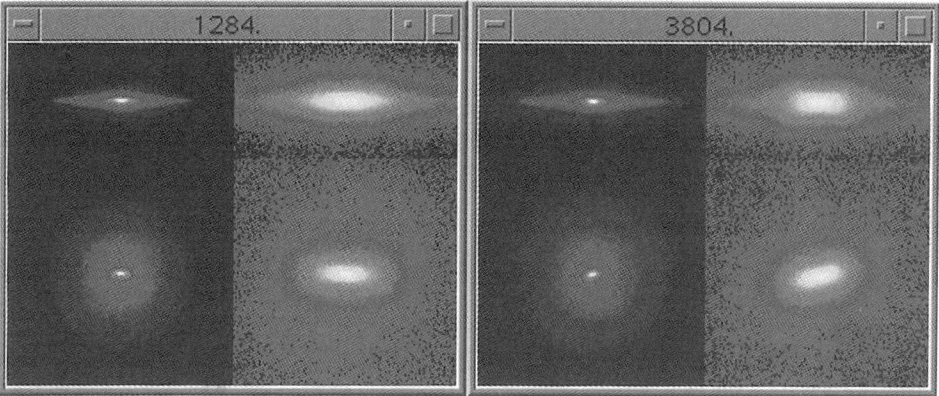

FIGURE 5. The now almost classical spontaneous formation of a peanut-shaped bar, seen regularly in collisionless *N*-body simulations of bars. The time in Myr is indicated at the top. *Left:* The initial stellar bar is shown face-on and end-on with a zoom on the central region. The orbital structure in the bar is such that 2/1 vertical resonances contribute to trigger a bending instability, not very conspicuous when the initial bar/disk is as thick as the one shown. *Right:* The long term evolution of the bar bending instability is a symmetric peanut-shaped bar-bulge. Note that the found time-scale in unperturbed simulations such as this one can be considerably reduced in strongly perturbed situations such as during galaxy interactions.

& Norman 1990; Hasan, Pfenniger & Norman 1993). It turns out that the accumulation of only a few percent of the mass within the Inner Lindblad Resonance is sufficient to modify a much larger region: because the elongated orbits sustaining the bar itself are strongly modified by a light central mass concentration, the whole bar can be destroyed by a well localized small dissipative perturbation. The result is not only that the bar can be axisymmetrized in the galaxy plane, but also (since bar dynamics is truly three-dimensional) that a nice spheroidal bulge results from the bar dissolution. The surprise here is that a small perturbation is amplified by the bar to a size which is a multiple of the original perturbed region.

For the reason that orbit shapes are only dependent on the mass distribution irrespective of the nature of the mass, the same bar destruction can be reached by any other means accumulating mass inside the Inner Lindblad Resonance. Another example will be shown below.

We have thus further fast mechanisms allowing to transform a Hubble type, from SB to SA, in rather short time intervals. Of course the time-scale for dissolution can be delayed for example if the gas infall rate is small.

4.5. *Multiple Bars*

For many years some astronomers considered bars as a disease, yet as we now know galaxies are not allergic to bars at all, and even appear sometimes addicted. Several galaxies are perfectly able to possess two or three of them simultaneously. Figure 6, kindly provided by Daniel Friedli, shows such a case. This illustrates how easily these structures form since the mutual perturbations of the bars do not prevent their co-existence, at least for a while. Numerical experiments show that the most favorable situation allowing these bars to survive the longest is when they rotate at different speeds, the inner bar being faster, and when the Inner Lindblad Resonance of the large bar coincides with the corotation of the inner bar. Such a resonance overlap minimizes the chaotic regions. Although it is theoretically thinkable that these bars rotate at the

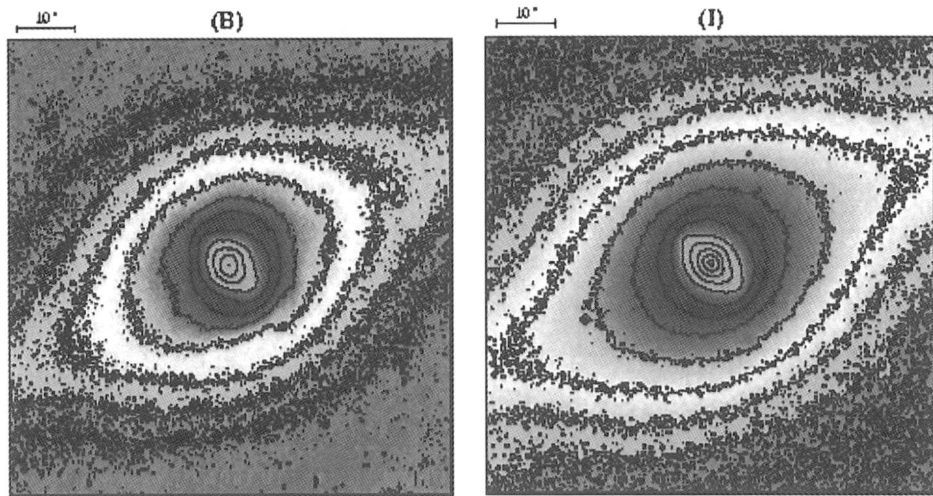

FIGURE 6. B and I views of the two bars inside NGC 5850 (Friedli et al. 1996). Such structures are best explained by embedded bars rotating at different speeds, the inner bar being faster. Such bars must exchange momentum and energy, so must be time-dependent with a limited lifetime. They provide an additional hint that galaxies are time-dependent structures.

same speed, in which case the inner bar should be perpendicular to the large one, most of the observed multiple bars do not show any preferred angle, so the interpretation of different pattern speeds appears the most plausible.

Despite their size differences, the mutual torques between embedded bars are not negligible, so these bars are unlikely to survive several Gyr. Simulations show also such structures for a limited time, so inevitably the question of the fate of these bars over the next Gyr must be addressed. As for bar destruction by accumulation of mass inside the Inner Lindblad Resonance, the observed future of such multiple bar systems in simulations is to dissolve into first a triaxial fat bulge (filling the zero-velocity surfaces given by Jacobi's integral, cf. Figure 3), and then to relax toward a classical spheroidal bulge indistinguishable from a bulge made differently (see Friedli & Martinet 1993; Friedli et al. 1996).

5. The Merger-Bar-Bulge Connection

5.1. *Environment*

Although a bar can grow spontaneously from its intrinsic small scale 'noise', an external finite perturbation can excite its growth even faster. Satellite tidal perturbations in marginally stable galactic disks are typical examples of such induced bar formation. The idea was explored already by Lindblad (1960a), and also studied by Noguchi (1988).

But after having excited a bar a galaxy satellite eventually penetrates the inner disk region by dynamical friction, and reaches the Inner Lindblad resonance of the recently triggered bar. To survive all this trip the galaxy satellite must be initially at least as dense as the bulge-bar region it penetrates, otherwise tidal stripping would dissolve the satellite before. In such conditions the bar can be destroyed into a spheroidal bulge if the satellite mass exceeds the $1 - 3\%$ critical mass ratio. Figure 7 shows such a scenario of bar destruction via satellite merging (Pfenniger 1992). In such experiments the surprise was that the time to change the apparent Hubble type from SB to SA was extremely

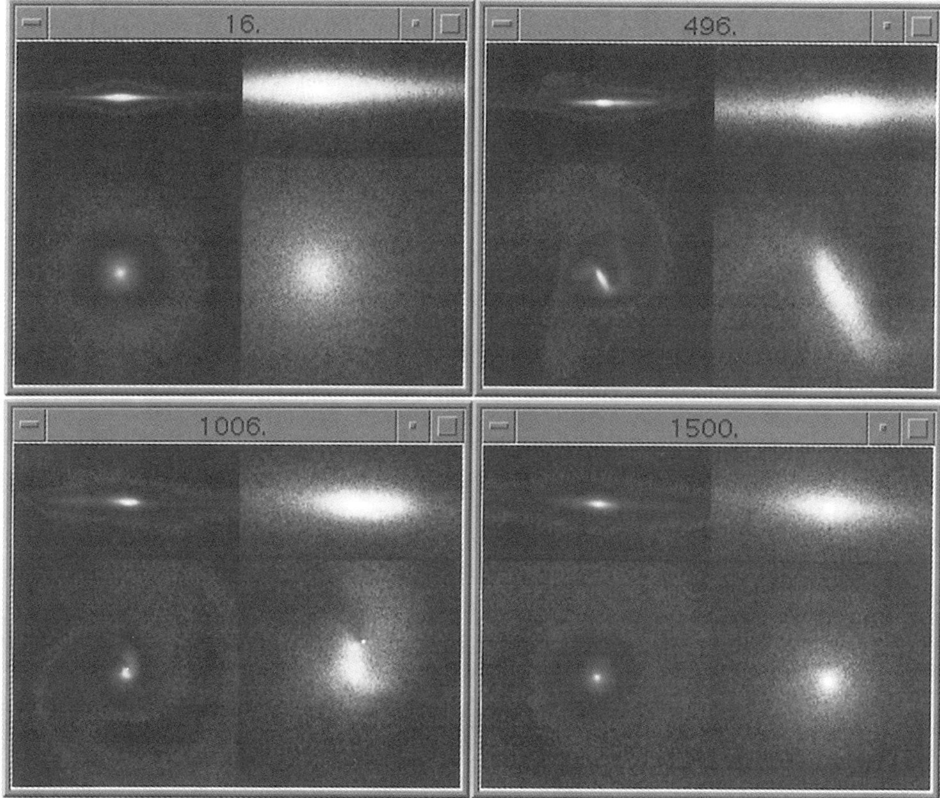

FIGURE 7. A N-body simulation showing the triggering of a bar and its subsequent destruction by a satellite weighting 10% of the galaxy mass (as in Pfenniger 1992). The time is indicated in Myr. The initial satellite, a point mass, is put on an inclined circular orbit slightly below the galaxy plane. At $t = 496\,$Myr the bar is well developed, at $t = 1006\,$Myr the satellite is close to the bar, while at $t = 1500$ the bar is destroyed into a symmetric spheroidal bulge. The only signature of accretion in the final frame is a slight disk inclination due to the deposited orbital angular momentum, and a disk inflation due to the deposited satellite energy.

short, of order of 20 Myr, essentially the time for the satellite to cross the Inner Lindblad resonance region and merge. The whole bar region readjusts quickly to the new central regime of resonances.

5.2. *Mild but Repeated Interactions, Harassment*

A galaxy may suffer several merging of small dwarf galaxies, not all reaching the centre. Therefore it was interesting to check which shape a whole disk adopts when stirred by the same amount of mass as before but by several satellites on uncorrelated orbits (Pfenniger 1993).

The result is that the disk inflates to lenticular shapes very reminiscent of S0's. Essentially all the *energy* deposited by the satellites serves to inflate the disk into a spheroidal bulge much larger than the disk scale-length since the satellites interact with the whole disk at several uncorrelated occasions. The whole process may take a couple of Gyr, depending on the satellite masses; around 5–10% mass accretion in this way is sufficient to inflate all the disk to large heights. This is roughly in agreement with the limit found by Tóth & Ostriker (1992), 4%, that the Galaxy could accrete without exceeding the

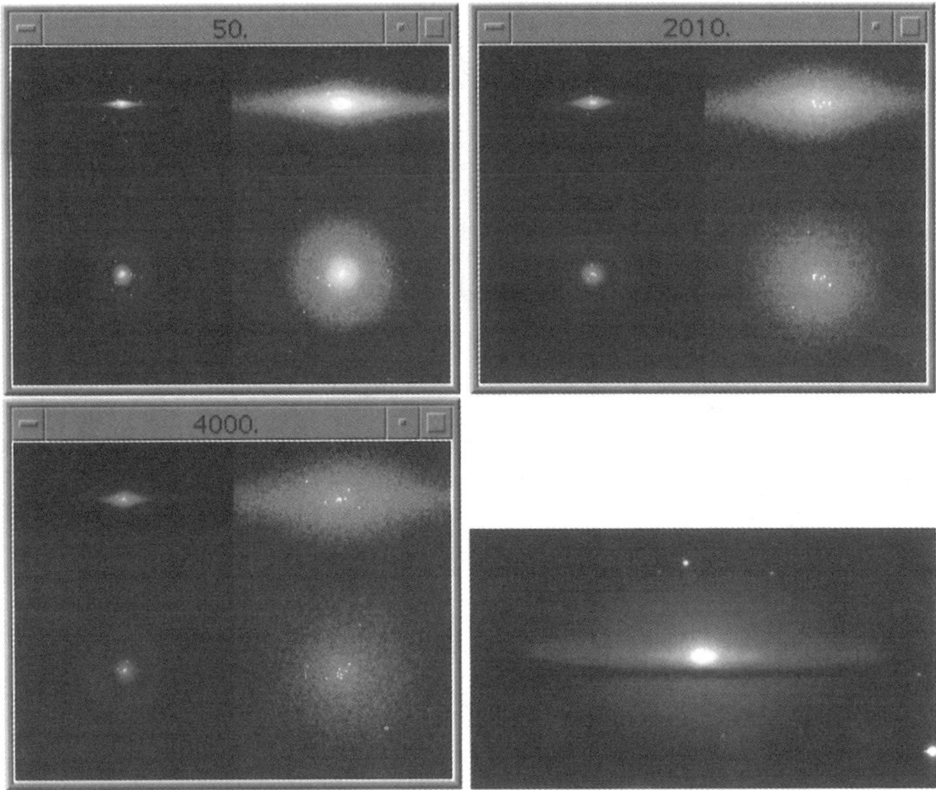

FIGURE 8. Transformation from a disk galaxy into a lenticular galaxy resembling the Sombrero (bottom right). After 4 Gyr of harassment by 10 point mass satellites weighting each 1% of the total mass and on random orbits, the initial disk is inflated such as to look like a lenticular (Pfenniger 1993). In the meanwhile the formed bar has been destroyed by the satellites.

known velocity dispersion. The final mass deposited by the satellites is almost negligible, but not the energy, therefore the final disk must be mostly made of potentially metal enriched disk material. The Hubble type is again modified toward earlier types over a fraction of the galaxy age.

6. Dynamical Models of the Inner Milky Way

Many models of the Milky Way have been produced in the recent years, but most suffer from the assumptions of strict axisymmetry and/or stability. On the contrary the recent N-body models including gas of Fux (1999) of the inner Milky Way are free of such assumptions. They can reproduce features seen in the HI and CO longitude–velocity diagrams in exquisite details (see Figure 9) at a very important condition: everything must be free of being time-dependent and not strictly symmetric. The N-body models, as all earlier unconstrained N-body models including gas, show a lot of transient features constantly regenerated, about like the spiral patterns produced by cream in rotating coffee.

Therefore the high degree of complexity existing in the observations of the Milky Way HI and CO are evidences that the Milky Way is substantially evolving still today. Dynamics can not be absorbed by symmetry assumptions when we want to interpret the

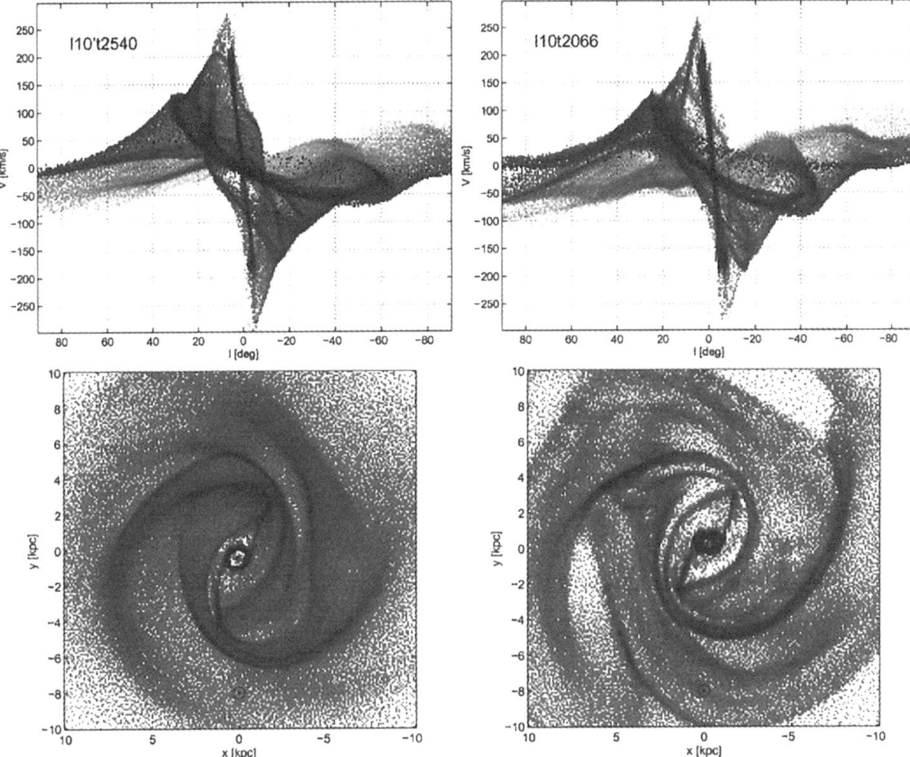

FIGURE 9. Two *N*-body models of the Milky Way by Fux (1997, 1999) for which a position can be found that satisfies well several observational constraints. The face-on distribution of the gas is shown in the lower frames. The position of the Sun is indicated by a ⊙ sign. The corresponding longitude–velocity diagrams are shown in the top frames. They reproduce many features seen in observations of HI or CO.

available observations. Time has come to realize for example that it is a waste of time to refine Oort's 'constants' when the underlying assumptions of axisymmetry and time independence included in these constants are not granted to the required level.

7. Secular Bulge Formation and the Hubble Sequence

It is useful to ponder about the consequences of the dynamical processes able to shape and transform galaxies, because the bulge feature is an important characteristic of the Hubble sequence monotonously increasing from the late to early types. Ideas of galaxy secular evolution related to dissolving bars and based on observer intuition were already expressed by Kormendy in his influential Saas-Fee Lectures (1982).

Since, as we have discussed, bulges may form after their surrounding disks, in relation to bars or not, it is quite obvious that once lifted into a bulge stars are unlikely to condense soon back to a disk. In other words inflating disk stars into a bulge is essentially an irreversible process, indicator of a sense of secular evolution in the Hubble sequence from late to early types. Once formed a bulge is essentially there forever.

A single irreversible process is not sufficient to draw the firm conclusion that most galaxies would really follow this evolutionary track. Yet other irreversible processes take place in spirals, and all indicate the same sense of evolution from late to early types. For example star formation, i.e. the transformation of gas into stars, and nucleosynthesis, i.e., the transformation of H, He into metals, both indicate that early-type galaxies have burned more gas, produced more stars, metals, and dust. Stellar populations too are, once formed, there forever.

Another indicator of irreversible evolution is just the depth of the potential well, proportional to the squared virial velocity. The more a gravitating system looses its energy, the higher its virial velocity. Early-type spirals rotate 3–4 times faster that late types, therefore early types must have lost 9–16 times more energy by unit mass that late types.

Another indicator of dynamical nature is the winding of spiral arms and the general degree of regularity displayed by a galaxy. The higher the degree of symmetry, the more rotational periods a galaxy requires to relax the irregularities. Early-type galaxies appear much more regular than late types, therefore look older.

Therefore, instead of the old views on a rapid early and synchronous formation of galaxies with a subsequent freezing of Hubble types, the consistent picture which emerges after $\approx$40 yr of modeling is that both secular and erratic brief episodes of dynamical evolution lead to the transformation of galaxy morphologies generally toward earlier Hubble types, at widely different speeds. Much depends on the environment, and indeed the denser the environment, the earlier galaxy types are found. Complex histories of galaxies appear inescapable when one accepts *all* the consequences implied by Newtonian mechanics.

8. Remarks and Conclusions

The studies of bars have taught us that:

1. The dynamics inside bars are markedly different from the dynamics inside axisymmetric disks or spheroids. Angular momentum is essentially *not* conserved inside bars. To be quantitative we checked in a N-body bar that the typical time for a particle to exchange 100% of its angular momentum with the bar back and forth is of order of a half rotational period (Pfenniger & Friedli 1991). Most orbits inside stellar bars spend of order of 30% of time being retrograde.

2. Bar dynamics is essentially three-dimensional and chaotic. Whenever a bar grows or dissolves, its effects concern the whole surrounding space, i.e., the bulge, the stellar disk, and the inner stellar and dark halo. Diffusion of stars in and out the bar region must also contribute to thickening the disk. Of course, other factors such as infalling satellites can also make thick disks.

3. Since bars are so widespread and probably not eternal, the heating of disks both in the radial and vertical directions due to bars is as certain as other predictions of classical mechanics.

4. We have seen several factors related to bars that produce small bulges growing from initially bulgeless disks. This is important since so many galaxies do not have yet a bulge. At least possibilities exist for these bulgeless galaxies to look later like their brighter, more experienced sisters.

5. Bulges much bigger than the disk scale length are unlikely made purely by a bar related process. Instead, accretion of galaxy satellites satellites is likely to be involved. Such accretion is able to produce the big bulges seen in lenticulars. As was the case for small bulges, big bulges do not necessarily form before their disks.

6. There is no unique way to make bulges. As for ellipticals, similar bulges may result from different histories. The important point to stress is that the age of the stars in today's bulges may have little relation with the age of formation of the bulge itself by dynamical heating.

7. The bulge sizes measure roughly the amount of complex past dynamical events (bar, accretion, ...). Dynamics still plays a determining role over the 10–15 Gyr following the proto-galaxy, because galaxies are generally insufficiently stable to be assumed so time-independent to neglect completely the secular effects of bars, spirals, interactions with the environment, and the dissipative component present in every galaxy.

8. Since galaxies are subject to fast dynamical transformations able to change markedly their morphologies, the Hubble types should no longer be seen as fixed. All the irreversible processes, including secular bulge growth, indicate that the Hubble sequence may be seen as a broad description of the aging of galaxies from late to early types.

This work has greatly benefited from the constructive atmosphere at Geneva Observatory, in particular in the group of galactic dynamics led by Louis Martinet, and from several stimulating stays at the Space Telescope Science Institute. Over the years the constant support from the Swiss National Science Foundation has also been essential.

REFERENCES

ARAKI, S. 1985, Ph. D. Thesis, Massachusetts Institute of Technology

BAHCALL, J.N., SONEIRA, R.M. 1980 *ApJS*, **44**, 73

BIENAYMÉ, O., ROBIN, A.C., CRÉZÉ, M. 1987 *A&A*, **180**, 94

BUREAU M., FREEMAN K.C. 1999 *AJ*, in press

COMBES, F., SANDERS, R.H. 1981 *A&A*, **96**, 164

COMBES, F., DEBBASCH, F., FRIEDLI, D., PFENNIGER, D. 1990 *A&A*, **233**, 82

CONTOPOULOS, G., GROSBØL, P. 1989 *A&A Rev*, **1**, 261

DEBATTISTA, V.P., SELLWOOD, J.A. 1998 *ApJ*, **493**, L5

DEHNEN, W., BINNEY, J. 1998 *MNRAS*, **294**, 429

FRIEDLI, D., MARTINET, L. 1993 *A&A*, **277**, 27

FRIEDLI, D., WOZNIAK, H., RIEKE, M., MARTINET, L., BRATSCHI, P. 1996 *A&AS*, **118**, 461

FUX, R. 1997 *A&A*, **327**, 983

FUX, R. 1999 *A&A*, in press

HASAN, H., NORMAN, C. 1990 *ApJ*, **361**, 69

HASAN, H., PFENNIGER, D., NORMAN, C. 1993 *ApJ*, **409**, 91

HOHL, F. 1971 *ApJ*, **168**, 343

KORMENDY, J. 1982, in *Morphology and Dynamics of Galaxies* (ed. L. Martinet & M. Mayor), p115. (Geneva Observatory)

LINDBLAD, P.O. 1960a *Stockholm Obs. Ann.* **21**, No. 3, 1

LINDBLAD, P.O. 1960b *Stockholm Obs. Ann.* **21**, No. 4, 1

MARTINET, L., PFENNIGER, D. 1987 *A&A*, **173**, 81

MATTEUCCI, F., BROCATO, E. 1990 *ApJ*, **365**, 539

MERRITT, D. 1999 *PASP*, **111**, 129

MILLER, R.H. 1964 *ApJ*, **140**, 250

NOGUCHI, M. 1988 *A&A*, **203**, 259

OLLÉ, M., PFENNIGER, D. 1998 *A&A*, **334**, 829

PFENNIGER, D. 1984 *A&A*, **134**, 373

PFENNIGER, D. 1985 *A&A*, **150**, 112

PFENNIGER, D. 1990 *A&A*, **230**, 55

PFENNIGER, D. 1991 *Dynamics of Disc Galaxies* (ed. B. Sundelius), p191. (Göteborg)

PFENNIGER, D. 1993, in *Galactic Bulges* (ed. H. Dejonghe & H.J. Habing), IAU Symp. **153**, p387. (Kluwer)

PFENNIGER, D., FRIEDLI, D. 1991 *A&A,* **252**, 75

PFENNIGER, D., NORMAN, C. 1990 *ApJ,* **363**, 391

RAHA, N., SELLWOOD, J.A., JAMES, R.A., KAHN, F.D. 1991 *Nature,* **352**, 411

SALUCCI P., PERSIC M. 1999 *A&A,* in press

SELLWOOD J.A., WILKINSON A. 1993 *Rep. Progr. Phys.,* **56**, 173

TÓTH, G., OSTRIKER, J.P. 1992 *ApJ,* **389**, 5

VON WEIZSÄCKER, C.F. 1951 *ApJ,* **114**, 165

Bars and Boxy/Peanut-Shaped Bulges: An Observational Point of View

By MARTIN BUREAU[1], K. C. FREEMAN[2]
AND
E. ATHANASSOULA[3]

[1]Sterrewacht Leiden, Postbus 9513, 2300 RA Leiden, The Netherlands

[2]Research School of Astronomy and Astrophysics, Institute of Advanced Studies,
The Australian National University, Mount Stromlo Observatory,
Private Bag, Weston Creek P.O., ACT 2611, Australia

[3]Observatoire de Marseille, 2 place Le Verrier, F-13248 Marseille Cedex 4, France

Prompted by work on the buckling instability in barred spiral galaxies, much effort has been devoted lately to the study of boxy/peanut-shaped bulges. Here, we present new bar diagnostics for edge-on spiral galaxies based on periodic orbits calculations and hydrodynamical simulations. Both approaches provide reliable ways to identify bars and their orientations in edge-on systems. We also present the results of an observational search for bars in a large sample of edge-on spirals with and without boxy/peanut-shaped bulges. We show that most such bulges are due to the presence of a thick bar viewed edge-on while only a few may be due to accretion. This strongly supports the bar-buckling mechanism for the formation of boxy/peanut-shaped bulges.

1. Introduction

Boxy/peanut-shaped bulges (hereafter referred to simply as boxy bulges) have, as their name indicates, excess light above the plane. They are thus easily identified in edge-on systems and display many interesting properties: their luminosity excess, an extreme three-dimensional structure, probable cylindrical rotation, etc. However, the main importance of boxy bulges resides in their incidence: at least 20-30% of all spiral galaxies possess a boxy or peanut-shaped bulge. They are thus essential to our understanding of bulge formation and evolution.

Early theories on the formation of boxy bulges were centered around accretion scenarios, where one or many satellites galaxies are accreted onto a preexisting bulge, and which lead to axisymmetric structures (e.g. Binney & Petrou 1985). However, such scenarios are restrictive, and it seems that the only viable path is the accretion of a small number of moderate-sized satellites. Thus, accretion probably plays only a minor role in the formation of boxy bulges. A more attractive mechanism is the buckling of a bar, due to vertical instabilities. This process can form boxy bulges even in isolated galaxies, and accounts easily for the fact that the fraction of boxy bulges is similar to that of (strongly) barred spirals. Soon after a bar is formed, it buckles and settles with an increased thickness, appearing boxy or peanut-shaped depending on the viewing angle (e.g. Combes *et al.* 1990). Hybrid scenarios, where a bar is excited by an interaction and then buckles, have also been suggested.

To test as directly as possible the bar-buckling hypothesis, we have developed reliable bar diagnostics for edge-on spirals (Bureau & Athanassoula 1999; Athanassoula & Bureau 1999), and have searched for bars in a sample of edge-on galaxies with and without boxy bulges (Bureau & Freeman 1999). This way, we can probe the exact relationship between bars and boxy bulges.

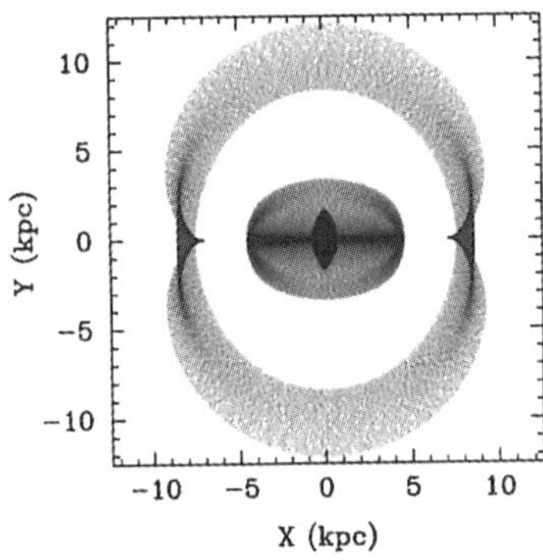
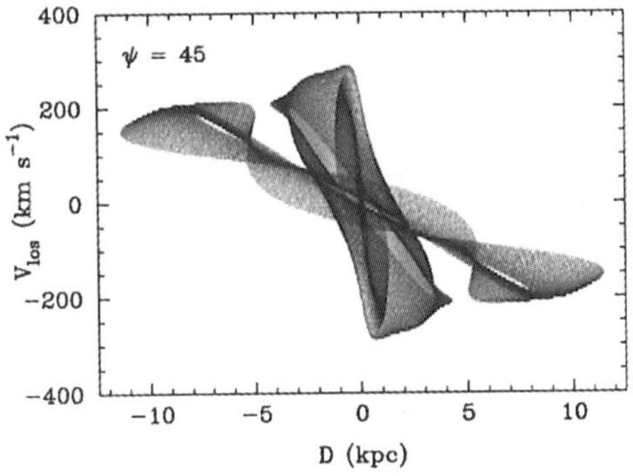

FIGURE 1. Surface density distribution (*left*) and projected PVD (*right*) of model 001, when considering the x_1, x_2, and outer 2:1 families of periodic orbits. In the left plot, the bar is horizontal, has a length of 5 kpc, and is viewed at an angle of $45°$ for the PVD.

2. Bar Diagnostics in Edge-On Spiral Galaxies: The Periodic Orbits Approach

There is no reliable photometric way to identify a bar in an edge-on spiral galaxy. However, Kuijken & Merrifield (1995) showed that an edge-on barred disk produces characteristic double-peaked line-of-sight velocity distributions which would not occur in an axisymmetric disk. Following their work, we also developed bar diagnostics based on the position-velocity diagrams (PVDs) of edge-on disks, which show the projected density of material as a function of line-of-sight velocity and projected position. The mass model we adopted has a Ferrers bar, two axisymmetric components yielding a flat rotation curve, and four free parameters. All our models are two-dimensional.

We first used the families of periodic orbits in our mass model as building blocks to model real galaxies (Bureau & Athanassoula 1999). Such an approach provides essential insight into the (projected) kinematics of spirals. We showed that the global appearance of a PVD can be used as a reliable tool to identify bars in edge-on disks. Specifically, the presence of gaps between the signatures of the various periodic orbit families follows directly from the non-homogeneous distribution of orbits in a barred galaxy. The two so-called forbidden quadrants of the PVDs are also populated because of the elongated shape of the orbits. Figure 1 shows the surface density and projected PVD of a typical model. The bar is viewed at an angle of $45°$ from the major axis and only the major families of periodic orbits are considered. The signatures of the x_1 (parallel to the bar) and x_2 (perpendicular to the bar) orbits are particularly important to identify the bar and constrain the viewing angle. Because of streaming, the parallelogram-shaped signature of the x_1 orbits reaches very high radial velocities when the bar is seen end-on and only relatively low velocities when it is seen side-on. The opposite is true for the x_2 orbits.

3. Bar Diagnostics in Edge-On Spiral Galaxies: Hydrodynamical Simulations

We also developed bar diagnostics using hydrodynamical simulations, targeting specifically the gaseous component of spiral galaxies (Athanassoula & Bureau 1999). The simulations are time-dependent and the gas is treated as ideal, isothermal, and non-viscous. We used the same mass model as above, without self-gravity, and modeled star formation and mass loss in a simplistic way. However, the collisional nature of the gas leads to better bar diagnostics than the periodic orbits approach.

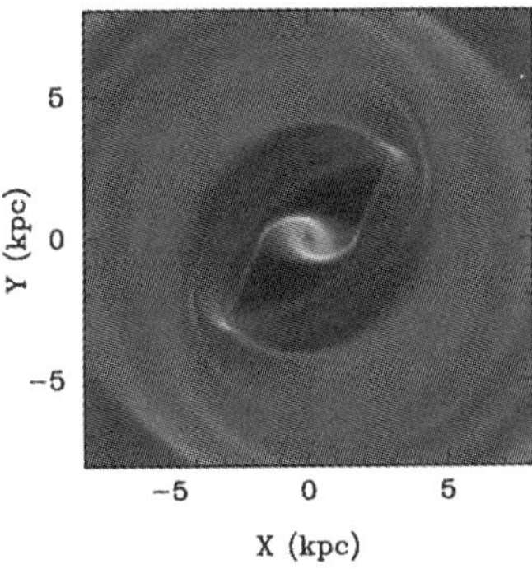
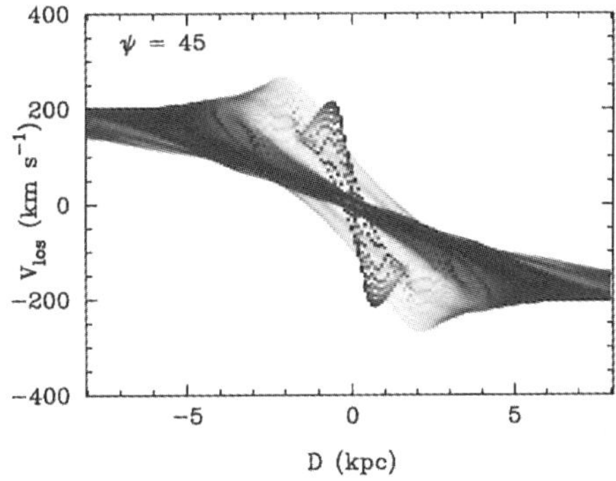

FIGURE 2. Surface density distribution (*left*) and projected PVD (*right*) of model 001, for the hydrodynamical simulations. In the left plot, the bar is now oriented diagonally and high densities are in white.

The main feature of the PVDs is a gap, present at all viewing angles, between the signature of the nuclear spiral (associated with x_2 orbits) and that of the outer parts of the disks. There is very little gas in x_1-like flows. This gap unmistakably reveals the presence of a bar in an edge-on disk. It occurs because the large scale shocks which develop in bars drive an inflow of gas toward the centers, depleting the outer bar regions. If a galaxy has no inner Lindblad resonance (ILR; or, equivalently, has no x_2 orbits), there is no nuclear spiral and the entire bar region is depleted. Then, the use of stellar kinematics is probably preferable to identify a bar. We will develop such diagnostics in a future paper. Figure 2 shows the gas density distribution and PVD for the same model as above, which has ILRs. Although not shown, the PVDs again vary significantly with the viewing angle, the signature of the nuclear spiral reaching its highest velocities when the bar is seen close to side-on. We also ran simulations covering a large fraction of the parameter space likely to be occupied by real galaxies. The PVDs can then be used to somewhat constrain the mass distribution and bar properties of observed systems.

4. The Nature of Boxy/Peanut-Shape Bulges

The PVDs produced are directly comparable to kinematic observations of edge-on spiral galaxies. In the hope of understanding the formation mechanism of boxy bulges, we searched for bars in a sample of 30 edge-on spirals with and without boxy bulges, using emission line long-slit spectroscopy (Bureau & Freeman 1999). The objects were selected from existing catalogs and 2/3 have probable companions. Of the 24 galaxies with a boxy bulge, 17 have extended emission lines and constitutes our main sample. The remaining 6 galaxies all have extended emission and form a control sample.

In the main sample, 14 galaxies display a clear bar signature in their PVD, and only 3 may be axisymmetric or have suffered interactions. None of the galaxies in the control sample shows evidence for a bar. This means that most boxy bulges are due to the presence of a thick bar viewed edge-on and only a few may be due to the accretion of external material. In addition, spheroidal bulges do appear axisymmetric. Thus, it seems that most boxy bulges are edge-on bars and that most bars are boxy or peanut shaped when viewed edge-on. However, the strength of this converse is limited by the small size of the control sample. To illustrate our data, we show the PVD of two galaxies in the main sample in Figure 3. Our association of bars and boxy bulges is supported by the anomalous emission line ratios observed in many objects. These galaxies display large

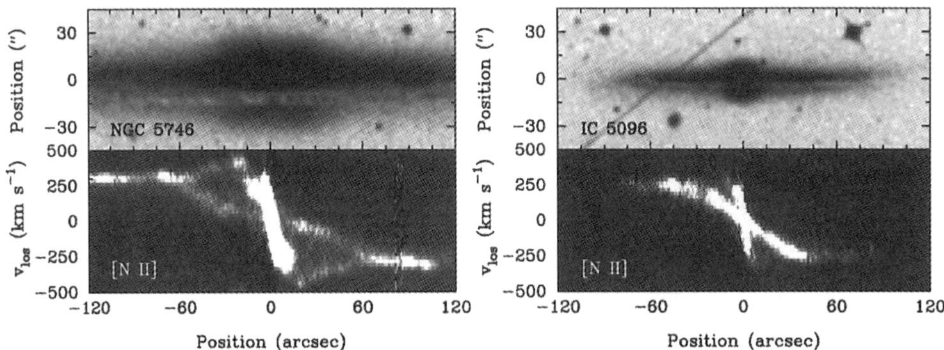

FIGURE 3. Image and ionised gas PVD (on the same scale and along the major axis) of two boxy galaxies. The bar signature in the bulge region is obvious in both cases.

Hα/[N II] ratios, often associated with shocks, and these ratios correlate with kinematical structures in the disks. Constraining the viewing angle to the galaxies with our models, the observations also appear to confirm the general prediction of N-body simulations, that bars are peanut-shaped when seen side-on and boxy-shaped when seen end-on.

Our results are consistent with the current knowledge on the bulge of the Milky Way and strongly support the bar-buckling mechanism for the formation of boxy/peanut-shaped bulges. However, we do not test directly for buckling, and other bar-thickening mechanisms and hybrid scenarios cannot be excluded. Nevertheless, it is clear that the influence of bars on the formation and evolution of bulges is primordial.

5. On-going Studies

The bar diagnostics we have developed open up for the first time the possibility of studying the vertical structure of bars observationally. To this end, we have obtained K-band images of all the sample galaxies. We have also obtained absorption line spectroscopic data to study the stellar kinematics, and a more in-depth investigation of line ratios will give us a better understanding of the large scale effects of bars in disks.

We would like to thank A. Bosma, A. Kalnajs, and L. Sparke for useful discussions at various stages of this work. We also thank J.-C. Lambert for computer assistance and G. D. Van Albada for the FS2 code.

REFERENCES

ATHANASSOULA, E., BUREAU, M. 1999 *ApJ,* submitted

BINNEY, J., PETROU, M. 1985 *MNRAS,* **214**, 449

BUREAU, M., ATHANASSOULA, E. 1999 *ApJ,* submitted

BUREAU, M., FREEMAN, K.C. 1999 *AJ,* submitted

COMBES, F., DEBBASCH, F., FRIEDLI, D., PFENNIGER, D. 1990 *A&A,* **233**, 82

KUIJKEN, K., MERRIFIELD, M.R. 1995 *ApJ,* **443**, L13

Boxy- and Peanut-Shaped Bulges

By R A I N E R   L Ü T T I C K E[1]
and R A L F - J Ü R G E N   D E T T M A R[1]

[1] Astronomisches Institut der Ruhr-Universität Bochum, Universitätsstr. 150, D-44780
Bochum, Germany

Our new statistical study of bulges of disk galaxies reveals a frequency of almost 50% being boxy-
or peanut-shaped. Therefore very common processes are required to explain this high fraction.
In an analysis of a possible relation between this internal structure and the environment of
galaxies with boxy/peanut-shaped bulge we find that on large scales there is no hint for a
connection. However, galaxies with boxy- or peanut-shaped bulges have more companions and
satellites and show more frequently interactions than a control sample. Thus we conclude that
the small-scale environment is important for the existence of such bulges. The most likely reason
responsible for the development of boxy/peanut-shaped bulges is a bar originating from galaxy
interaction in stable disks or by an infalling satellite.

1. Introduction

Boxy- and peanut-shaped (hereafter referred to simply as boxy or b/p) bulges are not
really as peculiar as it seemed in the past, and very common processes are required to explain their high frequency. At present several mechanisms for their origin are discussed.
Binney & Petrou (1985) and Whitmore & Bell (1988) suggest that these structures result
from material accreted from infalling satellite companions (soft merging). An alternative mechanism for forming boxy bulges are instabilities or resonances animated by bars
(Combes *et al.* 1990; Raha *et al.* 1991). N-body simulations for stars in barred potentials have demonstrated that this theory and observational evidence are consistent (in
particular from gas kinematics, e.g. Kuijken & Merrifield 1995). Within this framework,
however, the question of what may cause the bar becomes even more important. Dynamically cold disks can produce through a global instability a bar, or galaxy interaction can
drive bar formation in stable disks (Freeman 1996). However, a bar can also originate
from an infalling satellite so that accretion and bar hypotheses for the formation of boxy
bulges could be seen in a unified picture (Mihos *et al.* 1995).

2. Statitics of Boxy- or Peanut-Shaped Bulges

In a complete sample of edge-on disk galaxies selected from the RC3 (de Vaucouleurs
et al. 1991) with $D_{25} > 2$ arcmin and inclination greater than $\sim 70^0$ ($\sim$1350 galaxies)
we characterized bulges by their degree of boxy/peanut shape (type 1 = peanut-shaped
[Figure 2], 2 = boxy-shaped, and 3 = similar to boxy-shaped), as non-boxy (/), or
unclassifiable (mostly due to insufficient inclination or stars in the foreground) using the
'Digitized Sky Survey' (DSS). In a first go this is done by eye. We are currently in
the process of reobserving a subsample with CCDs in order to apply more quantitative
methods (Figures 1 and 2). The main result from the photographic material is that 46%
of all classifiable galaxies have a boxy- or peanut-shaped bulge: 4% type 1, 17% type 2,
and 25% type 3 (Lütticke & Dettmar 1999).
The frequency distributions of galaxies with boxy- or peanut-shaped bulges and barred
galaxies (all face-on disk galaxies in the RC3) binned by morphological type show the
same general dependences (S0 - Sd) with a maximum at Sc and a minimum at S0.

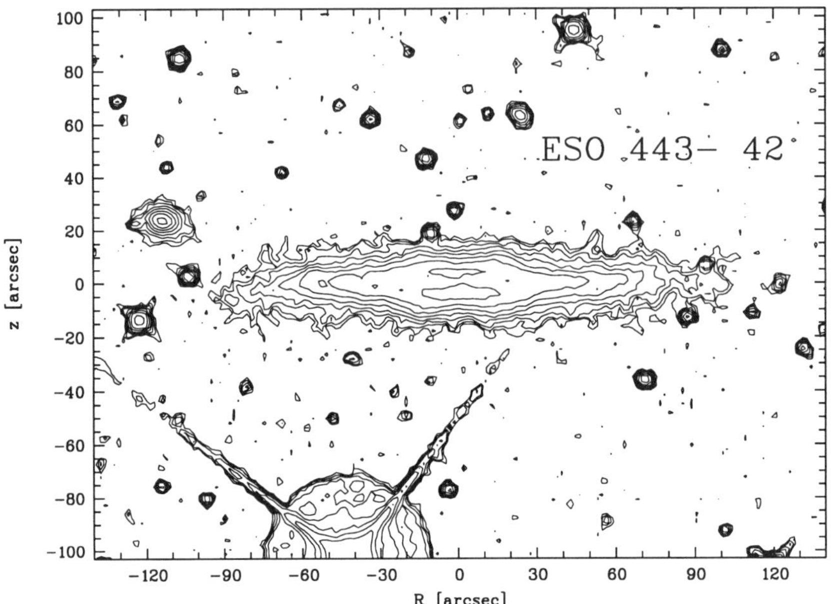

FIGURE 1. Contour plot of ESO 443- 42 from DSS

Parameter	b/p Bulges (type 1+2)	Non-b/p Bulges
NGT [galaxies/sq degree] [1]	3 ± 1	3 ± 1
D_{nb} [arcmin] [2]	24 ± 2.5	28 ± 3
D_{snb} [arcmin] [3]	34.5 ± 3	36 ± 3
D_{nb}^{10} [degree] [4]	1.4 ± 0.1	1.6 ± 0.1
ρ [galaxies/megaparsec^3] [5]	1.0 ± 0.2	0.75 ± 0.2

[1]: Number of galaxies (in Lauberts & Valentijn 1989).
[2], [3], [4]: Projected distance to the nearest, second nearest,
and 10th nearest neighbour (Lauberts & Valentijn 1989).
[5]: Local density (Tully 1988).

TABLE 1. Density Parameters.

3. Boxy/Peanut Bulges and the Influence of Environment

The mean density parameters for galaxies with (type 1 + 2) and without boxy- or peanut-shaped bulges are within the errors nearly the same (Table 1).

The analysis of galaxies in the Virgo- and Ursa Major-Cluster shows that in these clusters the fraction of boxy- or peanut-shaped bulges is the same as for the complete galaxy sample (Table 2).

The result of an investigation of the environment of galaxies with prominent boxy- or peanut-shaped bulges (type 1; N=26) is that the mean number of companions ($\Delta v <$ $1000 km/s$; NASA/IPAC Extragalactic Database) in a projected radius of 5 x D_{25} (RC3)

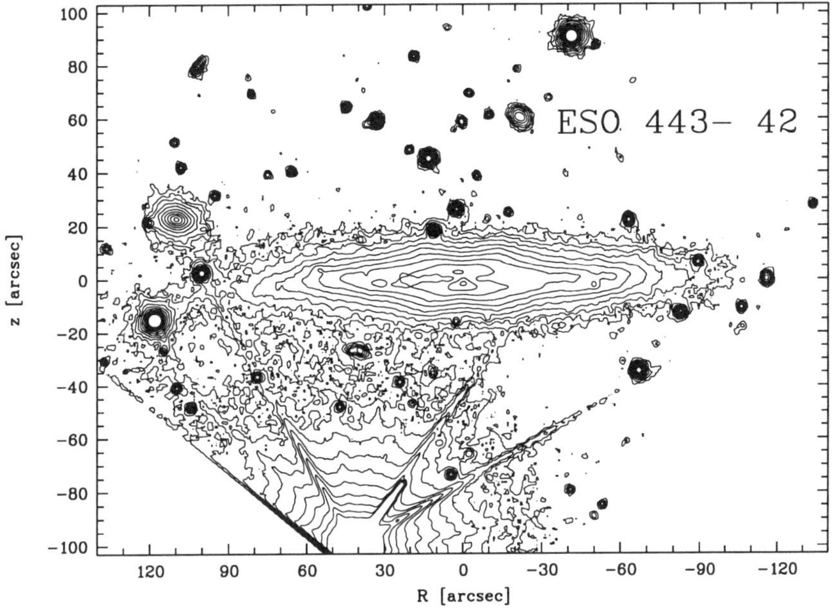

FIGURE 2. CCD follow-up in comparison (R-band CCD image, 0.9m Dutch/ESO)

Cluster	Fraction of b/p Bulges (all types)	Fraction of non-b/p Bulges
Virgo-Cluster [1]	44%	56%
Ursa Major-Cluster [2]	46%	54%

[1]: number of investigated galaxies: N = 18 ;   [2]: N = 13

TABLE 2. Clusters.

around the center of a host galaxy is 1.2 ± 0.3, and for half of this radius, 0.5 ± 0.2. In contrast, the mean number is 0.6 ± 0.2 resp. 0.2 ± 0.1 in a sample of galaxies without boxy bulges.

35% of the galaxies with boxy- or peanut-shaped bulges have at minimum one companion inside 2.5 x D_{25}, but in the control sample this is only 15%. From visual counts using DSS we obtain for the number of companions inside the $2.5 \times D_{25}$-radius 2.6 ± 0.4 for galaxies with, and 1.5 ± 0.3 for galaxies without, boxy- or peanut-shaped bulges. About 85% of the candidate companions have a diameter smaller than a third of the diameter (D_{25}) of the host galaxy. A very surprising result comes from the statistics of boxy bulges (N = 24) in isolated spiral galaxies as identified by Zaritsky *et al.* (1993, 1997). These galaxies in the field have the advantage that gravitional forces of nearby massive galaxies can be neglected. The only possible interaction can happen with a small companion so that its influence can be investigated by comparison with isolated galaxies without satellites. We find that 90% of the classifiable galaxies with satellites have boxy or peanut-shaped bulges. In contrast, there is no galaxy from the control sample without

Sample	$w=0$	$w=1$	$w=2$	$w=3$	$w=4$	I
Hubble Deep Field	203	2	20	27	39	0.96
Medium Deep Survey	444	19	37	8	4	0.26
Shapley-Ames Catalogue	821	22	65	0	3	0.18
Galaxies with type 1 b/p bulge	13	5	7	1	0	0.85 ± 0.2
Galaxies without b/p bulge	22	3	1	0	0	0.2 ± 0.1

TABLE 3. Interaction Index.

satellites showing a boxy- or peanut-shaped bulge.

The normalized interaction index defined by van den Bergh *et al.* (1996) $I = \sum_{i=1} \frac{w_i}{N}$ with $w=0$: number of objects showing no tidal distorstion, $w=1$: possible tidal effects, $w=2$: probable tidal effects, $w=3$: possible mergers, and $w=4$: almost certain mergers shows definitely a higher value for the sample of boxy- or peanut-shaped bulges (type 1) than for the control sample. The boxy- or peanut-shaped sample has a value similar to the value of the Hubble Deep Field, while the control sample is comparable to the Medium Deep Survey and the Shapley-Ames Catalogue (Table 3).

4. Conclusions

Our statistical study suggests that boxy- or peanut bulges preferentially occur in disk galaxies with companions. The favorable model scenarios for the development of boxy bulges are therefore resonances at a bar triggered by galaxy interaction or by an infalling satellite in an otherwise stable disk.

A sufficient mass concentration in the center of the bulge could finally cause the bar to dissolve. This proposed sccnario would let galaxies evolve from SA over SB to SA, with bulges growing through a boxy phase either by disk instabilities, accreted material, or by both. Along the Hubble sequence, galaxies would eventually evolve from Sd to S0/Sa. More recent results from N-body simulations support this evolutionary view of the morphology. However, currently it can not totally be excluded that some boxy- or peanut-shaped structures result from bars produced by dynamically cold disks through a global instability or directly from material accreted from infalling satellite companions (soft merging).

This work is partly supported by Deutsche Forschungsgemeinschaft and, in addition to using the DSS, is based on observations obtained at ESO/La Silla and DSAZ/Calar Alto (Spain).

REFERENCES

BINNEY, J., PETROU, M. 1985 *MNRAS*, **214**, 449

COMBES, F., DEBBASCH, F., FRIEDLI, D., PFENNIGER, D. 1990 *A&A*, **233**, 82

DE VAUCOULEURS, G., DE VAUCOULEURS, A., CORWIN, H.G. JR., BUTA, R.J., FOUQUÉ, P. 1991 *Third Reference Catalogue of Bright Galaxies*. (Springer-Verlag, New York). (RC3)

FREEMAN, K.C. 1996, in *Barred Galaxies* (ed. R. Buta R., D.A. Crocker & B.G. Elmegreen), ASP Conf. Ser. **91**, p1. (ASP)

KUIJKEN, K., MERRIFIELD, M.R. 1995 *ApJ*, **433**, L13

LAUBERTS, A., VALENTIJN, E.A. 1989 *The Surface Photometry Catalogue of the ESO-Uppsala Galaxies*. (ESO)

LÜTTICKE, R., DETTMAR, R.-J. 1999, in *The Magellanic Clouds and Other Dwarf Galaxies* (ed. T. Richtler & J.M. Braun), in press. (Shaker)

MIHOS, J.C., WALKER, I.R., HERNQUIST, L., DE OLIVEIRA, C.M., BOLTE, M. 1995 *ApJ*, **447**, L87

RAHA, N., SELLWOOD, J.A., JAMES, R.A., KAHN, F.D. 1991 *Nature*, **352**, 411

TULLY, R.B. 1988 *Nearby Galaxies Catalog.* (Cambridge). (NGC)

VAN DEN BERGH, S., ABRAHAM, R., ELLIS, R.S., TANVIR, N.R., SANTIAGO, B.X., GLAZE-BROOK, K.G. 1996 *AJ*, **112**, 359

WHITMORE, B.C., BELL, M. 1988 *ApJ*, **324**, 741

ZARITSKY, D., SMITH, R., FRENK, C.S., WHITE, S.D.M. 1993 *ApJ*, **405**, 464

ZARITSKY, D., SMITH, R., FRENK, C.S., WHITE, S.D.M. 1997 *ApJ*, **478**, 39

A New Class of Bulges

By **RAINER LÜTTICKE**[1]
and **RALF-JÜRGEN DETTMAR**[1]

[1] Astronomisches Institut der Ruhr-Universität Bochum, Universitätsstr. 150, D-44780
Bochum, Germany

Inspecting a sample of edge-on galaxies selected from the RC3 (de Vaucouleurs *et al.* 1991)
with D_{25} >2 arcmin ($\sim$1350 galaxies) on the 'Digital Sky Survey' we have identified a class of
approximately 20 disk galaxies with prominent, large, and boxy bulges. These bulges all show
irregularities and asymmetries which are suggestive of them being formed just recently and not
yet dynamically settled. We will present some examples and first results from CCD follow-up
observations.

While the large frequency of boxy- or peanut-shaped bulges in disk galaxies (nearly 50%) is
best explained by the response of the stellar disk to a bar potential, we propose soft-merging of
companions as the most likely scenario for the evolution of this new class of thick boxy bulges.

1. Introduction

Statistics of boxy- and peanut-shaped (b/p) bulges in edge-on galaxies show (Shaw
1987, Dettmar 1989) that such bulges are not really that peculiar as it seemed in the
past and very common processes are required to explain the high frequency. At present
several mechanisms for their origin are discussed. Binney & Petrou (1985) and Whitmore & Bell in their paper on IC 4767 (1988) suggested that these structures may result
from material accreted from infalling satellite companions (soft merging). An alternative mechanism for forming boxy bulges are instabilities or resonances animated by bars
(Combes *et al.* 1990; Raha *et al.* 1991). N-body simulations for stars in barred potentials
have demonstrated that with regard to the shape of bulges this theory and observational
evidence are consistent.

2. A New Blass of Bulges

While there is increasing evidence (in particular from gas kinematics, e.g., Kuijken &
Merrifield, 1995) that the 'bar resonance' is indeed at work in most cases, it was pointed
out earlier (Dettmar & Barteldrees 1990, Dettmar 1996) that some large boxy bulges can
not be directly explained this way. They rather show irregularities and asymmetries that
hint at an origin from a more recent merger event. Prototypes are the objects NGC 1055
(Shaw, 1993) and IC 4745 (Dettmar & Barteldrees, 1990).

From a new and complete survey of edge-on galaxies we have compiled a list of objects
that fall into this category of hosting disturbed, large, and boxy bulges. In the following
we will briefly describe the survey and present the object catalogue with some examples
including some first images from our CCD follow-up.

3. The Survey

In a complete sample of edge-on disk galaxies selected from the RC3 (de Vaucouleurs
et al. 1991) with D_{25} >2 arcmin ($\sim$1350 galaxies) we characterized bulges by their degree
of boxy or peanut shape, as non-boxy, or unclassifiable using the 'Digitized Sky Survey'.
In a first go this is done by eye. We are currently in the process of reobserving a

Object	RA (2000)	DEC (2000)	Morph. type[1]	Source [2]	Comment
ESO 13- 12	01 07	-80 18	S0a	DSS	
NGC 1030	02 40	+18 02	S?	CCD	likely S0
NGC 1055	02 42	+00 27	Sb	DSS	prominent bulge well studied in literature
NGC 1589	04 31	+00 52	Sab	DSS	
UGC 3458	06 26	+64 44	Sb	DSS	
ESO 494- 22	08 06	-24 49	Sa	DSS	affected by stars due to position behind MW
NGC 3573	11 11	-36 52	S0a	DSS	
ESO 506- 3	12 22	-25 05	Sab	CCD	
ESO 322-100	12 49	-41 27	S0	CCD	
ESO 383- 5	13 29	-34 16	Sbc	DSS	
ESO 21- 4	13 33	-77 51	S0a	DSS	affected by stars due to position behind MW
NGC 5719	14 41	-00 19	Sab	DSS	
UGC 9759	15 11	+55 21	S?	DSS	~ S0a; faint galaxy; confirmed by CCD follow-up
ESO 514- 5	15 19	-23 49	Sa	DSS	
UGC 10205	16 07	+30 06	Sa	CCD	VV 624; prominent bulge known in the literature
IC 4745	18 42	-64 56	Sab	DSS	
IC 4757	18 44	-57 10	S0a	CCD	
IC 4767	18 48	-63 24	S0	CCD	large peanut-shaped bulge
ESO 142- 19	19 33	-58 07	Sb	DSS	
NGC 6848	20 03	-56 05	Sa	DSS	
NGC 7183	22 02	-18 55	S0	DSS	

[1]: From RC3 (de Vaucouleurs *et al.* 1991).
[2]: DSS = detected in a complete sample of edge-on disk galaxies selected from the RC3 with $D_{25} > 2$ arcmin; CCD = own CCD-observations.

TABLE 1. Catalogue of Disturbed, Large and Boxy Bulges.

subsample with CCDs in order to apply more quantitative methods. The main result from the photographic material (DSS) is that 46% of all classifiable galaxies have a boxy or peanut-shaped bulge (Lütticke & Dettmar 1999, and these proceedings).
15 galaxies of the DSS sample and additional 6 galaxies from CCD-observations show such extreme 'boxy' isophotes. Their large bulges have extra features such as twists of the isophotes, large scale asymmetries, and other irregularities. In the case of IC 4745 the asymmetries are even reflected in the rotation curve (Dettmar & Barteldrees 1990). We have catalogued these galaxies in Tab. 1 and show some examples in Figures 1 – 3 (another example is presented in Dettmar & Lütticke, 1999 and these proceedings). These figures give an example for one of the DSS scans (Figure 1) as well as first results from our CCD follow-up observations (Figures 2 and 3). In all cases for which CCD imaging could be obtained the suspected peculiarities identified on DSS scans were confirmed.

4. Discussion

From a comparison of the relative sizes of the boxy bulge and the bar structure as deduced from the rotation curve (e.g., NGC 1055) it becomes clear that 'bar resonances'

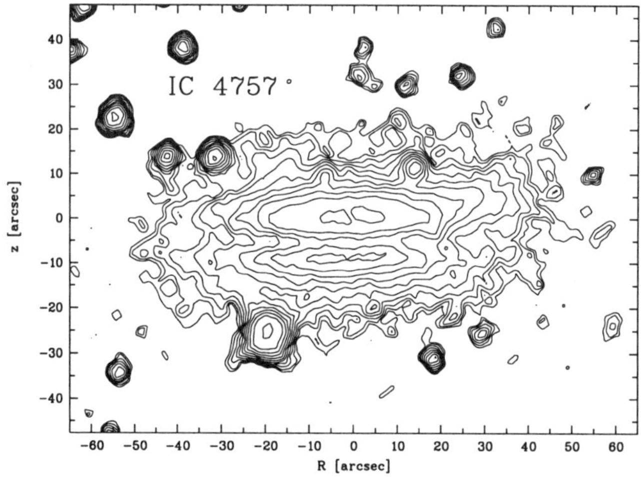

FIGURE 1. Contour plot of IC 4757 from DSS.

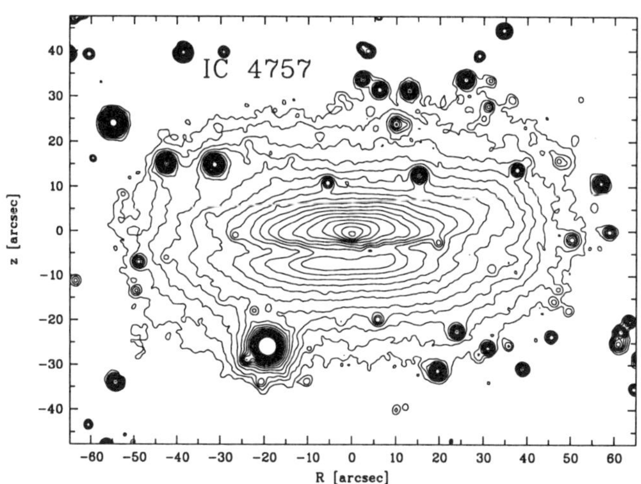

FIGURE 2. R-band CCD image of IC 4757 obtained with the ESO/MPI 2.2 m on La Silla.

can not explain these huge boxy bulges. In addition, we observe in many cases large scale asymmetries and/or small scale substructure which both can be considered as indicators for interaction events. This adds a new morphological feature into a possible merger sequence: interaction and mergers with the smallest satellites cause a bar instabilty which leads to a boxy- or peanut-shaped bulge, while soft mergers with more massive satellites result in thick boxy bulges.

The CCD follow-up observations will allow us to discuss colors as an additional indicator for the past history of these exceptional objects, and we are planning to derive kinematical data, too.

This work is partly supported by Deutsche Forschungsgemeinschaft and, in addition

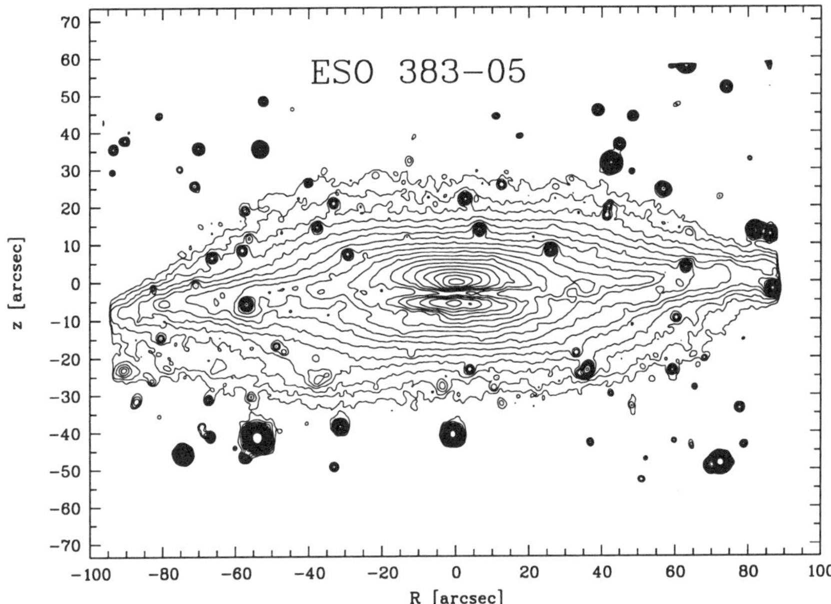

FIGURE 3. R-band CCD image of ESO 383-05 obtained with the ESO/MPI 2.2 m on La Silla.

to using the DSS, is based on observations obtained at ESO/La Silla and DSAZ/Calar Alto (Spain).

REFERENCES

BINNEY, J., PETROU, M. 1985 *MNRAS*, **214**, 449

COMBES, F., DEBBASCH, F., FRIEDLI, D., PFENNIGER, D. 1990 *A&A*, **233**, 82

DETTMAR, R.-J. 1996, in *Unsolved problems of the Milky Way* (ed. L. Blitz & P.J. Teuben), IAU Symp. **169**, p335. (Kluwer)

DETTMAR, R.-J. 1989, in *The World of Galaxies* (ed. H.G. Corwin & L. Bottinelli), p229. (Springer, New York)

DETTMAR, R.-J., BARTELDREES, A. 1990, in *Bulges of Galaxies* (ed. D. Terndrup & B. Jarvis), p255. (ESO)

DETTMAR, R.-J., LÜTTICKE, R. 1999, in *The Third Stromolo Symposium: The Galactic Halo* (ed. Gibson, B.K., Axelrod, T.S. & Putman, M.E.), in press. (ASP)

DE VAUCOULEURS, G., DE VAUCOULEURS, A., CORWIN, H.G. JR., BUTA, R. J., FOUQUÉ, P. 1991 *Third Reference Catalogue of Bright Galaxies.* (Springer-Verlag, New York). (RC3)

KUIJKEN, K., MERRIFIELD, M.R. 1995 *ApJ*, **433**, L13

LÜTTICKE, R., DETTMAR, R.-J. 1999, in *The Magellanic Clouds and Other Dwarf Galaxies* (ed. T. Richtler & J.M. Braun), in press. (Shaker)

RAHA, N., SELLWOOD, J.A., JAMES, R.A., KAHN, F.D. 1991 *Nature*, **352**, 411

SHAW, M. 1987 *MNRAS*, **229**, 691

SHAW, M. 1993 *A&A*, **280**, 33

WHITMORE, B.C., BELL, M. 1988 *ApJ*, **324**, 741

The Role of Secondary Bars in Bulge Formation

By HASHIMA HASAN

Office of Space Science, NASA Headquarters, Washington DC 20546, USA
Space Telescope Science Institute, 3700 San Martin Dr., Baltimore MD 21218, USA

An analysis of stellar orbits in a doubly barred galaxy shows that the effect of a secondary bar is to destabilize the orbits, the process being accompanied by the appearance of vertical resonances which would enable stars to leave the galactic plane and move into the bulge. This phenomenon could contribute to bulge formation. Results of the orbital analysis are presented and their significance discussed.

1. Introduction

The role of a secondary bar in shaping the morphology of a galaxy and its possible contribution to bulge formation is an issue which is currently largely unexplored. With more powerful observing techniques beginning to become available, a new look at galaxies which had been classified as unbarred shows that several of them possess a primary bar and some even show secondary bars (Mulchaey *et al.* 1997). If secondary bars are more prevalent than previously supposed, it is conceivable that they play a role in the secular evolution of galaxies much in the same way as do central mass concentrations (e.g. Hasan & Norman 1990, Sellwood & Moore 1999, Merritt 1998.) Nested gaseous bars have been produced in N-body simulations (Friedli & Martinet 1993; Heller & Shlosman 1994) suggesting that a system of embedded bars may be effective in transporting gas to the galactic center (Pfenniger & Norman 1990, Shlosman *et al.* 1989), thus influencing galactic evolution. An intuitive insight into the evolutionary process may be gained by examining the stellar dynamics in such systems. Maciejewski and Sparke (1997) have demonstrated that nonchaotic multiply periodic particle orbits can exist in doubly barred galaxies.

A preliminary computation of stellar orbits in doubly barred galaxies was reported earlier by Hasan (1996), where it was shown that the growth of a secondary bar causes orbit destabilization. I present more details here and discuss the results in the context of bulge building.

2. Formulation

A two dimensional formulation of the problem is considered, in which the galactic potential is approximated by a disk component modelled as a Plummer sphere, and two bars each of which is represented as a Ferrers bar. We consider orbits in the galactic plane corotating with the bars (both bars are considered to have the same pattern speed).

For normalization purposes it is found convenient to fix the primary bar semi-major axis a_1 at 9 kpc, and the total mass $M_T = 4.67 \times 10^{10} M_\odot$ (Hasan & Norman 1990). For all cases studied this normalization results in a pattern speed $\Omega_p \sim 15$ km/sec/kpc so that $R_{cr}/a_1 = 1$. The parameters fixed for all calculations are: the primary bar mass, $M_{b_1}/M_T = 0.3$, its semi-minor axis, $b_1 = 4.05$ kpc, secondary bar major axis, $a_2 = 1.8$ kpc and the length scale of the central component, $A_c/a_1 = 0.5$. The Jacobi constant is fixed so that a star can reach a maximum distance corresponding to the edge of the minor axis of the primary bar.

Two scenarios were studied: (1) The semi-minor axis of the secondary bar was fixed to $0.45a_2 = 0.81$ kpc and the seconadry bar mass varied. (2) The secondary bar mass fixed at $0.05M_T$ and semi-minor axis of the secondary bar varied so that $b_2/a_2 = 0.45, 0.35, 0.25$.

3. Bar Dissolution, Bulge Formation, and Resonances

For each case studied, surfaces of section were obtained, examined for orbital stability, and physical insight gained into processes contributing to galactic evolution. Stable direct (x_1) orbits which support the bar and which manifest themselves as continuous elliptic curves in surfaces of section, fill most of the available phase space in the presence of a small ($M_{b_2}/M_T = 0.05$) secondary bar. Increasing the secondary bar mass causes the appearance of resonances and leads eventually to orbital stochasticity. The orientation and width of the secondary bar also affect stochasticity and consequent weakening and dissolution of the primary bar. A thinner, more compact secondary bar causes greater stochasticity than does a less dense one. Perpendicular secondary bars lead to a richer phase-space structure, representing orbits that develop complicated shapes, while still generally aligned with the primary bar. Some of the orbits supporting the primary bar re-orient themselves to align with the secondary bar, thus leading to a 'boxier' shape.

An example of the surface of section is given in Figure 1, where we clearly see that a perpendicular secondary bar has a more dramatic effect on the surface-of-section structure than a parallel bar. The direct orbits are plotted in the right hand side of each figure. The islands in the surfaces of section represent resonance orbit families. Of particular significance is the minor family which appears towards the center of the figure for the case where the secondary bar is perpendicular to the primary bar. This 'banana' orbit family also appeared in the case of spherical central masses (Hasan & Norman 1990, Hasan *et al.* 1993), and it was found that these orbits would enable stars to leave the galactic plane. The 'banana' orbits do not appear for a parallel secondary bar till its mass is close to 11% of the primary bar, and have disappeared for a perpendicular secondary bar more massive than about 9% of the primary bar. As the orbits supporting the primary bar start taking more complicated shapes, the primary bar starts weakening and will dissolve if the secondary bar becomes quite massive ($M_{b_2}/M_T = 0.15$).

Unlike the case of a spherical central mass (e.g. Hasan & Norman 1990), increasing the mass of the secondary bar has a marked effect on the retrograde orbits. These orbits do not support the bar and thus are not of importance to its evolution. They may, however, provide an alternative path for dust and gas to circulate within the galaxy. This may be a question worth investigating.

The effect of different secondary bar parameters on orbital behavior may be understood by borrowing concepts from the epicyclic theory. Frequency curves were computed for the two bar orientations considered here. Not surprisingly, for the parallel secondary bar, two Inner lindblad Resonances (ILRs) are present, thus making a larger volume in phase space available for x_1 orbits than for an anti-aligned secondary bar. As the secondary bar mass is increased, the inner ILR moves inward and the outer ILR moves outward, reducing the phase space available for x_1 orbits for both the parallel secondary bar and the primary bar and leading to a destruction of both. An anti-aligned secondary bar is likely to maintain its shape while the primary bar is weakened.

The rate of mass accumulation in the secondary bar will determine how much of the dissolution process will contribute to bulge formation. If mass accumulation slows or stops when $M_{b_2}/M_T = 0.07$, dissolution of the primary bar will slow down, but bulge building will continue. Since in reality secondary bars are neither parallel nor perpendicular to the primary bar at all times, but are rotating at a slower pattern speed

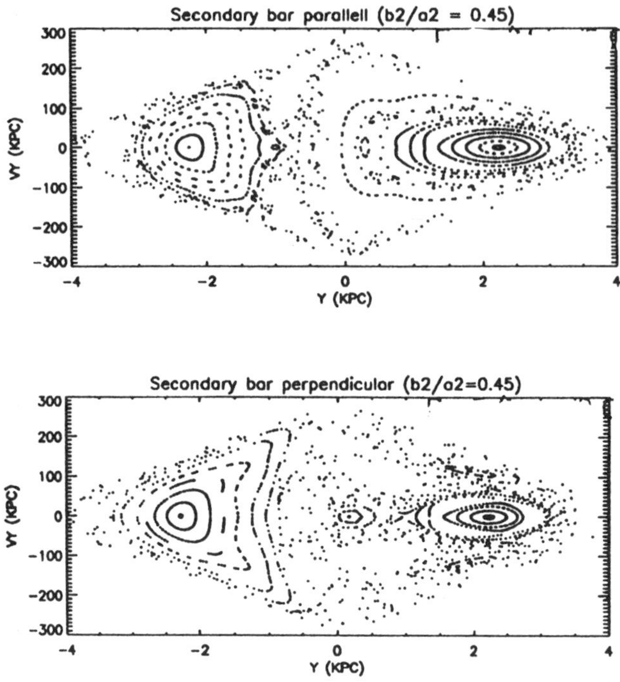

FIGURE 1. Surfaces of section for a secondary bar mass $M_{b_2}/M_T = 0.07$, when the secondary bar is (1) parallel (top), and (2) perpendicular (bottom) to the primary bar. The right hand side of each figure represents direct orbits. The smooth almost elliptical curves represent the x_1 orbits which support the primary bar. The secondary bar causes resonances which manifest themselves as islands on the surface of section, stochastic orbits which cause regions of irregularity in the surfaces of section, and 'banana' orbits which appear as a series of closed curves towards the center of the figure for the perpendicular bar.

than the primary bar, the actual mass ratios computed here should only be considered as very approximate figures. Furthermore, it is difficult to say if the secondary bar will maintain its shape, be destroyed together with the primary bar, or lead to more complex nuclear structures. Observationally, more and more complex structures are being seen in galaxies (e.g. Carollo *et al.* 1997), for which there is no good theoretical description. The main purpose of the simplistic computations presented here is to gain physical insight into dynamical processes within the galactic nucleus and to lead the way to more detailed investigations.

3.1. *Discussion*

An understanding of processes responsible for bulge building is far from complete as evidenced by the papers and discussions in this conference. Combining the results presented here with earlier ones (e.g. Hasan *et al.* 1993, Combes *et al.* 1990, Sellwood & Wilkinson 1993), it is becoming clear that processes taking place in the central regions of barred galaxies impact bulge formation. A dense, compact object whether in the form of a sphere or a small rotating bar, will cause orbital stochasticity accompanied by bulge building. The denser and more compact the object is, the quicker the bar will dissolve. However, bulge formation may take place for only a fraction of the bar dissolution time. For more diffuse central objects, the bar dissolution process may be slower and the bulge formation period longer. Indeed, the bar may only weaken and not be completely destroyed. Detailed N-body simulations of this process would shed more light on this phenomenon.

REFERENCES

CAROLLO, C.M., STIAVELLI, M., DE ZEEUW, P.T., MACK, J. 1997 *AJ*, **114**, 2366

COMBES, F., DEBASCH, F., FRIEDLI, D., PFENNIGER, D. 1990 *A&A*, **233**, 82

FRIEDLI, D., MARTINET, L. 1993 *A&A*, **277**, 27

HASAN, H. 1996, in *Barred Galaxies* (ed. R. Buta, D.A. Crocker & B.G. Elmgreen), ASP Conf. Ser. **91**, p464. (ASP)

HASAN, H., NORMAN, C. 1990 *ApJ*, **361**, 69

HASAN, H., PFENNIGER, D., NORMAN, C. 1993 *ApJ*, **409**, 91

HELLER, C.H., SHLOSMAN, I. 1994 *ApJ*, **424**, 84

MACIEJEWSKI, W., SPARKE, L. 1997 *ApJ*, **484**, L117

MERRITT, D. 1998 *Comments on Astrophysics*, **19**

MULCHAEY, J.S., REGAN, M.W., KUNDU, A. 1997 *ApJS*, **110**, 299

PFENNIGER, D., NORMAN, C. 1990 *ApJ*, **363**, 391

SELLWOOD, J., MOORE, E.M. 1999 *ApJ*, in press

SELLWOOD, J., WILKINSON, A. 1993 *Rep. Prog. Phys.*, **56** , 173

SHLOSMAN, I., FRANK, J. AND BEGELMANN, M.C. 1989 *Nature*, **338**, 45

Radial Transport of Molecular Gas to the Nuclei of Spiral Galaxies

By **KAZUSHI SAKAMOTO**[1,2], **S. K. OKUMURA**[1], **S. ISHIZUKI**[1], AND **N. Z. SCOVILLE**[2]

[1]Nobeyama Radio Observatory, Nagano 384-1305, JAPAN

[2]Radio Astronomy, California Institute of Technology, MS105-24, Pasadena CA91125, USA

The NRO/OVRO imaging survey of molecular gas in 20 spiral galaxies is used to test the theoretical predictions on bar-driven gas transport, bar dissolution, and bulge evolution. In most galaxies in the survey we find gas condensations of 10^8–10^9 $M_\odot$ within the central kiloparsec, the gas masses being comparable to those needed to destroy bars in numerical models. We also find a statistically significant difference in the degree of gas concentration between barred and unbarred galaxies: molecular gas is more concentrated to the central kiloparsec in barred systems. The latter result supports the theories of bar-driven gas transport. Moreover, it constrains the balance between the rate of gas inflow and that of gas consumption (i.e., star formation, etc.), and also constrains the timescale of the possible bar dissolution. Namely, gas inflow rates to the central kiloparsec, averaged over the ages of the bars, must be larger than the mean rates of gas consumption in the central regions in order to cause and maintain the higher gas concentrations in barred galaxies. Also, the timescale for bar dissolution must be longer than that for gas consumption in the central regions by the same token.

1. Introduction

Radial transport of gas in galactic disks likely plays an important role in the formation and evolution of bulges. There are two aspects in the effect of gas transfer to bulges, in both of which stellar bars are involved. First, theories predict that bars efficiently transport interstellar gas to the nuclei of spiral galaxies, providing star forming material to the bulge regions. Second, simulations have shown that the gas accumulation at a galactic center changes the gravitational potential and eventually destroys the bar (c.f., a review by Pfenniger in these proceedings). Bulges may grow through this process by gaining stars from disks.

Observationally, evidence for bar-driven gas transport and for bar dissolution has been limited compared to the large amount of theoretical work. The pieces of observational evidence supporting the bar-driven gas transport are the estimation of gas inflow rates in two barred galaxies using CO and NIR observations and dynamical models (Quillen *et al.* 1995; Regan & Vogel 1997), shallower metallicity gradients in barred than unbarred galaxies (Zaritsky *et al.* 1994; Martin & Roy 1994), and larger Hα luminosities in the nuclei of barred galaxies presumably due to larger amount of gas in barred nuclei (e.g., Ho *et al.* 1995). In order to further investigate the relation between bars, gas, and bulges, it is important to observe gas in many galaxies.

The NRO/OVRO CO imaging survey mapped the distribution of molecular gas in the central kiloparsecs of 20 ordinary nearby spirals using the millimeter arrays of the two observatories (Sakamoto *et al.* 1998, 1999). The 20 northern spiral galaxies were selected on the basis of inclination (face-on), lack of significant dynamical perturbation, and reasonable single-dish CO flux to allow high-resolution observations. No selection was made on starburst, nuclear activity, far-infrared luminosity, and galaxy morphologies. The sample contains 10 barred (SB+SAB) and 10 unbarred (SA) spirals with the mean distance of 15 Mpc and with luminosities $\sim L^*$. Our aperture synthesis observations have

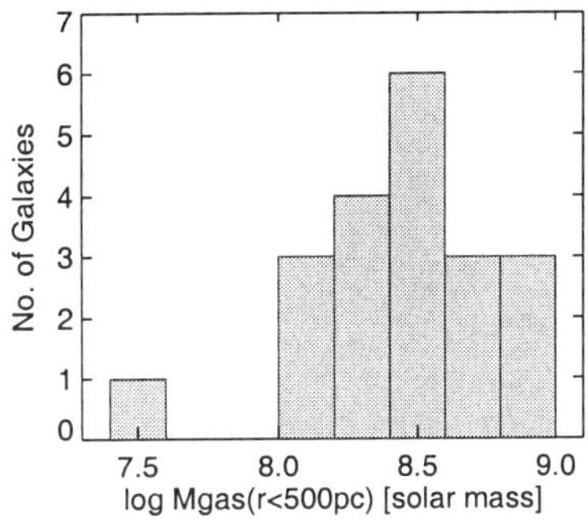
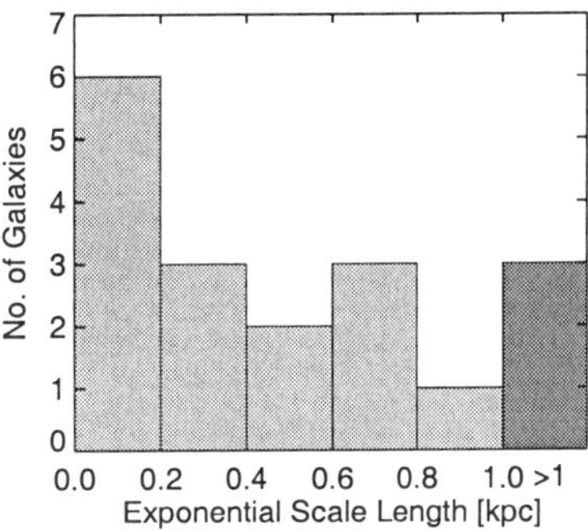

FIGURE 1. (Left) Molecular gas masses within the central kiloparsec derived from CO emission. (Right) Scale lengths of CO radial distributions in the galactic centers.

a mean resolution of $4''$ ($= 300$ pc at 15 Mpc) and recovered most (70 ± 14 %) of the single-dish flux. We use the data to set constraints on the above theoretical predictions.

2. Central Gas Condensations

Most galaxies in our sample show strong condensations of CO at their centers. Figure 1 shows the histogram of CO-derived masses of molecular gas within the central kiloparsec. The central gas masses are mostly in the range of 10^8–10^9 $M_\odot$. It thus seems not unusual for a large gas-rich galaxy to have a condensation of such a large amount of gas at the center. The gas condensations generally have radial profiles sharply peaking toward the galactic centers, when observed with sub-kiloparsec resolutions. The distribution of radial scale lengths of CO is also in Figure 1. The central scale length is defined as the radius at which a radial profile falls to $1/e$ of its maximum value, and is not affected much by the missing flux (15 % error at most). It is apparent that most galaxies have sub-kiloparsec scale lengths in the nuclear regions. The gas condensations are thus not simple extensions of outer exponential disks, which usually have scale lengths larger than a few kpc. It is interesting to note that the highest mass of the gas condensations, 10^9 $M_\odot$, is comparable to the mass needed to destroy bars in simulations.

3. Higher Gas Concentrations in Barred Galaxies

In order to quantify the degree of gas concentration in disk galaxies, we compare in Figure 2 the gas surface densities averaged in the central kiloparsec with those averaged over the optical galactic disks (i.e., $R < R_{25}$). The former are calculated from our data and the latter are calculated from the single-dish mapping data of the FCRAO survey (Young *et al.* 1995). The ratio of the two surface densities, $f_{\rm con} \equiv \Sigma_{\rm gas}^{R<500\rm pc}/\Sigma_{\rm gas}^{R<R_{25}}$, is an indicator of gas concentration to the central kiloparsec. The concentration factor $f_{\rm con}$ shows more than 100-fold variation in our sample.

Barred (i.e., SB+SAB) and unbarred (SA) galaxies are plotted with different symbols in Figure 2. It is apparent that barred galaxies have higher concentration factors than unbarred galaxies. The difference is statistically significant according to the Kolmogorov-Smirnov test; the probability to observe the difference of $f_{\rm con}$ in Figure 2 would be 0.007 if there were no difference between the two classes of galaxies.

We note that the conversion factor from CO flux to mass of molecular gas, $X_{\rm CO}$, is unlikely to produce the difference of $f_{\rm con}$ between barred and unbarred galaxies. The ratio

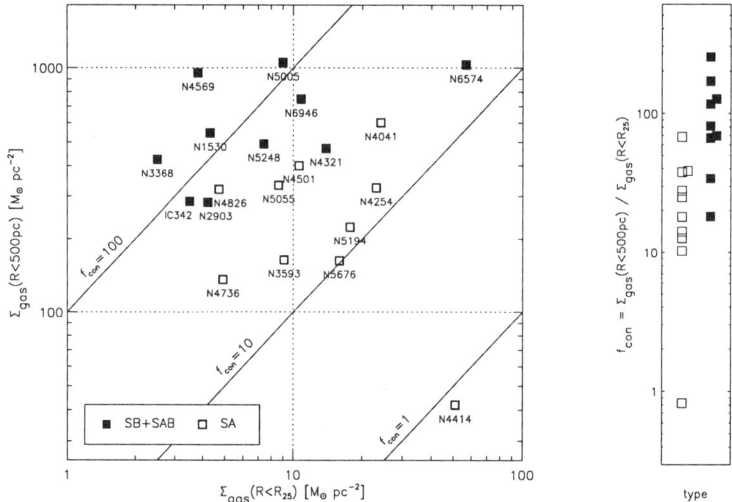

FIGURE 2. (Left) Surface densities of molecular gas averaged within the central kiloparsec are compared to those averaged over the optical galactic disks. The ratio of the central-to-disk averaged surface densities is an index of gas central concentration. Galaxies in the upper-left part of the panel have higher ratios, i.e., higher gas concentrations. (Right) Distributions of the surface density ratios (i.e., concentration factors f_{con}) for barred and unbarred galaxies.

f_{con} is independent of X_{CO} if it has the same form of radial distribution in galaxies, e.g., $X_{CO}(r)$ being either a (const.), ae^{br}, or ar^b with the same radial scale b. The multiplier a can be different from one galaxy to another without changing f_{con}. A systematic difference in the radial profile of X_{CO} between barred and unbarred galaxies may exist if the conversion factor scales with metallicity. However, the shallower metallicity gradients in barred galaxies would make the apparent CO concentrations lower in barred galaxies. Thus the correction for metallicity would only enhance the difference of f_{con} between the two types of galaxies. No other cause is known to create a systematic difference in radial profiles of X_{CO} between barred and unbarred systems.

We used classifications in the Third Reference Catalogue to distinguish barred and unbarred galaxies. It is possible that the classifications based on optical images missed small nuclear bars or misidentified open spiral arms as a bar. However, nuclear bars naturally have smaller power of gas transport, and spiral arms masquerading a bar create a global nonaxisymmetry in the gravitational potential as a bar does. Thus the optical classification is a qualitative index of the strength of nonaxisymmetry in mass distribution and in gravitational potential. We conclude therefore that galaxies with larger nonaxisymmetries (called 'barred' galaxies) have higher gas concentrations than galaxies with smaller nonaxisymmetries (i.e., 'unbarred' galaxies) †.

4. Implications for the Bar-Dissolution Scenario

The higher gas concentrations in barred galaxies are most likely due to radial transport of gas in the barred potentials. However, the transport of gas is not a sufficient condition to cause and maintain the higher gas concentrations in barred galaxies. It is also necessary

† The f_{con} in *unbarred* galaxies are low but ≥ 1. Possible causes for this, other than bars that had been destroyed, are weak nonaxisymmetries in the galaxies, viscous accretion of gas, and centrally peaked distribution of stars producing gas. Bar dissolution is thus *not required* here.

that the molecular gas funneled to the galactic centers remains there in molecular form *and* that the bars responsible for the gas transport remain. These requirements set constraints on the relation between the rates of gas inflow and gas consumption, and also on the timescale for the possible bar dissolution.

First, the total amount of gas funneled to the center of a barred galaxy must be larger than the total amount of stars formed in the same region, because otherwise the higher gas concentration in the barred galaxy can not be sustained. Dividing the total masses by the age of the bar, the above relation translates to the condition that the time-averaged rate of gas inflow must be larger than that of star formation. One may be able to estimate the time-averaged rate of star formation from an ensemble-average of star formation rates in the centers of barred galaxies, thereby setting a lower limit to the mean gas inflow rate. If there are other ways of losing molecular gas, such as an outflow due to starburst and accretion to active nucleus, then the lower limit becomes higher.

The second condition we can deduce is that the timescale of gas consumption in the central regions is longer than that of the possible bar dissolution. In other words, if bars are to be destroyed by the gas inflow of 10^8–10^9 $M_\odot$ to the central kiloparsec and if the bar dissolution is much quicker than the gas consumption in the central regions, then we would see currently unbarred but previously barred galaxies with high central gas concentrations that destroyed the bars. The lack of such galaxies (i.e., unbarred spirals with $f_{con} \geq 100$) allows us to set the above condition on the timescale of bar dissolution.

Quantitative evaluation of the above conditions is hampered by the difficulty in accurately estimating star formation rates in galactic centers. The current star formation rates crudely estimated from $H\alpha$ in the centers of the sample galaxies are ~ 0.1–1 $M_\odot \mathrm{yr}^{-1}$, which set a lower limit to the mass inflow rate. The consumption time of the gas concentrations is 10^8–10^{10} yrs. The lower value does not contradict with the predicted timescale of bar dissolution, which is comparable to the dynamical time or a few 10^8 yrs. If the higher value is the case in many spirals, then the bar dissolution must take longer time than predicted, or will not happen for the 10^8–10^9 $M_\odot$ gas concentrations.

It seems worthwhile to compile more data of gas concentration and star formation to tighten the above constraints on the mass transfer in galactic disks and on the fate of stellar bars. The index of gas concentration f_{con} may be useable as a tool to find out unbarred galaxies that were barred and galaxies with young bars: the former must have higher concentration factors for unbarred galaxies and the latter must have lower factors for barred galaxies. Observations of such galaxies would tell us about evolution of disks, bars, and bulges.

Stimulating conversations at the workshop with Drs. Norman, Pfenniger, Hasan, Regan, and Wada are acknowledged. K.S. was supported by JSPS grant-in-aid.

REFERENCES

Ho, L.C., Filippenko, A.V., Sargent, W.L.W. 1997 *ApJ*, **487**, 591

Martin, P., Roy, J-R. 1994 *ApJ*, **424**, 599

Quillen, A.C., et al. 1995 *ApJ*, **441**, 549

Regan, M.W., Vogel, S.N., Teuben, P.J. 1997 *ApJ*, **482**, L143

Sakamoto, K., Okumura, S., Ishizuki, S., Scoville, N.Z. 1998, in *The Central Regions of the Galaxy and Galaxies*, IAU Symp. **184**, p215.

Sakamoto, K., Okumura, S., Ishizuki, S., Scoville, N.Z. 1999 *ApJ*, submitted

Young. J.S., et al. 1995 *ApJS*, **98**, 219

Zaritsky, D., Kennicut, R.C., Huchra, J.H. 1994 *ApJ*, **420**, 87

Dynamical Evolution of Bulge Shapes

By MONICA VALLURI

Department of Physics and Astronomy, Rutgers University, 136 Frelinghuysen Road,
Piscataway, NJ 08854-8019, USA

Figure rotation substantially increases the fraction of stochastic orbits in triaxial systems. This increase is most dramatic in systems with shallow cusps showing that it is not a direct consequence of scattering by a central density cusp or black hole. In a recent study of stationary triaxial potentials (Valluri & Merritt 1998) it was found that the most important elements that define the structure of phase space are the two-dimensional resonant tori. The increase in the fraction of stochastic orbits in models with figure rotation is a direct consequence of the destabilization of these resonant tori.

The presence of a large fraction of stochastic orbits in a triaxial bulge will result in the evolution of its shape from triaxial to axisymmetric. The timescales for evolution can be as short as a few crossing times in the bulges of galaxies and evolution is accelerated by figure rotation. This suggests that low luminosity ellipticals and the bulges of early type spirals are likely to be predominantly axisymmetric.

1. Introduction

It is now widely believed that the effects of central black holes and cusps on the dynamics of triaxial galaxies are well understood: the box orbits which form the back bone of triaxial elliptical galaxies become chaotic due to scattering by the divergent central force (e.g. Gerhard & Binney 1985). The scattering of these orbits then results in the evolution of the triaxial galaxy to an axisymmetric one whose dynamics is dominated by well behaved families of regular orbits. Thus most studies of elliptical galaxies still focus on the nature of the regular orbits. Recent investigations of the structure of phase space in triaxial ellipticals have shown that phase space is rich in regular and chaotic regions even in the absence of black holes and steep cusps.

Studying the effects of central black holes on galaxies has taken on renewed importance because of the discovery that many if not most bulge dominated galaxies have central black holes. The existence of central black holes as the end products of the QSO and AGN phenomena is justified by energetic arguments. But less is known about the interplay between the growth of a black hole and the shape of its host galaxy. Most models for the fueling of QSO and AGN require a high degree of triaxiality to transport fuel to the center and to simultaneously transport angular momentum outwards (Rees 1990). Understanding the interplay between black hole growth and galaxy shape is one motivation for studying the behavior of orbits in triaxial potentials.

There have been several studies of the effect of figure rotation on the orbits of stars in triaxial galaxies. Most studies have focused on the behavior of the periodic orbits in the plane perpendicular to the rotation axis. Some authors (Martinet & Udry 1990) found that increasing figure rotation resulted in a decrease in the phase space occupied by the unstable x_3 family and consequently a reduction in the overall chaos. Others (Udry & Pfenniger 1988, Udry 1991) found that increasing figure rotation had negligible effect on the stochasticity of orbits in 3-dimensional models. More recently it has been shown (Tsuchiya *et al.* 1993) that orbits of all 4 major families in a perfect ellipsoidal model (completely integrable when stationary) became stochastic when figure rotation is added. Rapidly rotating triaxial bars can be almost completely regular (Pfenniger &

136

Friedli 1991) although more slowly rotating bars and bars with high central concentrations generally contain a large fraction of stochastic orbits that eventually destroy the bars (Norman *et al.* 1996, Sellwood & Moore 1999).

We use the frequency analysis technique (Laskar 1990) to study the behavior of orbits in a family of triaxial density models with figure rotation. The models have a density law that fits the observed luminosity profiles of ellipticals and the bulges of spirals and is given by Dehnen's law

$$\rho(m) = \frac{(3-\gamma)M}{4\pi abc} m^{-\gamma}(1+m)^{-(4-\gamma)}, \qquad 0 \leq \gamma < 3 \qquad (1.1)$$

where

$$m^2 = \frac{x^2}{a^2} + \frac{y^2}{b^2} + \frac{z^2}{c^2}, \qquad a \geq b \geq c \geq 0 \qquad (1.2)$$

and $M = 1$ is the total mass. The parameter γ determines the slope of the central density cusp and a, b, c are the semi-axes of the model. In some cases we also introduced a central point mass M_h representing a nuclear black hole. The figure rotates about its short axis and the degree of figure rotation can be small (as in the case of giant ellipticals) or reasonably large as in the case of bulges. The co-rotation radius R_Ω is parameterized in units of the half-mass radius of the model and ranges from $R_\Omega = 25$ (slowly rotating) to $R_\Omega = 3$ (rapidly rotating). Frequency analysis was restricted to $\sim 10^4$ orbits in each model. Orbits were launched from the equi-effective-potential surface corresponding to the half-mass radius (thus all orbits have the same Jacobi Integral, $E_J = E - \frac{1}{2}|\Omega \times \mathbf{r}|^2$). The initial conditions for the orbits were selected in two different ways to study orbits from all four major families.

2. Frequency Mapping and Resonant Tori

Laskar's (1990) frequency analysis technique is based on the idea that regular orbits have 3 isolating integrals of motion which are related to 3 fundamental frequencies. A filtered Fourier transform technique can be used to accurately determine these 3 frequencies (ω_x, ω_y, ω_z). While stochastic orbits do not really have fixed frequencies, quantities resembling frequencies which measure their local behavior can be used to determine how they diffuse in frequency space. Regular orbits come in three types: (1) Orbits in regions that maintain their regular character in spite of departures of the potential from integrable form; (2) orbits associated with stable resonant tori; (3) orbits associated with stable periodic orbits, or 'boxlets'.

The use of frequency mapping has shown that even in the case of weakly chaotic systems, it is the 'resonant tori' that provide the skeletal structure to regular phase space (Valluri & Merritt 1998). Frequency mapping provides the simplest method for finding resonant tori. They are families of orbits which satisfy a condition: $l\omega_x + m\omega_y + n\omega_z = 0$ with (l, m, n) integers. Such orbits are restricted to 2-dimensional surfaces in phase space and we refer to them as *thin orbits*. Thin boxes are the most generic box orbits in non-integrable triaxial potentials. They avoid the center because they are two-dimensional surfaces. They generate families of 3-D boxes whose maximum thickness is determined by the strength of the central cusp or black hole (Merritt 1999). The closed periodic boxlet orbits lie at the intersection of two or more resonance zones. High order resonances also exist for tube orbit families. Unlike the well known thin tube families around the long and short axes, thin resonant tubes are often surrounded by unstable regions, making it difficult to find them without a technique like frequency mapping.

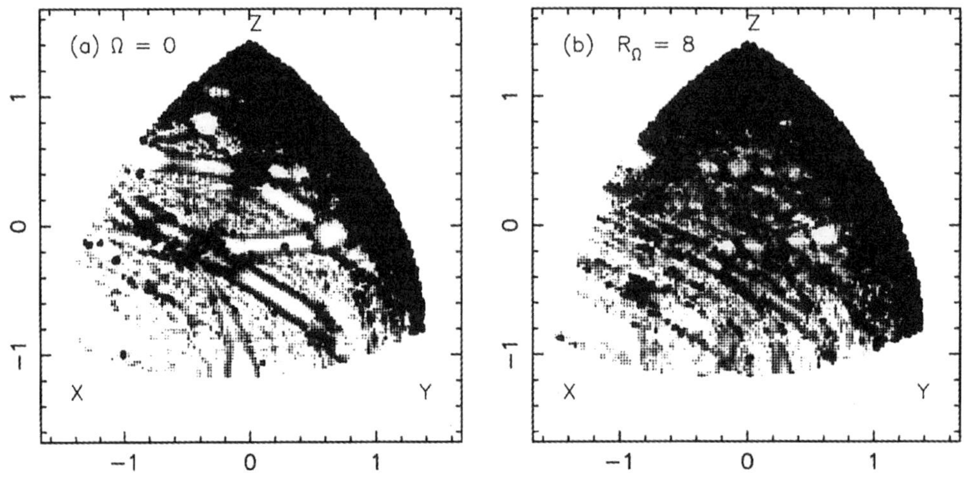

FIGURE 1. (a) The initial-condition-space diffusion map of box-like orbits in the non-rotating model. (b) Like (a) for orbits started on the equi-effective-potential surface of a slowly rotating model with $R_\Omega = 8$. Features on the maps are described in the text

3. Results: Destruction of the Resonant Tori

A box or boxlet orbit reverses its sense of progression around the rotation axis every time it reaches a turning point. In a rotating frame this means that the path described during the prograde segment of the orbit is not retraced during the retrograde segment. This 'envelope doubling' is a consequence of the Coriolis forces on the two segments being different (de Zeeuw & Merritt 1983). Envelope doubling effectively thickens the thin box orbits driving them closer to the center. This results in a narrowing of the stable portion of the resonance layer and renders a large fraction of the orbits stochastic. The degree of 'thickening' increases with increasing figure rotation and results in a corresponding rise in the fraction of stochastic box-like orbits.

Figure 1 (a) shows a plot of a quantity measuring the diffusion rates of 10^4 orbits started at rest at the half-mass equi-potential surface in a non-rotating triaxial model with central cusp slope $\gamma = 0.5$. Only one octant of the surface is plotted. The grey scale is proportional to the logarithm of the diffusion rate: the dark regions indicate initial conditions corresponding to stochastic orbits, the white regions correspond to regular orbits. Figure 1 (b) shows the same set of orbits started from the equi-effective-potential surface of a model with $R_\Omega = 8$. Rotation results in the broadening of the unstable regions with a resultant narrowing of the stable (white) regions. It also gives rise to new unstable and stable resonances which are seen in Figure 1 (b) as dark striations within the white regions. The increase in the number of resonances and their broadening results in greater overlap of nearby stochastic layers eventually leading to the onset of global stochasticity (e.g. Chirikov 1979).

Contrary to the finding of Tsuchiya *et al.* (1993) we find that figure rotation has a strong destabilizing effect on inner-long axis tubes. The low angular momentum z-tubes and the outer x-tubes also become more stochastic. The high angular momentum z-tubes are much less affected. The increased stochasticity of tube orbits can be attributed largely to the increase in the width of the stochastic layers associated with the resonant tube orbit families. We emphasize that for the tube orbits it is the destabilization of resonant tubes and not scattering by divergent central forces that determines their stability.

4. Conclusions

It is a popular misconception that in the presence of figure rotation box orbits in a triaxial elliptical will loop around the center due to Coriolis forces thereby reducing stochasticity. We find that on the contrary stochasticity increases with increasing figure rotation primarily because the thin box orbits and resonant tubes, which play a crucial role in structuring phase space, are broadened and destabilized by the 'envelope doubling' effect.

Models for the fueling of AGN and QSOs require triaxial central potentials which aid accretion onto a black hole, but the same black holes would tend to destroy triaxiality. Low luminosity ($M_B > -19$) ellipticals and the bulges of spirals are expected to evolve into axisymmetric shapes on time scales much shorter than the age of the Universe (Valluri & Merritt 1998). If the peanut-shaped bulges in nearby galaxies are in fact triaxial they are probably dynamically young or are composed of only tube-like orbits.

I thank David Merritt for useful discussions. This work was supported by NSF grants AST 93-18617 and AST 96-17088 and NASA grant NAG 5-2803 to Rutgers University.

REFERENCES

CHIRIKOV, B. 1979 *Rep. Prog. Phys.*, **52**, 263

DE ZEEUW, P.T., MERRITT, D. 1983 *ApJ*, **267**, 571

GERHARD, O.E., BINNEY, J. 1985 *MNRAS*, **216**, 467

LASKAR, J. 1990 *Icarus*, **88**, 266

MARTINET, L., UDRY, S. 1990 *A&A*, **235**, 69

MERRITT, D. 1999, in preparation

MERRITT, D., VALLURI, M. 1996 *ApJ*, **471**, 82

NORMAN, C.A., SELLWOOD, J.A., HASAN, H. 1996 *ApJ*, **462**, 114

REES, M.J. 1990 *Nature*, **249**, 817

SELLWOOD, J.A., MOORE, E. 1998 *ApJ*, **510**, 125

TSUCHIYA, T., GOUDA, N., YAMADA, Y. 1993 *Prog. Th. Phys*, **89**, 793

UDRY, S. 1991 *A&A*, **245**, 99

UDRY, S., PFENNIGER, D. 1988 *A&A*, **198**, 135

VALLURI, M., MERRITT, D. 1998 *ApJ*, **506**, 686

Two–Component Stellar Systems: Phase–Space Constraints

By **L U C A   C I O T T I**[1,2]

[1]Osservatorio Astronomico di Bologna, via Zamboni 33, 40126 Bologna, ITALY

[2]Scuola Normale Superiore, Piazza dei Cavalieri 7, 56126 Pisa, ITALY

In the context of studying the properties of the mutual mass distribution of the bright and dark matter in bulges (or elliptical galaxies), the properties of the analytical phase–space distribution function (DF) of two–component spherical self–consistent stellar systems (where one density distribution follows the Hernquist profile, and the other a $\gamma = 0$ model, with different total masses and core radii [H0 models]) are here summarized. A variable amount of radial Osipkov–Merritt (OM) orbital anisotropy is allowed in both components. The necessary and sufficient conditions that the model parameters must satisfy in order to correspond to a model where each one of the two distinct components has a positive DF (the so–called model consistency) are analytically derived, together with some results on the more general problem of the consistency of two–component $\gamma_1 + \gamma_2$ models. The possibility to add in a consistent way a black hole at the center of radially anisotropic γ-models is also discussed. In the particular case of H0 models, it is proved that a globally isotropic Hernquist component is consistent for any mass and core radius of the superimposed $\gamma = 0$ halo. On the contrary, only a maximum value of the core radius is allowed to the $\gamma = 0$ component when a Hernquist halo is added. The combined effect of halo concentration and orbital anisotropy is successively investigated. It is suggested that the observed centrally steep density profiles of bulges (and ellipticals) can be a natural consequence of the underlying dark matter distribution, if this is distributed similarly to, e.g., the 'universal' profile of Navarro, Frenk & White (1997).

1. Introduction

In the study of stellar dynamical models the fact that the Jeans equations have a physically acceptable solution is not a sufficient criterion for the validity of the model: the essential requirement to be met is the positivity of the DF of each distinct component. A model satisfying this minimal requirement is called a *consistent* model. In order to recover the DF of spherical models with anisotropy, the OM technique has been developed (Osipkov 1979; Merritt 1985), and numerically applied (see, e.g., Ciotti & Pellegrini 1992, CP92; Carollo, de Zeeuw, & van der Marel 1995; Ciotti & Lanzoni 1997, CL97). In the OM framework, a simple approach in order to check the consistency of spherically symmetric, multi–component models (avoiding the recovering of the DF itself), is described in CP92. It is now accepted that a fraction of the mass in galaxies is made of a dark component, whose density distribution – albeit not well constrained by observations – differs from that of the visible one (see, e.g., Bertin *et al.* 1994; Carollo *et al.* 1995; Buote & Canizares 1997; Gerhard *et al.* 1998). Moreover, there is an increasing evidence of the presence of massive black holes (BHs) at the center of most (if not all) elliptical galaxies (see, e.g., Harms *et al.* 1994; van der Marel *et al.* 1997; Richstone 1998). Unfortunately, only a few examples of two–component systems in which both the spatial density and the DF are analytically known are at our disposal, namely the Binney–Evans model (Binney 1991; Evans 1993), and the two–component Hernquist model (HH model, Ciotti 1996, C96). It is therefore of interest that the DF of H0 models with OM anisotropy is completely expressible in analytical way (Ciotti 1999, C99). This family of models is made by the superposition of a density distribution following the

Hernquist profile (Hernquist 1990), and another density distribution following the $\gamma = 0$ profile [see eq. (3.5)], with different total masses and core radii. OM orbital anisotropy is allowed in both components. Strictly related to the last point above, is the trend shown by the numerical investigations of CP92, i.e., the difficulty of consistently superimposing a centrally-peaked distribution to a centrally-flat one. More specifically, CP92 showed numerically that King (1972) or quasi–isothermal density profiles can not be coupled to a de Vaucouleurs (1948) model, because their DFs run into negative values near the model center. From this point of view, the C96 work on HH models is complementary to the investigation of CP92: in the HH models the two density components are both centrally peaked, and their DF is positive for all the possible choices of halo and galaxy masses and concentrations (in the isotropic case). The implications of these findings have not been sufficiently explored. For example, one could speculate that in the presence of a centrally peaked dark matter halo, elliptical galaxies or bulges with flat cores should be relatively rare, or, *vice versa*, that a galaxy or a bulge with a central power–law density profile cannot have a dark halo that is too flat in the center. In fact observational results on the bulges of spirals (Carollo & Stiavelli 1998), and on the central surface brightness profiles of elliptical galaxies (see, e.g., Jaffe *et al.* 1994; Møller, Stiavelli, & Zeilinger 1995; Lauer *et al.* 1995), as well as high–resolution numerical simulations of the formation of dark matter halos (Dubinsky & Carlberg 1991; Navarro, Frenk, & White 1997) seem to point in this direction. In C99, I explore further the trend emerged in CP92 and in C96, considering the analytical DFs of the H0 models and determining the structural and dynamical limitations imposed to them by dynamical consistency.

2. The Consistency of Multi–Component Systems

For a multi–component spherical system, where the orbital anisotropy of each component is modeled according to the OM parameterization, the DF of the density component ρ_k is given by:

$$f_k(Q_k) = \frac{1}{\sqrt{8}\pi^2}\frac{d}{dQ_k}\int_0^{Q_k}\frac{d\varrho_k}{d\Psi_T}\frac{d\Psi_T}{\sqrt{Q_k - \Psi_T}}, \quad \varrho_k(r) = \left(1 + \frac{r^2}{r_{a,k}^2}\right)\rho_k(r), \qquad (2.1)$$

where $\Psi_T(r) = \sum_k \Psi_k(r)$ is the total relative potential, $Q_k = \mathcal{E} - L^2/2r_{a,k}^2$, and $0 \leq Q_k \leq \Psi_T(0)$. $\mathcal{E}$ and L are respectively the relative energy and the angular momentum modulus per unit mass, r_a is the anisotropy radius, and $f_k(Q_k) = 0$ for $Q_k \leq 0$. If *each* f_k is non-negative over all the accessible phase–space, the system is 'consistent'. In C92 it was proved that:

Theorem: A necessary condition (NC) for the non-negativity of f_k given in eq. (2.1) is:

$$\frac{d\varrho_k(r)}{dr} \leq 0, \quad 0 \leq r \leq \infty. \qquad (2.2)$$

If the NC is satisfied, a 'strong' (SSC) and a 'weak' sufficient condition (WSC) for the non-negativity of f_k are respectively:

$$\frac{d}{dr}\left[\frac{d\varrho_k(r)}{dr}\frac{r^2\sqrt{\Psi_T(r)}}{M_T(r)}\right] \geq 0, \quad \frac{d}{dr}\left[\frac{d\varrho_k(r)}{dr}\frac{r^2}{M_T(r)}\right] \geq 0, \quad 0 \leq r \leq \infty. \qquad (2.3)$$

Some considerations follow looking at the previous conditions. The first is that the violation of the NC is connected only with the radial behavior of ρ_k and the value of $r_{a,k}$, and so this condition applies independently of any other interacting component added to the model. Even when the NC is satisfied, f_k can be negative, due to the radial behavior of the integrand in eq. (2.1), which depends on the total potential, on the particular $\rho_{\rm k}$, and on $r_{\rm a,k}$; so, a range of permitted values of $r_{\rm a,k}$ satisfying the NC must be discarded. Naturally, the true critical anisotropy radius is always larger than or equal to that given by the NC, and smaller than or equal to that given by the SSC (WSC). To summarize: a model failing the NC is *certainly* inconsistent, and a model satisfying the SSC (WSC) is *certainly* consistent; the consistency of a model satisfying the NC and failing the SSC (WCS) can be proved only by direct inspection of the DF.

3. Results and Conclusions

Both density distributions defining the H0 models belong to the family of the γ-models (Dehnen 1993):

$$\rho(r) = \frac{3 - \gamma}{4\pi} \frac{M\, r_{\rm c}}{r^\gamma (r_{\rm c} + r)^{4-\gamma}}, \quad 0 \leq \gamma < 3, \tag{3.1}$$

where M is the total mass and $r_{\rm c}$ a characteristic scale–length. The main results obtained in C99 can be summarized as follows:

(1) The NC, WSC, and SSC that the model parameters must satisfy, in order to correspond to an H0 system for which the two physically-distinct components have a positive DF, are analytically derived using the method introduced in CP92. Some conditions are obtained for the wider class of two–component $\gamma_1 + \gamma_2$ models (of which the H0 models are a special case). In particular, it is shown that the DF of the γ_1 component in isotropic $\gamma_1 + \gamma_2$ models is nowhere negative, independently of the mass and concentration of the γ_2 component, whenever $1 \leq \gamma_1 < 3$ and $0 \leq \gamma_2 \leq \gamma_1$. As an interesting application of this result, it follows that a black hole of any mass can be consistently added at the center of any isotropic member of the γ-models family, when $1 \leq \gamma < 3$. Two important consequences follow. The first is that the consistency of isotropic HH (or H+BH) models proved in C96 using an 'ad hoc' technique is not exceptional, but a common property of a large class of two–component γ-models: for example, also isotropic two–component Jaffe (Jaffe 1983, $\gamma = 2$ in eq. [3.4]) or Jaffe+BH models can be safely assembled. The second is that in two–component isotropic models, the component with the steeper central density distribution is usually the most robust against inconsistency.

(2) It is shown that an analytical estimate of a minimum value of $r_{\rm a}/r_{\rm c}$ for one–component γ-models with a massive (dominant) BH at their center can be explicitly found. As expected, this minimum value decreases for increasing γ.

(3) It is shown that the analytical expression for the DF of H0 models with general OM anisotropy can be found in terms of elliptic functions; the special cases in which each one of the two density components are embedded in a dominant halo are also discussed.

(4) The region of the parameter space in which H0 models are consistent is explored using the derived DFs: it is shown that, at variance with the H component, the $\gamma = 0$ component becomes inconsistent when the halo is sufficiently concentrated, even in the isotropic case. This is an explicit example of the negative result found by CP92 described in the Introduction.

(5) The combined effect of halo concentration and orbital anisotropy is finally investigated. The trend of the minimum value for the anisotropy radius as a function of the halo concentration is qualitatively similar in both components, and to that found for HH models in C96: a more diffuse halo allows a larger amount of anisotropy. A qualitatively new behavior is found and explained investigating the DF of the $\gamma = 0$ component in the halo–dominated case for high halo concentrations. It is analytically shown that there existsa small region in parameter space where a sufficient amount of anisotropy can

compensate for the inconsistency produced by the halo concentration on the structurally analogous – but isotropic – case.

(6) As a final remark, it can be useful to point out some general trends that emerge when comparing different one- and two-component models with OM anisotropy, such as those investigated numerically in CP92 and CL97, and analytically in C96 and C99. The first common trend is that OM anisotropy produces a negative DF outside the galaxy center, while the halo concentration affects mainly the DF at high (relative) energies. The second is that the possibility to sustain a strong degree of anisotropy is weakened by the presence of a very concentrated halo. The third is that in two–component models, in the case of very different density profiles in the central regions, the component with the flatter density is the most 'delicate' and can easily be inconsistent: particular attention should be paid when constructing such models.

REFERENCES

BERTIN, G., ET AL. 1994 *A&A*, **292**, 381

BINNEY, J. 1981 *MNRAS*, **196**, 455

BUOTE, D.A., CANIZARES, R.C. 1997 *ApJ*, **474**, 650

CAROLLO, C.M., ET AL. 1995 *ApJ*, **441**, L25

CAROLLO, C.M., DE ZEEUW, P.T., VAN DER MAREL, R.P. 1995 *MNRAS*, **276**, 1131

CAROLLO, C.M., STIAVELLI, M. 1998 *AJ*, **115**, 2306

CIOTTI, L. 1996 *ApJ*, **471**, 68 (C96)

CIOTTI, L. 1999 *ApJ*, in press (C99)

CIOTTI, L., LANZONI, B. 1997 *A&A*, **321**, 724 (CL97)

CIOTTI, L., PELLEGRINI, S. 1992 *MNRAS*, **255**, 561 (CP92)

DE VAUCOULEURS, G. 1948 *Ann. d'Ap.*, **11**, 247

DEHNEN, W. 1993 *MNRAS*, **265**, 250

DUBINSKI, J., CARLBERG, R.G. 1991 *ApJ*, **378**, 496

EVANS, N. 1993 *MNRAS*, **260**, 191

HARMS, R.J., ET AL. 1994 *ApJ*, **435**, L35

HERNQUIST, L. 1990 *ApJ*, **536**, 359

GERHARD, O., JESKE, G., SAGLIA, R.P., BENDER, R. 1998 *MNRAS*, **295**, 197

JAFFE, W. 1983 *MNRAS*, **202**, 995

JAFFE, W., ET AL. 1994 *AJ*, **108**, 1567

KING, I. 1972 *ApJ*, **174**, L123

LAUER, T.R., ET AL. 1995 *AJ*, **110**, 2622

MERRITT, D. 1985 *AJ*, **90**, 1027

MØLLER, P., STIAVELLI, M., ZEILINGER, W.W. 1995 *MNRAS*, **276**, 979

NAVARRO, J.F., FRENK, C.S., WHITE, S.D.M. 1997 *ApJ*, **490**, 493

OSIPKOV, L.P. 1979 *Pis'ma Astron. Zh.*, **5**, 77

RICHSTONE, D.O. 1998, in *The Central Region of the Galaxy and Galaxies* (ed. Y. Sofue), IAU Symp. **184**, in press

VAN DER MAREL, R.P., ET AL. 1997 *Nature*, **385**, 610

NGC 2146: A Firehose-Type Bending Instability?

By EVGENY GRIV AND M. GEDALIN

Department of Physics, Ben-Gurion University of the Negev, Beer-Sheva 84105, Israel

The 'firehose' instability in central disks is discussed. This instability may arise in the centers of galaxies where the stars move in thin, practically non-rotating disks. N-body simulations described here predict the existence of a new type of structure – small-scale $\sim h$ out-of-plane bends of newly formed OB stars – in the central regions of spiral galaxies with high star formation rates.

1. Introduction

As shown by gravitational N-body simulations and observations of highly flattened giant galaxies including the Milky Way, the central parts of these systems at distances of, say, $r \lesssim 0.7\text{-}1$ kpc from the center rotate slowly, and their local circular velocities of regular galactic rotation become less than (or comparable to) the residual (random) velocities. In such a thin, practically nonrotating ('pressure-supported') central disk, a typical star moves along the bending, perpendicular to the equatorial plane layer, under the action of two forces which act in opposite directions: the destabilizing centrifugal force, F_c, and the restoring gravitational attraction, F_g. Obviously, fierce instabilities of the buckling kind developing perpendicular to the plane may not be avoided if $F_c > F_g$. The latter condition is nothing else than the well-known condition of the so-called firehose electromagnetic instability in collisionless plasmas. The source of free energy in the instability is the intrinsic anisotropy of a velocity dispersion ('temperature'). It seems reasonable that this is a natural mechanism for building a snake-shaped radio structure which has recently been observed in the central region of the spiral starburst galaxy NGC 2146 with the VLA at an angular resolution of $2'$ (Zhao *et al.* 1996).

Apparently, the firehose-type bending instability of a sufficiently thin stellar disk has been predicted by Toomre (1966) by using a simplified theory based on moment equations. This instability was also discovered by Kulsrud *et al.* (1971) with a more accurate kinetic theory. Fridman & Polyachenko (1984) have discussed the role of the instability in explaining the existence of maximum oblateness in elliptical galaxies and the formation of the bulges of disk-shaped galaxies of stars. Combes & Sanders (1981) and then Raha *et al.* (1991) found the firehose-type bending instability to be a precursor of galactic bulge formation in the central, almost non-rotating regions of warm, planar N-body disks which initially developed planar bars.

2. N-body Simulations of the Firehose-Type Bending Instability

In order to investigate the dynamics of the central region, N-body simulations of the firehose-type bending instability as *a precursor* of galactic bulge formation are presented for NGC 2146. Use of current computers has enabled us to make long simulation runs using a sufficiently large number of particles in the direct summation code, $N = 20\,000 - 40\,000$. In contrast to all previous N-body simulations of bending instabilities, we show how bending structures may be longer-lived in real starburst galaxies than in the (previous) computer models.

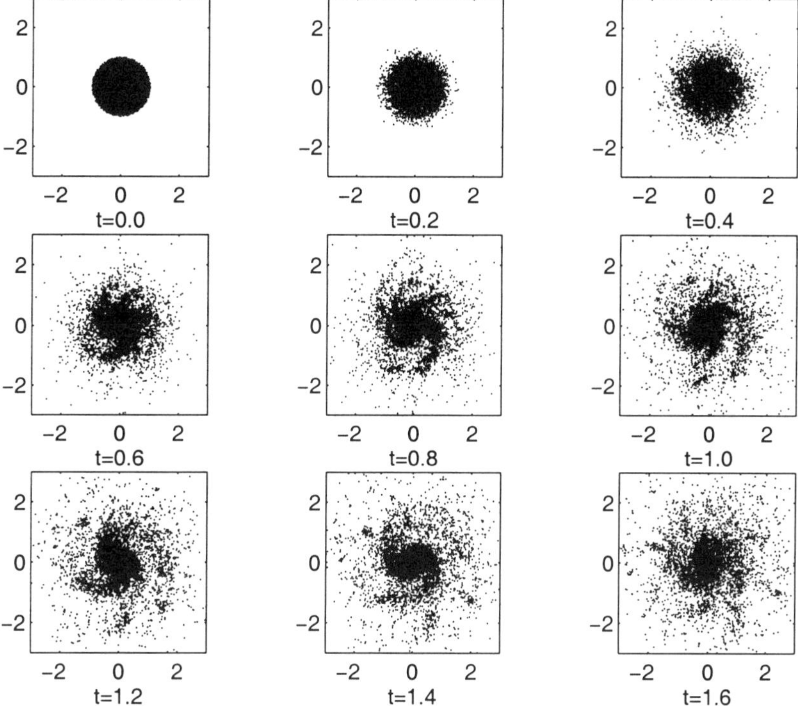

FIGURE 1. The time evolution (face-on view) of a three-dimensional disk of stars ($N = 30'000$). The system is violently unstable with respect to short-lived $m = 2-4$ spiral modes (or sometimes $m = 1$) of the Jeans type developing in the equatorial plane, where m is the azimuthal mode number (i.e., the number of spiral arms). A time $t = 1$ was taken to correspond to a single revolution of the initial disk. The sense of disk rotation was taken to be counter-clockwise. At first (at $t \approx 0.6$) a multi-armed, moderately tightly wound spiral structure is developed in the plane of the system. It is interesting to notice that in a sample of 654 spiral galaxies (Elmegreen & Elmegreen 1989), two-armed 'grand design' galaxies like M51 and the Sc-shaped Whirlpool galaxy in Canes Venatici are roughly a factor of 6 times rarer than such many-armed 'flocculent' galaxies as NGC 613, an SBb galaxy in Sculptor. Then, after ~ 1 rotation, a prominent massive bar forms; the $m = 1$ instability shifts the point with highest density from the center of mass. The underlying potential in a large fraction of spiral galaxies, e.g., in the spiral galaxies M101 and NGC 1300, is now believed to have this lop-sided form; such a deviation is due to the one-arm Jeans instability developing in the plane of the system under study (Griv & Chiueh 1998). Note that in the single-arm galaxy NGC 4378, the spiral arm can be traced over most 1.25 revolutions. We suggest that the structures observed in our N-body simulations originate from the collective-type modes of practically collisionless galactic models — the classical Jeans-type modes as seen in Figure 1 and firehose-type bending modes (see Figure 2).

At the start of the N-body integration, our simulation initializes identical particles on a set of 100 rings with a circular velocity V of galactic rotation in the r, φ plane; the system is isolated in vacuum. Consider a uniformly rotating model disk of stars with a surface mass density variation given by

$$\Sigma_0(r) = \Sigma(0)\sqrt{1 - r^2/R^2}, \qquad (2.1)$$

where $\Sigma(0)$ is the central surface density, and R is the radius of the initial disk. As a solution of a time-independent collisionless Boltzmann equation, to ensure initial equilibrium,

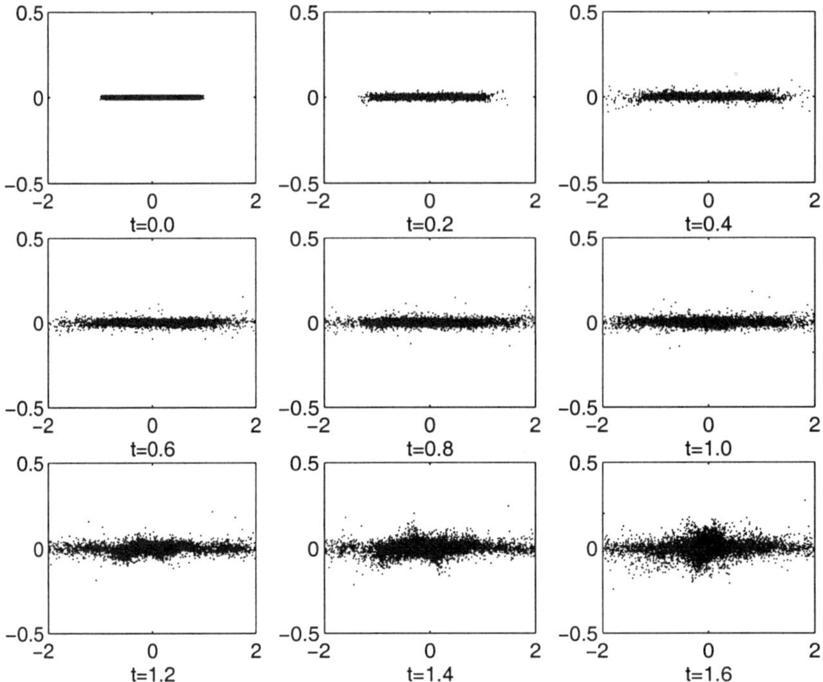

FIGURE 2. Higher-resolution plots of the central parts (edge-on view) for the simulation shown in Figure 1. Within the framework of our model, newly formed, dynamically young stars form the very thin disk ($h \ll R$). At the start of simulations the disk thickness was chosen $h = 0.025R$. Slight corrections have been applied to the resultant velocities of the model stars so as to include the weak effect of the finite thickness of the disk to the gravitational potential. One can see (at $t \approx 1.2$) the rapid development of the snake-shaped, firehose-type bending instability in the central, practically non-rotating portions of the system. This instability is a transient one, and develops on a timescale of single rotation of the system. The linear dimension of the bending along the z-axis is comparable to the disk thickness and much less than the radius of the outermost parts of the system. The usual name for this instability of collective systems – firehose-type instability – recalls the fact that the unstable motion of a hose, when the flow velocity of water inside becomes very high, is essentially of the same nature. As a result of the instability, a 'box-shaped' or sometimes 'peanut-shaped' bar structure is developed. The projected densities of the central bar resemble the bulge light distribution measured by the *COBE* satellite in the Milky Way. Furthermore, a significant fraction of edge-on spiral galaxies (and therefore presumably of all spirals) show boxy or peanut-shaped isophotes in the bulge region. An isophotal plot of the central region of NGC 2146, like that of Figure 4a of Zhao *et al.* (1996), looks remarkably similar to this Figure. We attribute the origin of a snake-shaped structure in NGC 2146 to the bending instability in the inner 'pressure-supported' parts of the galaxy. It is important to note that the bending perturbations begin to grow at later stages of the evolution when the stellar layer becomes dynamically hot enough, $c_r > c_T$, in the r, φ plane. In turn, the layer becomes hot as a result of the Jeans instability of non-axisymmetric (spiral) gravity perturbations (Griv & Peter 1996).

the uniform angular velocity to balance the zero-velocity dispersion disk,

$$\Omega_0 = \pi \sqrt{G\Sigma(0)/2R}, \qquad (2.2)$$

was adopted (Griv & Chiueh 1998). Then the position of each particle was slightly perturbed by applying a pseudo-random number generator. For the uniformly rotating disk, the Maxwellian-distributed random velocities with radial c_r and azimuthal c_φ dispersions in the plane $z = 0$ according to the well-known Toomre's criterion,

$$c_T \equiv \frac{3.36 G \Sigma_0}{\kappa} = 0.341 \Omega_0 \sqrt{R^2 - r^2}, \tag{2.3}$$

was added ($c_r = c_\varphi$) to the initial circular velocities $V = r\Omega_0$, where $\kappa(r)$ is the ordinary epicyclic frequency. It is crucial to realize that in this case, according to Lin & Lau (1979), Morozov (1981), and Griv & Peter (1996), initially the disk is Jeans-stable against the small-scale axisymmetric (radial) perturbations but unstable against the relatively large-scale non-axisymmetric (spiral) perturbations. The initial vertical velocity dispersion was chosen equal to $c_z = 0.3 c_r$. Finally, the angular velocity Ω_0 was replaced by (Griv & Chiueh 1998)

$$\Omega = \left\{ \Omega_0^2 + \frac{1}{r\Sigma(r)} \frac{\partial}{\partial r} \left[\Sigma(r) c_r^2(r) \right] \right\}^{1/2}. \tag{2.4}$$

The simulations (Figures 1 and 2) clearly confirm the qualitative picture and, moreover, are in fair quantitative agreement with the theory. A theoretical prediction is confirmed that the instability is driven by an excess of plane kinetic energy of random motions of stars, when the ratio of the dispersion of radial velocities of stars in the plane c_r to the velocity dispersion in the perpendicular direction c_z is large enough, $c_r \gtrsim 0.6 c_z$. In other words, the instability occurs if the thickness of the stellar disk $h \propto c_z$ is small enough. The extent to which our results on the stability of the disk can have a bearing on observable spiral galaxies with a high star formation rate in the central parts is discussed as well. In particular, the discovery of the snake-shaped structure in central parts of the starburst galaxy NGC 2146 made by Zhao *et al.* raises the question of whether this feature, which we tentatively relate to the firehose-type instability feature, is common in these objects or whether this galaxy is a special case. We suggest that these alternatives can be distinguished with sensitive, high-frequency observations of other nearby starburst galaxies seen almost edge-on, for example, IC 2531, Maffei 2, NGC 3079, NGC 3628, NGC 3666, NGC 4627, NGC 4631, NGC 4700, and NGC 4945 using the VLA with an angular resolution of $\sim 1''$. At present, the N-body experiments described here are meant to predict the existence of a new type of structures — small-scale $\sim h$ out-of-plane bends of newly formed OB stars — in the central regions of spiral galaxies with a high star formation rate (see Griv & Chiueh 1998 for a discussion).

REFERENCES

COMBES, F., SANDERS, R.H. 1981 *A&A*, **96**, 164

ELMEGREEN , B.G., ELMEGREEN, D.M. 1989 *ApJ*, **342**, 677

FRIDMAN, A.M., POLYACHENKO, V.L. 1984 *Physics of Gravitating Systems*, **1**. (Springer, New York)

GRIV, E., PETER, W. 1996 *ApJ*, **469**, 84

GRIV, E., CHIUEH, T. 1998 *ApJ*, **503**, 186

KULSRUD, R.M., MARK, J.W.K., CARUSO, A. 1971 *Ap. Sp. Sci.*, **14**, 52

LIN, C.C., LAU, Y.Y. 1979 *SIAM J. Appl. Math.*, **29**, 352

MOROZOV, A.G. 1981 *Soviet Astron.*, **25**, 421

RAHA, N., SELLWOOD, J.A., JAMES, R.A., KAHN, F.D. 1991 *Nature*, **352**, 411

TOOMRE, A. 1966, in *Geophys. Fluid Dyn.*, **66-46**, 111

ZHAO, J.H., ANATHARAMAIAH, K.R., GOSS, W.M., VIALLEFOND, F. 1996 *ApJ*, **472**, 54

Bulge Formation: The Role of the Multi-Phase ISM

By MARCO SPAANS

Harvard-Smithsonian Center for Astrophysics, 60 Garden Street, Cambridge MA 02138, USA

Star formation in bulges has likely been a rather efficient process. An efficient formation of stars depends strongly on the presence of metallic atoms and molecules. These species provide the necessary cooling for the ambient medium to sustain star formation. In order to assess the epoch and timescales for bulge-formation, it is therefore important to investigate the structure of the multi-phase ISM as a function of redshift and the formation of stars in such a medium. Calculations are presented which incorporate feedback effects and the thermal and chemical balance of interstellar gas. Predictions are made for the star formation histories of spheroids of various masses, and compared to similar estimates for disks.

1. Introduction

Star formation (SF) is a local phenomenon which must find its explanation in the stability and fragmentation of dense molecular clouds. Studies in our own Galaxy have focussed on the structure of dense proto-stellar cores and the chemical and thermal balance of star-forming regions. These studies lend indirect support to a Schmidt (1959) law, but emphasize the need to include explicitly the structure of the multi-phase ISM to model accurately the most important heating and cooling processes. A large unknown in these investigations is the role of feedback. Supernova explosions and stellar radiation associated with the process of SF influence the global physical structure of the interstellar gas which supports this process. A detailed discussion of the structure of the ISM and the importance of feedback effects for different kinds of stellar systems can be found in Norman & Spaans (1997), Spaans & Norman 1997 and Spaans & Carollo (1997).

2. Short Model Description

The theoretical background for the evolution of the multi-phase ISM in primordial galactic structures is fully described in Spaans & Carollo (1998), where the employed numerical methods and a discussion of the spatial and temporal resolution of the code can be found as well.

The models include three stellar and three gaseous components. The gas phases include cold molecular clouds, the warm neutral/ionized medium, and the hot tenuous interiors of supernova bubbles. The phases are assumed to be in pressure equilibrium and their chemical and thermal balance is computed explicitly. The stellar components are divided according to their final evolutionary stages into massive stars of more than 11 $M_\odot$, which explode as type II supernovae, and low-mass stars, which are assumed to lose their material in a planetary nebula phase. The low-mass stars will become 0.6 $M_\odot$ white dwarfs (Weidemann & Koester 1983). Stars with masses below 0.6 $M_\odot$ do not evolve during the lifetime of a galaxy. The third class of stars comprises the stellar remnants in the form of white dwarfs, neutron stars, and black holes. A fraction of the white dwarfs gives rise to type Ia supernovae in the merging CO-dwarf picture.

Although no dynamics are included explicitly, there is mass exchange between the cold molecular and hot tenuous phases due to cloud evaporation. The density dependence for mass evaporation is given by $\propto E^{6/5} n_{\mathrm{h}}^{-4/5}$, with E the supernova energy and n_{h} the density of the hot phase (McKee & Ostriker 1977).

Following Larson (1991) the SF rate is calculated with a Schmidt-law applied to some volume in a galaxy with a mass M

$$\frac{\partial M_{\mathrm{cm}}}{\partial t} = \alpha_{\mathrm{SF}} \left(\frac{n_{\mathrm{cm}}}{n_{\mathrm{cm}}^0}\right)^b, \tag{2.1}$$

where the label 'cm' indicates the cold molecular phase, n_{cm} is a number density, $n_{\mathrm{cm}}^0 = 40$ cm^{-3} and $b = 1 - 2$. The coefficient α_{SF} is normalized to the SF rate as observed in molecular gas in the Solar vicinity

$$\alpha_{\mathrm{SF}} = \frac{M}{M_{\mathrm{g}}} \qquad \mathrm{M}_\odot \mathrm{yr}^{-1}, \tag{2.2}$$

with M normalized to the total body of gas M_{g} in the Milky Way (see below). For our own Galaxy as a whole this corresponds to an instantaneous rate of ~ 3 M$_\odot$ yr^{-1}. The assumed value of b is 1.3.

The density of the cold molecular phase follows from application of the Field, Goldsmith, & Habing (1969) formalism to the thermal balance of a multi-phase ISM. The angular momentum of star-forming clouds is not included beyond the accuracy of the local Schmidt-law applied to the cold molecular component. As such, we cannot address the question of cloud fragmentation, and the IMF is an input parameter. A time-independent Salpeter IMF $\phi(m) \propto m^{-2.35}$ is assumed. Stellar masses between 0.1 M$_\odot$ and 40 M$_\odot$ are considered. It follows that approximately 12% of the newly formed stellar mass is incorporated into massive stars with a lifetime of no more than 2×10^7 years.

3. Results

3.1. *The Multi-Phase ISM*

The main results regarding the structure of the ISM will be illustrated for a proto-galactic spheroid with a total mass of 2×10^{11} M$_\odot$. The pressure is proportional to the square of the column density from Archimedes' law. The nine panels of Figure 1 show how multiple solutions for the density and temperature can be realized as a function of metallicity Z and background star formation rate $\dot{S}$. For a metallicity of more than a few percent of Solar, the abundance of molecules like OH, H$_2$O and CO has become large enough to form and cool a molecular phase. It is at this point that a phase transition to an ISM with multiple components takes place (Field, Goldsmith & Habing 1969).

The background SF rates determine the rate at which the medium is enriched, and the amount of radiation (O/B stars) and kinetic (supernovae) energy which is injected into the ambient gas. It turns out that once a multi-phase ISM has been established, the injected radiation energy and the thermalized component of the supernova ejecta can be effectively cooled away through atomic and molecular line emission of H, H$_2$, [CII], [CI], [OI], and CO (see also Gerritsen 1997). Nevertheless, large local fluctuations exist in the heating rate, which influence the subsequent formation of stars. The bulk motions initiated by supernova shock fronts can lead to outflows and significant loss of (enriched) interstellar material.

3.2. *Star Formation Histories*

Figure 2 shows a collection of SF histories for spheroids, disks and dwarf galaxies. The curves were obtained by convolution with the mass distribution for a standard cold dark matter model with a bias parameter of unity. Although quantitatively the curves depend

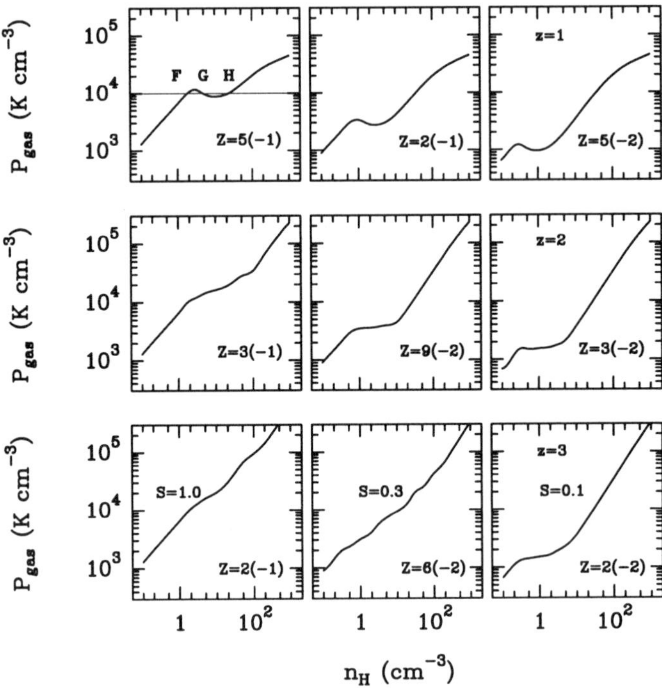

FIGURE 1. The redshift evolution of a 2×10^{11} M$_\odot$ proto-galactic disk.

on the specific assumptions and approximations, three qualitatively different evolutionary scenarios can be distinguished.

Dwarf galaxies have shallow gravitational potentials and a few supernova explosions are sufficient to unbind their ISM. This implies that they do not retain their produced metals very efficiently. Therefore, the average metallicity in a dwarf galaxy increases only slowly with time. This leads to a delayed peak in their cosmological SF history until $z \approx 0.5$.

Proto-galactic disks experience their peak in SF at a redshift of $z \approx 1-2$. Even though disks are massive enough to retain their ISM, they can develop supernova driven outflows due to their flat geometry. As a supernova expansion front reaches one scale height in the atmosphere of a spiral it will break out, and vent the hot, metal-rich gas behind the shock front (De Young & Heckman 1994). The propagation of the front occurs on a time scale which is shorter than the cooling time of the hot phase. Therefore, more robust molecular clouds cannot form in time and the metal-rich gas is blown out (and re-accreted on a free-fall time of the order of 1 Gyr).

In contrast to disks and dwarfs, outflows in spheroids require more massive starbursts, and these objects can therefore form stars at a high rate and at the same time retain their metal-enriched ISM. This leads to a SF peak for spheroids at $z \approx 3 - 4$. Of course, this conclusion requires the efficient formation of such objects at high redshift (note however that the required time to form a proto-galactic bulge of 3×10^9 M$_\odot$ is only of the order of 10^8 yr).

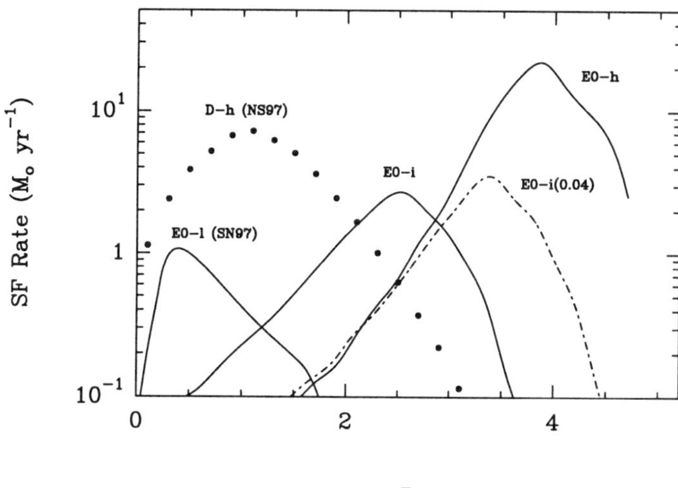

FIGURE 2. Star formation histories for spheroids (E0-h, E0-i), disks (D-h), and dwarf (E0-l) galaxies. The label E0-i(0.04) indicates a baryonic to dark matter ratio of 0.04, whereas the other models have a ratio of 0.1. The labels h, i and l indicate total masses of 10^{11}, 10^{10} and 10^9 $M_\odot$, respectively.

4. Conclusions and Discussion

The epoch of efficient SF is initiated by a transition to a multi-phase ISM at a typical metallicity of ~ 0.03 $Z_\odot$. While the SF history peaks at $z \sim 0.5$ for dwarf galaxies and at $z \sim 1$-2 for massive disks, this peak occurs at $z = 3$-4 for massive spheroidals.

The calculations presented here attempt to incorporate the chemical and thermal balance of the ISM, and to ascertain the efficiency of SF. Ultimately, such an approach needs to be implemented in a hydrodymanical code suited for cosmological applications. As they stand, the presented results already indicate that the multi-phase structure of the ISM plays an essential role in the evolution of the cosmic SF rate in the universe.

REFERENCES

DE YOUNG, D.S., HECKMAN, T.M. 1994 *ApJ*, **431**, 598

FIELD, G.B., GOLDSMITH, D., HABING, H.J. 1969 *ApJ*, **155**, 49

GERRITSEN, J.P.E. 1997, Ph.D. Thesis, University of Groningen

LARSON, R.B. 1974 *MNRAS*, **166**, 385

MCKEE, J.F., OSTRIKER, J.P. 1977 *ApJ*, **218**, 148

NORMAN, C.A., SPAANS, M. 1997 *ApJ*, **480**, 145

SCHMIDT, M. 1959 *ApJ*, **129**, 243

SPAANS, M., NORMAN, C.A. 1997 *ApJ*, **483**, 87

SPAANS, M., CAROLLO, C.A. 1997 *ApJ*, **482**, 93

SPAANS, M., CAROLLO, C.A. 1998 *ApJ*, **502**, 640

WEIDEMANN, V., KOESTER, D. 1983 *A&A*, **121**, 77

Global Evolution of a Self-Gravitating Multi-Phase ISM in the Central Kpc Region of Galaxies

By KEIICHI WADA[1,2] AND COLIN A. NORMAN[1]

[1] Johns Hopkins University, 3400 N. Charles Street, Baltimore MD 21218, USA

[2] National Astronomical Observatory, Mitaka, 181, Japan

Using high resolution, two-dimensional hydrodynamical simulations, we investigate the evolution of a self-gravitating multi-phase interstellar medium in the central kiloparsec region of a galactic disk. We find that a gravitationally and thermally unstable disk evolves, in a self-stabilizing manner, into a globally quasi-stable disk that consists of cold ($T < 100$ K), dense clumps and filaments surrounded by hot ($T > 10^4$ K), diffuse medium. In the quasi-stable phase where cold and dense clouds are formed, the effective stability parameter, Q, has a value in the range 2-5. The dynamic range of our multi-phase calculations is $10^6 - 10^7$ in both density and temperature. Phase diagrams for this turbulent medium are analyzed and discussed. We also succeeded in modeling star formation in the multi-phase ISM with 2 pc resolution. Massive stars formed in the dense, cold clouds are tracked for their life time, and finally explode as SNe. The filament-like structure of the cold gas is stable for the SNe, although bubbles of the hot gas ($T > 10^6$ K) are formed. We observed recurrent burst-like SNe production.

1. Introduction

We model the multi-phase and inhomogeneous interstellar medium (ISM) in the inner region of a galactic disk including fundamental physical processes crucial for understanding star formation, global and local dynamics of the ISM in galaxies, and aspects of galaxy formation such as feedback. Most numerical simulations of the ISM and of star formation in galaxies have assumed simpler ISM models, e.g. an isothermal or nearly-isothermal equation of state, and either a smooth medium or discrete clouds.

In the present paper, we report on the structure and global evolution of multi-phase ISM models in two-dimensions, taking into account self-gravity of the gas, galactic rotation, radiative cooling, and heating due to UV background radiation, SN explosions and stellar wind. We use an Eulerian hydro-code without periodic boundary conditions. Our code allow us to handle over seven orders of magnitude for density and temperature in a $\sim$ kpc-scale region around the galactic center with $\sim$ pc-scale resolution.

2. Numerical Method and Models

We solve the hydrodynamical equations and Poisson equation with an Euler mesh code (AUSM, see Liou & Steffen 1993). We use 1024^2 Cartesian grid points covering a 2 kpc $\times$ 2 kpc region. Therefore, the spatial resolution is 1.95 pc. The second-order leap-frog method is used for the time integration. We adopt implicit time integration for the cooling term.

The initial condition are an axisymmetric and rotationally supported disk with the Toomre stability parameter $Q = 1.2$ over the whole disk. Random density and temperature fluctuations are added to the initial disk. These fluctuations are less than 1% of the unperturbed values. The initial temperature is set to 10^4 K over the whole region.

We assume the gaseous disk is rotating in an axisymmetric, time-independent stellar potential. The mass of the gas disk is about 10% of the stellar mass.

We take into account two feedback effects of massive stars on the gas dynamics, namely stellar winds and supernova explosions. We first identify grids which satisfy criteria for star formation. The criteria are a surface density threshold for star formation $(\Sigma_g)_{i,j} >$ Σ_c and a critical temperature $(T_g)_{i,j} < T_c$ below which star formation is allowed. In these simulations we have chosen $T_c < 50K$ and $\Sigma_c = 5 \times 10^4 M_\odot$ pc^{-2}. We do not assume any global criteria for gravitational instability. The stars radiate energy due to stellar winds during their lifetime, which is approximately$\sim 10^7$ yr. When the star explodes as supernova, an energy of 10^{51} ergs is injected into the mesh where the particle is located. The cooling procedure is not used for such meshes, but the meshes adjoining the supernova mesh are treated normally. With our code, the 2-D evolution of blastwaves caused by supernovae in an inhomogeneous medium with global rotation is followed explicitly. Therefore, we can follow consistently the thermal and dynamical evolution of the ISM around the star forming regions and the associated supernovae remnants and superbubbles.

3. Morphology and Evolution of the ISM

The most prominent feature of the resulting quasi-stable structure is its filamentary appearance (Figure 1). The high density clumps are embedded in less dense filaments ($\Sigma_g \sim 10^{3-4} M_\odot$ pc^{-2} and $T_g < 100$ K). The characteristic size of the highest density clouds ($\Sigma_g \geq 10^5 M_\odot$ pc^{-2}) is about 5-50 pc. The temperature of the very low density gas in the voids sometimes reaches 10^6 K, due to shock heating. The energy source of this heating is the turbulent, random motion of the gas, at velocities $\sim 10 - 100$ km s^{-1}.

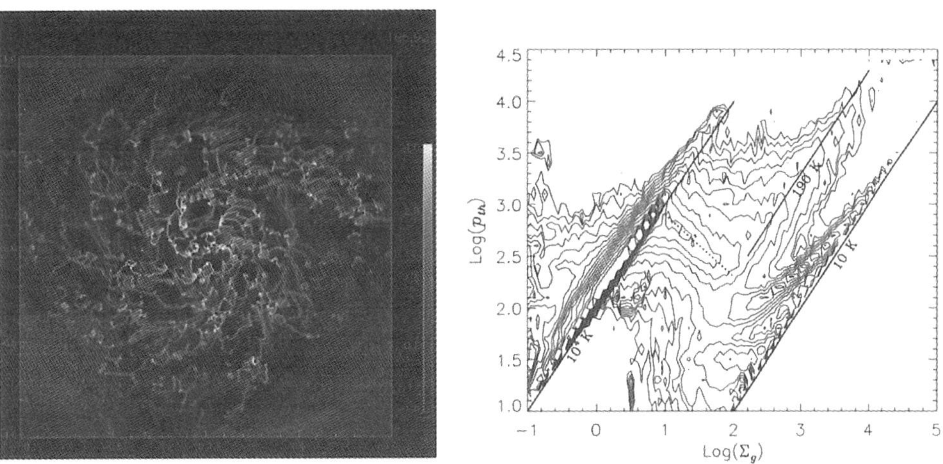

FIGURE 1. (left) Density map of a model without star formation at $t = 166$ Myr. The density range is 10^{-1} to $10^4 M_\odot$ pc^{-2}

FIGURE 2. (right) Pressure ($\Sigma_g T_g$) vs. density (Σ_g) phase diagram for the model shown in Figure 1. Three diagonal lines show the gaseous temperature 10^4 K, 190 K, and 10 K. The dotted line means thermally unstable gas.

Figure 2 is a phase diagram ($p_{\text{th}} \equiv \Sigma_g T_g$ vs. Σ_g) for the model shown in Figure 1. The left and right diagonal lines correspond to states in which the gas temperature is 10^4

and 10 K. There are also two less prominent 'ridges' between the hot and cold phases, i.e. $(p_{\mathrm{th}}, \Sigma_g) = (3, 1) - (2.5, 2)$ and $(2.5, 2) - (4, 4)$. One ridge corresponds to filaments with temperature $T_g \sim 200$ K. The second ridge indicated by a dotted line has 'negative γ', where γ is the adiabatic index. The gas in this state is thermally unstable. In spite of this instability the gas component does not have temporal variations. This is not a paradox, if we consider the kinematic or turbulent pressure of the gas as well as the thermal pressure. As can be seen in this phase diagram, the system cannot be described as a simple two- or three-phase media with pressure equilibrium between the phases.

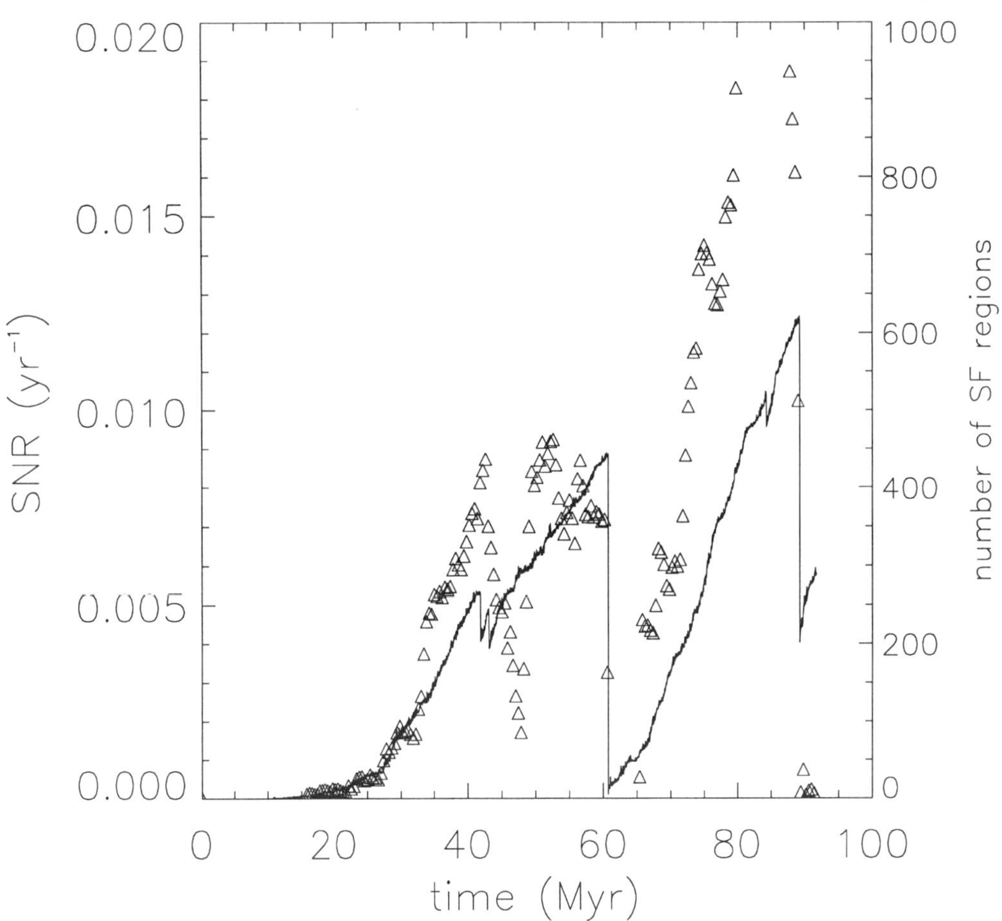

FIGURE 3. Time evolution of the SN rate (triangles) and the number of the star forming regions (solid line).

Although the initial Q value is 1.2 (i.e. the disk is stable for axisymmetric modes), the effective Q value becomes much less than unity after a few Myr, because the cooling time is very short. The gas temperature decreases from the initial value ($\sim 10^4$ K) to an equilibrium temperature ($\sim 10^2$ K) within 10^5 yr. Since the initial density is higher in the central region of the disk than in the outer region, gravitational instabilities begin in the inner few 100 pc and develop outward. At $t \sim 10$ Myr ($\sim$ one rotational period at $R \sim 200$ pc), the gravitational instabilities in the central 500 pc are already in a non-linear phase, where the maximum density contrast is about 10^6. At $t \sim 25$ Myr, the instability grows non-linearly over the whole disk, and many clumps, filaments and low density voids are formed. At this stage, Q_{eff} increases above the marginal stability value for the whole disk. This means that the system is globally self-stabilized. We found that the stabilization is mostly due to the increase of the velocity dispersion of the gas.

About 80% of the mass is in the cold ($T_g < 100$ K) phase. As expected, on the other hand, the volume filling factors of the hot, warm and cold gas indicate that the hot gas ($T_g \geq 9000$ K) occupies a volume $10^{2.5}$ times larger than the cold gas. The volume filling factor for the gas in the temperature range $11000 \leq T_g < 10^5$ K increases from 0 to 30% until $t = 30$ Myr. After 30 Myr, the volume filling factor of each component does not change appreciably, and this also shows that the system reaches a quasi-equilibrium state.

It is notable that the multi-phase, quasi-stable ISM is formed from a highly gravitationally and thermally unstable state without heating due to SNe. The diffuse UV heating does not contribute to the global stabilization. Shock heating is a dominant heating mechanism for $T_g > 10^4$ K in this system. Shearing of the disk induced by differential rotation and gravitational perturbation from clumps are the main energy source for the shock heating.

We found in a model with star formation that the SNe cannot significantly change the multi-phase structure described above. The quasi-stationary, filamentary structure of the cold gas is almost the same as seen in the model without star formation, although very hot and diffuse gas ($T > 10^6$ K) is produced due to the SNe. It is hard to produce huge holes (> 100 pc) in a region occupied with many dense clumps/filaments. This is because the radiative cooling at the dense region is very effective. The SNe and blast waves, however, trigger new star formation at the dense clouds around them. Figure 3 shows the time evolution of the SN rate and the number of the star forming sites. Recurrent burst-like SNe production is clear. The period of the burst is about 10 Myr, which is the life time of massive stars. That is, one burst of SNe produces new massive stars, and then they will explode after about 10 Myr.

4. Summary

We have shown that a globally stable, multi-phase ISM is formed as a natural consequence of the non-linear evolution of a self-gravitating gas disk. The gas is in so many and so various phases represented by a wide range of density and temperature that multi-phase is probably an inadequate description. The gas has properties more like a phase-continuum. The density ranges over seven orders of magnitude from $10^{-1} - 10^6 M_\odot$ pc^{-2}, and the temperature extends from $10 - 10^8$K. We found that the ISM cannot be expressed simply by the two-phase or three-phase model (e.g. McKee & Ostriker 1977; Ikeuchi, Habe, & Tanaka 1984; Norman & Ikeuchi 1989; Norman & Ferrara 1996), because the gas dynamic processes, such as turbulence, global rotation, shear motion and shocks, are at least as important as thermal processes. Our high resolution Eulerian hydrodynamic code allows us to handle gas dynamics on a kiloparsec scale with a few parsec resolution regardless of the gas density. Possible star forming sites where the gas density is very high ($n > 10^4$ cm^{-3}) and the temperature is less than 100 K can be identified clearly, even if the disk is globally stable. The evolution of supernova remnants in an inhomogeneous and rotating media is fully followed, and supernovae themselves can trigger new star formation in surrounding dense clouds.

Numerical computations were carried out on VPP300/16R at the Astronomical Data Analysis Center of the National Astronomical Observatory, Japan.

REFERENCES

IKEUCHI, S., HABE, A., TANAKA, Y.D. 1984 *MNRAS*, **207**, 909

LIOU, M., STEFFEN, C. 1993 *J. Comp. Phys.*, **107**, 23

MCKEE, C.F., OSTRIKER, J.P. 1977 *ApJ*, **218**, 148

NORMAN, C.A., IKEUCHI, S. 1989 *ApJ*, **345**, 372

NORMAN, C.A., FERRARA, A. 1996 *ApJ*, **467**, 280

SPAANS, M., NORMAN C. 1997 *ApJ*, **483**, 87

Part 5

BULGE PHENOMENOLOGY

This section collects a few contributions which do not address specifically any of the main themes of the workshop, but do provide state-of-the-art information on several additional constraints that need to be considered when building a self-consistent picture for bulge formation:

Möellenhof finds that the slope of the bulge surface brightness distribution increases with Hubble type and bulge radius; Pompei stresses that the trixiality is a long-standing feature (evident in the old bulge stellar population); Bertola *et al.* discuss the geometrical and kinematical decoupling between bulges and disks in two Sa galaxies, and suggest that perhaps the disk may represent a second event in the formation history of bulge-dominated spirals; Corsini *et al.* argue for an internal origin for the dynamically-hot ionized gas they detect in a massive S0 bulge; Maeda and collaborators discuss ASCA observation of the Galactic bulge, and argue that its X-ray emission arises from an optically thin thermal plasma; Fabbiano summarizes integrated X-ray observations of several early-type spirals and detailed observations of the bulge of M31, and suggests that several processes may contribute to the X-ray emission from spheroids.

The last three papers address the phenomenological connection between 'central activity' and hosts: Urry *et al.* describe an HST survey of ≈ 100 BL Lac objects up to a redshift $z \sim 1.4$; Verdoes *et al.* present the nuclear properties of a sample of 21 Fanaroff-Riley type-I radio-ellipticals; Cappellari *et al.* use HST/FOS spectroscopy to argue that the nuclear UV-spikes found in many spheroids may have different physical origins and duty-cycles. An unbiased census of central black hole masses and kind/intensity of central activity as a function of host-properties and redshift is still awaited to clarify what is the cause and what the effect between central black holes properties, accretion and fueling efficiencies and mechanisms, and structure/evolution of the hosts.

Bulge-Disk Decomposition of Spiral Galaxies in the NIR

By CLAUS MÖLLENHOFF

Landessternwarte, Königstuhl 12, 69117 Heidelberg, Germany

A method for fitting the near-infrared surface brightness distribution of spiral galaxies by two-dimensional disk- and bulge-functions is presented. First results for a sample of 40 spirals are shown.

1. Introduction

An important tool for galaxy research is the study of the surface brightness (SB) distribution. For spiral galaxies the determination of the scale length of the exponential disk has a long traditition (e.g. Courteau 1996). However, the errors in these results are still rather large (Knapen & van der Kruit 1991).

For a better understanding of spiral galaxies it is necessary to study the structure of both disk and bulge as well. In order to separate non-axisymmetric structures as bars or triaxial bulges from the axisymmetric disk, two-dimensional fits are advantageous (e.g. de Jong 1996). In the following I present a generalization of a nonlinear direct fit method to the two-dimensional SB distribution of near-infrared (NIR) images of spiral galaxies.

2. NIR Data

The aim of this project is the study of the distribution of the mass-carrying evolved stars in spiral galaxies of different Hubble types. For this purpose, NIR observations are advantageous since they have much less perturbations due to dust or especially bright young stars.

The observations were performed during several runs at the 2.2m telescope of the German-Spanish observatory on Calar Alto, Spain. The detector was the MAGIC-NIR-camera with a NICMOS chip of $(0.67'')$ 256×256 pixels, for a total field of view of $\approx 3' \times 3'$. The total exposure times were 9 min on the objects in each filter J, H, K. The exposures were chopped in many (typically 48) short integrations (with a sky exposure in between each science exposure).

The complete sample comprises ≈ 100 galaxies with $B_T < 12$ and Hubble types Sa to Sc, and 100 SBa to SBc galaxies. Here I consider a first subsample of 40 galaxies with low inclination and without a strong bar.

3. Two-Dimensional Surface Fits

Two-dimensional SB functions for disk and bulge were fitted simultaneously to the observed flux distribution. For the flux of the inclined disk we assumed an exponential radial density law

$$F_d(r) = I_d exp(-r/r_d) \qquad (3.1)$$

where I_d is the central flux density and r_d the radial scale length. The inclination leads to the elliptical geometry

$$r^2 = (xcos\phi_d + ysin\phi_d)^2/q_d{}^2 + (ycos\phi_d - xsin\phi_d)^2 \qquad (3.2)$$

where ϕ_d is the position angle (PA) of the major axis versus North and $q_d = b/a$ is the axis ratio. Thus there are 4 free fit parameters for the disk.

For the bulge flux we assume a generalized de Vaucouleurs radial density law

$$F_b(r) = I_e exp(-b_e[(r/r_e)^\beta - 1])   \tag{3.3}$$

with the effective flux density I_e and the half light radius r_e. Correspondingly we have

$$r^2 = (xcos\phi_b + ysin\phi_b)^2/q_b{}^2 + (ycos\phi_b - xsin\phi_b)^2   \tag{3.4}$$

with ellipticity q_b and the position angle ϕ_b. The exponent β determines the slope of the projected bulge SB distribution and varies between ≈ 0.15 to 1.2 (for the de Vaucouleurs profile $\beta = 0.25$). If the constant b_e is replaced by a function of β

$$b_e = 1.9986/\beta - 0.327,   \tag{3.5}$$

then r_e is the half-light radius and I_e the flux at r_e (Caon *et al.* 1993). Thus we have 5 free fit parameters for the bulge. The sum flux function $F_{fit} = F_d + F_b$ is fitted simultaneously to the observed two-dimensional SB distribution under the condition that

$$\chi^2 = \sum_{i,k} (F_{obs}(i,k) - F_{fit}(i,k))^2 \times Weight(i,k)^2 = Min.   \tag{3.6}$$

This leads to a nonlinear system of equations for the 9 free parameters which has to be solved iteratively. To this purpose the *Levenberg-Marquardt*-method was adopted.

It is necessary to consider the effects of atmospheric seeing on the galaxy image, especially for the bulge fit. If enough stars are present in the image, the (mean) two-dimensional point spread function (PSF) was determined and applied to the fit function during the iteration process. If not enough stars were present, the PSF was approximated by a Gaussian profile. In this case the full width half maximum (FWHM) of the Gaussian profile was considered as a 10th independent variable and was fitted together with the galaxy parameters.

Some care had to be taken to find the real minimum of χ^2: In a first step, only the bulge parameters were iterated for the inner zone, and the disk parameters were kept fixed; in the outer zone, the opposite was initially done to obtain a reasonable guess for the disk parameters. In a second step, the results from the first step were iterated by the simultanous fit of disk and bulge to the two-dimensional image.

Figure 1 (left) shows the J image of NGC 3147. It can easily be seen that the bulge has a different PA compared to the disk; obviously the bulge is triaxial. This requires a two-dimensional fit. The residual image $F_{obs} - F_{fit}$ (Figure 1, right panel) shows all structures which could not be described by the adopted elliptical geometry for disk and bulge: asymmetries in the inner disk region, spiral arms, etc. Figure 2 shows the quantitive results in form of horizontal and vertical cuts through galaxy, fit-model, and residual.

4. Limits and Errors

The fit procedure and its stability were tested extensively with artificial galaxies, including photon noise and seeing convolution. The relevant errors in this procedure are not the very small statistical errors from the χ^2-minimization, but are the systematic errors, e.g. , non-optimal sky-subtraction, non-uniformness of the sky, irregular perturbations of the light distribution in the galaxy. By comparing the results from independent fits in the different NIR bands, and from applying a slightly different sky subtraction, the errors on the fit parameters are estimated to be $< 15\%$.

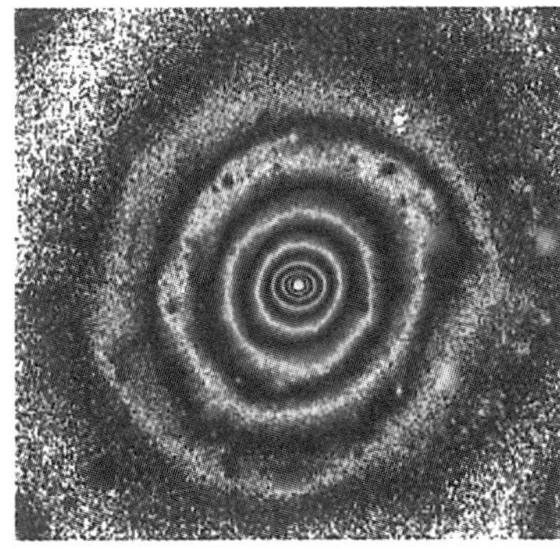 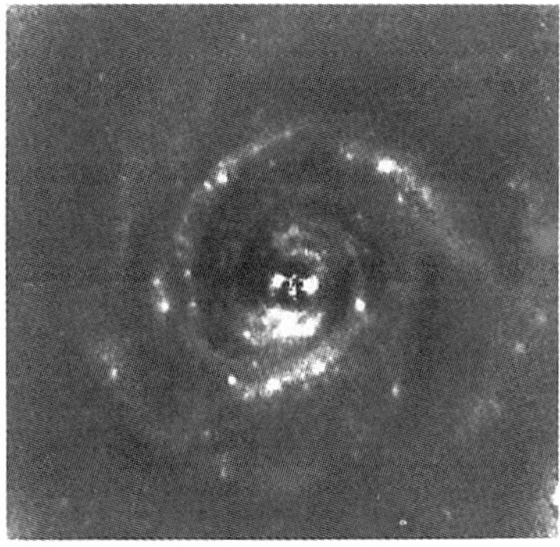

FIGURE 1. *Left:* J image of NGC 3147. The field is $3' \times 3'$; North is up. *Right:* Residual-image after subtraction of a two-dimensional bulge-disk model from the image of NGC 3147. The displayed flux depth of the residual-image is less than $\approx 3\%$ of that of the observed image.

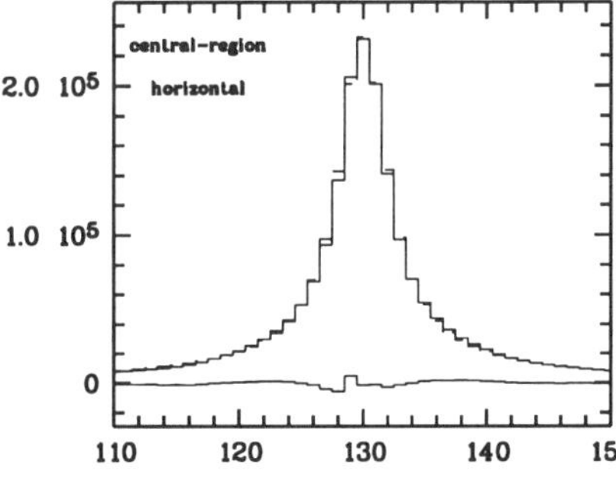 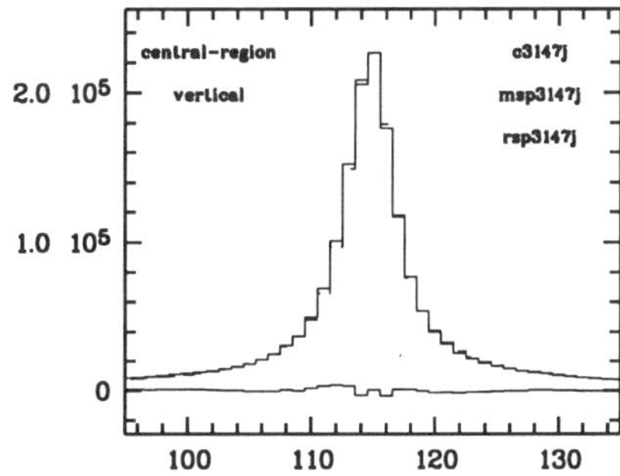

FIGURE 2. Horizontal (left) and vertical cuts (right) through the center of NGC 3147, the bulge-disk model, and the residual-image (solid lines at the bottom). The solid lines of the original galaxy and the dashed lines of the model are hardly distinguishable and demonstrate quantitatively the good quality of the fits.

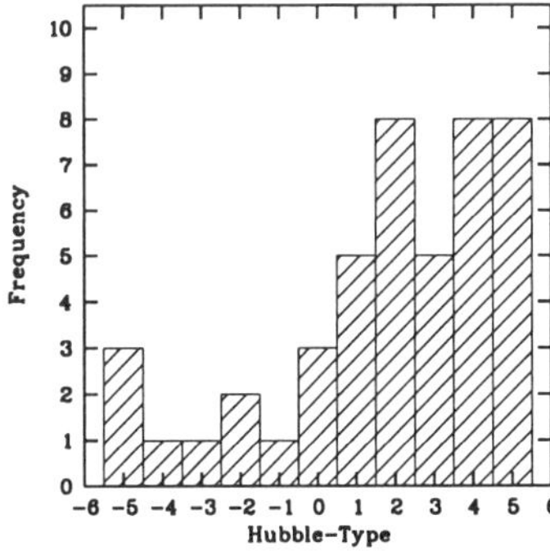 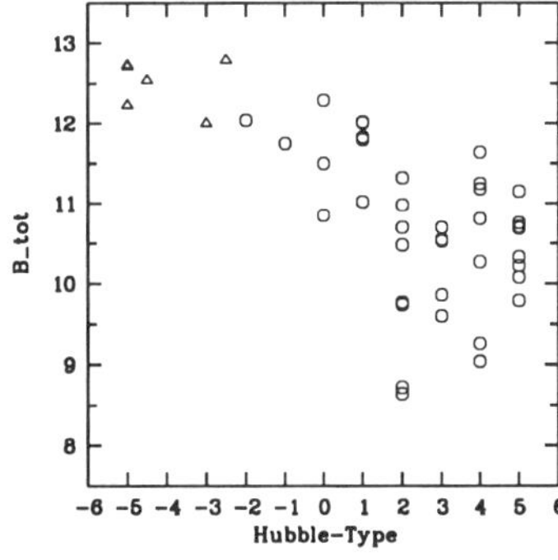

FIGURE 3. *Left:* Distribution of the different Hubble types in our sample. For comparison reasons 5 elliptical galaxies ($T = -3, -4, -5$) have been included. *Right:* Total B magnitudes of our sample ($\circ = spirals$, $\Delta = ellipticals$).

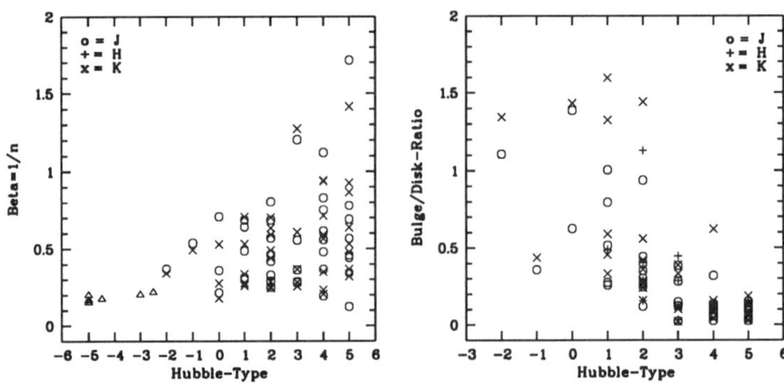

FIGURE 4. *Left*: The bulge exponent $\beta = 1/n$ increases with Hubble type, with an increasing scatter for late spirals (o $= J$, $+ = H$, $\times = K$ for spirals, $\Delta = R$ for ellipticals). *Right*: The bulge-to-disk ratio decreases with Hubble type

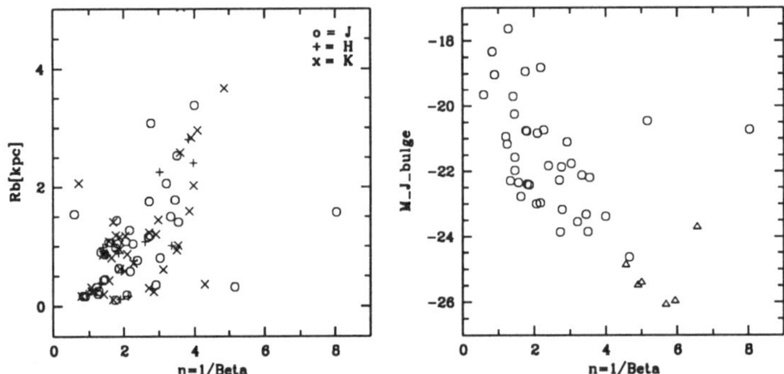

FIGURE 5. *Left*: The linear effective bulge radius $R_b = r_e$ correlates with the bulge exponent $\beta = 1/n$. *Right*: The absolute J-luminosity of the bulges correlates strongly with the bulge exponent $\beta = 1/n$, early-type bulges are brighter and flatter (o $=$ spirals, $\Delta =$ ellipticals).

5. Results

Some of the most important fit parameters for the 40 spiral galaxies are presented in Figures 3, 4, and 5. There is a good consistency of the fit parameters from the different colors. The bulge exponent β increases with Hubble type; however, there is an increasing scatter towards the late-type spirals (Figure 4, left panel; cf. Andredakis *et al.* 1995). A somewhat weaker correlation is found between bulge-to-disk ratio and Hubble type (Figure 4, right panel); correlations between β and bulge radius and between β and bulge luminosity are also detected (Figure 5).

REFERENCES

ANDREDAKIS, Y.C., PELETIER, R.F., BALCELLS, M. 1995 *MNRAS*, **275**, 874

CAON, N., CAPACCIOLI, M., D'ONOFRIO, M. 1993 *MNRAS*, **265**, 1013

COURTEAU, S. 1996 *ApJS*, **103**, 363

DE JONG, R.S. 1996 *A&AS*, **118**, 557

KNAPEN, J.H., VAN DER KRUIT, P.C. 1991 *A&A*, **248**, 57

The Triaxial Bulge of NGC 1371

By EMANUELA POMPEI[1,2] F. MATTEUCCI[1]

AND

I. J. DANZIGER[3]

[1]Universitá di Trieste, Dipartimento di Astronomia, Via G.B. Tiepolo 11, 34100 Trieste, Italy

[2]Osservatorio Astronomico di Torino, Via dell'Osservatorio 20, 10025 Pino Torinese (TO)

[3]Osservatorio Astronomico di Trieste, Via G.B. Tiepolo 11, 34100 Trieste, Italy

We present here an optical and near-infrared (NIR) photometric study of the bulge of NGC 1371, an Sa galaxy in the Fornax cluster. The galaxy hosts a nuclear bar, from which two spiral arms depart, and a triaxial bulge and it is the most peculiar object in a sample of 17 isolated spiral galaxies studied here. The triaxial shape and the bar are apparent also in the H band, i.e. where the emission from the old ($t > 10^7$ yr) stellar population peaks (Grauer & Rieke, 1998). The implications of our findings for bulge formation and bar secular evolution models are discussed.

1. Introduction

Bulge morphology has often been compared to that of elliptical galaxies, both of which were initially thought to be axisymmetric. Later it was discovered that elliptical galaxies are triaxial (see for instance de Zeeuw 1989; Bender 1988; and references therein) and soon afterward also triaxial bulges in spiral galaxies were found (Kormendy 1982; Zaritsky & Lo 1986; Bertola 1989, 1991; Shaw 1993; Varela *et al.* 1996). The radial surface brightness profile of a triaxial bulge usually follows a classic $r^{1/4}$ law and the distribution of triaxial bulges in barred and unbarred galaxies is similar (Pompei 1998), so in principle triaxiality and barred potentials are unrelated. It should be noted however that early-type galaxies host the strongest bars, which are currently supposed to have formed a long ($t > 10^8$ yr) time ago (Noguchi 1996). These bars are also responsible for strong gaseous inflows shortly after their formation, while the inflow decreases with increasing bar age. Finally a clear trend between triaxiality and morphological type appears, with early-type galaxies hosting a greater percentage of triaxial bulges relative to late-type galaxies. Observations of bulge stellar populations (Idiart *et al.* 1996; Jablonka 1997; Trager *et al.* 1998) have revealed that bulges are very similar to elliptical galaxies and presumably are very old ($t > 10^9$ yr). These conclusions stand for galaxies with morphological type up to Sc. Since triaxiality is observed to be a long standing characteristic of bulges, it can be inferred that whatever process was responsible for forming a triaxial bulge, it must have happened shortly after the epoch of bulge formation, unless the process merely rearranges the matter distribution without generating significant star formation. There are three possible explanations for the formation of a triaxial bulge, namely merging (Binney & Petrou 1985), bar secular evolution (Friedli & Benz 1993, 1995) or external torque (May *et al.* 1985). The last one may be justified only in the case of a very faint and dissipative collapse, the second is strongly dependent on the modalities of the interaction bar/bulge and on the gas inflow timescale (Noguchi 1996), while the first often needs fine tuning of the parameters. In order to gain some insights into which of these processes may be responsible for triaxial bulges, we have undertaken a photometric study in the optical and NIR of 17 isolated spiral galaxies, with no nuclear activity and no known bar. In the next section we will discuss the observations and data reduction and analysis, while in Section 3 we discuss our findings and give our conclusions.

TABLE 1. Observational set-up

Telescope	Detector	Scale Factor	Filters	Seeing
2.2m	IRAC2b	0.507 ($''/pix$)	J, H, K$'$	0.8$''$
1.54m D	CCD	0.39 ($''/pix$)	B, V	1.1$''$

2. Observations, Data Reduction and Analysis

All the observations were performed at the European Southern Observatory, La Silla, during the nights 28-29 November 1996 (NIR) with the 2.2 m telescope and the IRAC2b detector, and 3-5 December 1996 (optical) with the Danish 1.54m telescope and a 2048x2048 CCD. The relevant parameters for both observing runs are indicated in Table 1.

2.1. *Optical Data Reduction*

After bias, dark current subtraction, flat fielding and cosmic ray removal, the frames were sky subtracted fitting a least squares plane to the sky background. The dispersion in the background sky level was about 0.5% of the sky level in B and 0.8% in V. The images were aligned using as reference points the centroids of nine stars and added together. The resulting frame was median filtered and the centre of the galaxy was calculated with a 0.2 pixel accuracy, fitting the nucleus with a gaussian profile. The data were calibrated to the standard photometric system using standard stars and the calibration errors are about 0.02 mag.

2.2. *NIR Data Reduction*

For each filter 6 frames of the galaxy and 6 frames of the sky were obtained, with the same exposure time, 10m. From each galaxy frame an average of the preceding and following sky frames was subtracted; the sky subtracted frames were flat fielded using dome flat fields. The processed frames were aligned using the centre of the galaxy as reference point and averaged into one single image, which was cleaned of remaining bad pixels using a bad pixel mask. The atmospheric extinction correction was performed using the mean atmospheric extinction of the observing site. The Galactic extinction correction was performed using the values from Rieke & Lebofsky (1985), while no correction for internal extinction has been made.

The package 'ellipse' in IRAF has been used to extract the surface brightness radial profile of the galaxy. The disk was fitted with a Freeman's law, a model image was constructed using the best fit parameters and subtracted from the original frames. The disk subracted images were analyzed for evidence of bulge triaxiality, i.e. tipped bulge isophotes and a difference in the major axis position angle of the bulge relative to that of the disk of at least 10°.

In Figure 1 we show the disk subtracted image of NGC 1371; a small bar from which two spiral arms depart and an ovoidal shape almost perpendicular to the bar are apparent. The disk position angle is $133^o \pm 3^o$, while the bulge position angle is $103^o \pm 3^o$, resulting in a twist angle between the two of 30°.

To check if the bulge is really triaxial, we chose to employ a geometrical model (Stark 1977), rather than reconstructing the surface density distribution from surface brightness data. This is because Gerhard & Binney (1996) demonstrated that for any inclination angle i $< 90^o$ there exists a family of densities that shows positive and negative regions (konus density). Any of these densites added to the galaxy will produce the same surface brightness distribution and will be completely invisible for any inclination angle less than

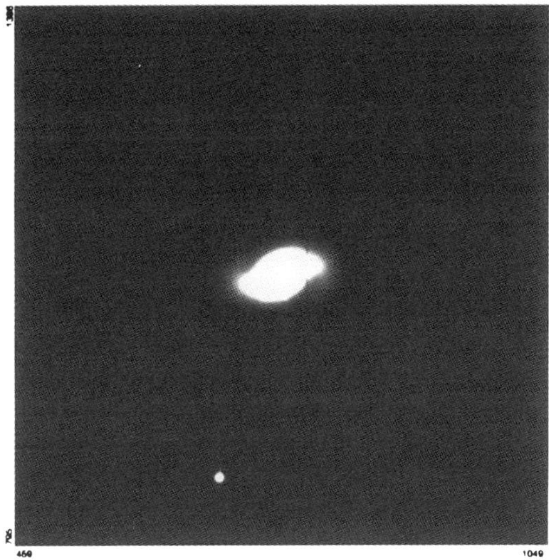

FIGURE 1. Bulge image (galaxy-exponential disk) of NGC 1371 in V.

TABLE 2. Euler Angles and True Axis' Lengths for NGC 1371 as given by Stark's Model.

ϕ	ψ	θ	x_{max}	y_{max}	z_{max}
39°	30°	46°	21.008	31.819	10.309

90°. Since the inclination angle of NGC 1371 is 42°, a geometrical model seems more reliable than an attempt to reconstruct the surface density distribution. In his model Stark demonstrates that a triaxial ellipsoid with costant luminosity on similar ellipsoids will show elliptical isophotes; then, from the shape of the isophotes it is possible to reconstruct the shape of the triaxial ellipsoid with a simple geometrical transformation that involves 3 Euler angles (ψ, θ, ϕ). If it is possible to assume that the ellipsoid has the minor axis in common with the disk, from the observations it is possible to measure 2 of the 3 Euler angles needed for the transformation (the twist angle ψ and the complement of the inclination angle θ) and the apparent axial ratio, so the ellipsoid can be modeled leaving the third angle as a free parameter. The best solution will be the one that gives back the apparent measured axial ratio. The application of this model to NGC 1371 gives solutions (i.e. all three axes with positive values) between $37.5^o \leq \phi \leq 67^o$ assuming as outermost radius for triaxiality $r_{max} = 30''$, i.e. the bulge effective radius, and as apparent axis ratio β=1.5. The angle for which we found the measured axis ratio is ϕ=$39^o \pm 1^o$. The true axis lengths are given in Table 2; as can be seen, all the axes have different lengths, confirming the true triaxial shape of the bulge.

3. Discussion and Conclusions

From the data presented in the previous section it is clear that:
(A) the bulge of NGC 1371 is triaxial;
(B) the triaxiality is evident also in the old stellar population (H band).

From item *(B)* it is possible to infer that triaxiality is a long standing feature inside spiral galaxies, so if we wish to explain it with a secular bar evolution process, where a triaxial bulge is the remnant of a double bar system (Friedli & Benz 1995), we have to assume that the bar has been destroyed a long time ago or that the effect of the bar is just rearranging the matter distribution inside the galaxy without changing in a significant way its star formation history. On the other side, triaxial bulges are observed both in barred and unbarred galaxies, so the hypothesis of bar destruction appears unlikely. This leaves us with merging or external torques as possible explanations for triaxial bulges. If the first hypothesis is preferred, then we should assume that the fraction of merging galaxies some 10^9 years ago is comparable to the fraction of triaxial systems observed today; moreover we should be able to explain the observed trend of triaxiality with morphological type. The main conclusion of the work is that, although we observe that a non-negligible percentage (from 10% to 30%) of spiral galaxies hosts a triaxial bulge, we still do not know the physical cause of this phenomenon. Observations of the same morphological type of galaxies at different redshift, a careful study of the environment of the host galaxy and a spectroscopic follow up of the promising candidates should help to solve the problem.

REFERENCES

BENDER, R. 1988 *A&A*, **193**, L7

BERTOLA, F., RUBIN, V.C., ZEILINGER, W.W. 1989 *ApJ*, **345**, L29

BERTOLA, F., VIETRI, M., ZEILINGER, W.W. 1991 *ApJ*, **374**, L13

BINNEY, J.J., PETROU, M. 1985 *MNRAS*, **214**, 449

DE ZEEUW, P.T., FRANX, M. 1989 *ApJ*, **343**, 617

FRIEDLI D., BENZ, W. 1993 *A&A*, **268**, 65

FRIEDLI D., BENZ, W. 1995 *A&A*, **301**, 649

GERHARD, O.E., BINNEY, J.J. 1996 *MNRAS*, **279**, 993

GRAUER, A.D., RIEKE, M.J. 1998 *ApJS*, **116**, 29

IDIART, T.P., DE FREITAS-PACHECO, J.A., COSTA, R.D.D. 1996 *AJ*, **112**, 254

JABLONKA, P. 1997, in *Galaxy Scaling Relations: Origins, Evolutions and Applications* (ed. L. Nicolaci, L. da Costa & A. Renzini), p140. (Springer)

KORMENDY, J. 1982 *ApJ*, **257**, 75

MAY, A., VAN ALBADA, T.S., NORMAN, C.A. 1985 *MNRAS*, **214**, 131

NOGUCHI, M. 1996 *ApJ*, **469**, 605

POMPEI, E. 1998, Ph.D. Thesis, Universitá degli Studi di Trieste

RIEKE, G.H., LEBOFSKY, M.J. 1985 *ApJ*, **288**, 618

SHAW, M. 1993 *MNRAS*, **261**, 718

STARK, A.A. 1977 *ApJ*, **213**, 368

VARELA, A.M., MUNOZ-TUNON, C., SIMMONEAU, E. 1996 *A&A*, **306**, 381

ZARITSKY, D., LO, K.Y. 1986 *ApJ*, **303**, 66

The Bulge-Disk Orthogonal Decoupling in Galaxies: NGC 4698 and NGC 4672

By FRANCESCO BERTOLA[1], E. M. CORSINI[1],
M. CAPPELLARI[1] J. C. VEGA BELTRÁN[2],
A. PIZZELLA[3], M. SARZI[1],
AND
J. G. FUNES S. J.[1]

[1]Dipartimento di Astronomia, Università di Padova, Vicolo dell'Osservatorio 5, I-35122
Padova, Italy

[2]Telescopio Nazionale Galileo, Osservatorio Astronomico di Padova, Vicolo dell'Osservatorio 5,
I-35122 Padova, Italy

[3] European Southern Observatory, Alonso de Cordova 3107, Casilla 19001, Santiago 10, Chile

We report the case of the geometrical and kinematical decoupling between the bulge and the disk of the Sa galaxy NGC 4698. The $R-$band isophotal map of this spiral shows that the bulge structure is elongated perpendicularly to the major axis of the disk. At the same time a central stellar velocity gradient is found along the major axis of the bulge. We also present the Sa galaxy NGC 4672 as being a good candidate for a spiral hosting a bulge and a disk that are orthogonally decoupled with respect to one other. This decoupling of the two fundamental stellar components suggests that the disk could represent a second event in the history of early-type spirals.

1. Introduction

NGC 4698 is classified Sa by Sandage & Tammann (1981) and Sab(s) by de Vaucouleurs *et al.* (1991; RC3). Sandage & Bedke (1994; CAG) present NGC 4698 as an example of the early-to-intermediate Sa type since it is characterized by a large central bulge and tightly wound spiral arms. In addition to a remarkable geometrical decoupling between the bulge and the disk (whose apparent major axes appear oriented in an orthogonal way upon simple visual inspection of galaxy plates; see Panels 78, 79 and 87 in CAG), a spectrum taken along the minor axis of the disk shows the presence of a stellar velocity gradient which could be ascribed to the bulge. For this reason NGC 4698 is a noteworthy case for the study of the formation processes of disks in spirals.

2. Results

The $R-$band isophotal map of NGC 4698 (Figure 1 left panel) shows the geometrical decoupling (position angle twist $\Delta\,\mathrm{PA} \simeq 90°$) between the bulge and the disk of this spiral galaxy. This phenomenon is visible in both the inner isophotes and the outermost isophote; the latter is characterized by two 'bumps', protruding perpendicularly to the disk major axis. The isophotes between $4''$ and $19''$ appear round in the plot. However, once an exponential disk is subtracted, they become elongated in a direction perpendicular to the disk major axis. In order to disentangle the contributions of the bulge and of the disk to the total light, we decomposed the surface-brightness radial profiles extracted along different axes of NGC 4698 into the sum of an $r^{1/4}$ bulge ($\mu_e = 19.6$ mag·arcsec^{-2}; $r_e = 11.3''$; $q = 1.1$) and an exponential disk ($\mu_0 = 19.2$ mag·arcsec^{-2}; $r_d = 32.2''$; $i = 60°$). The bulge is found to be the dominant component inside $10''$ along the galaxy major axis, and inside $14''$ along the minor axis. The axial ratio of the bulge is found

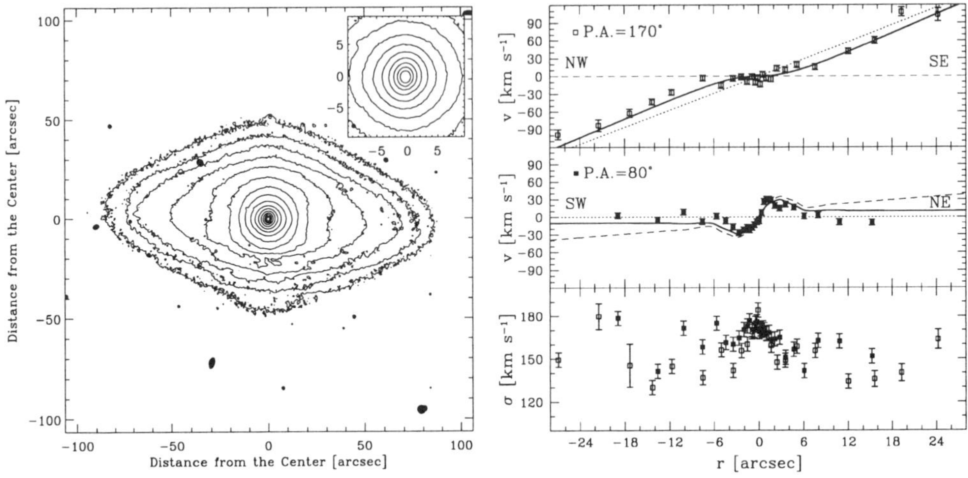

FIGURE 1. *Left panel:* $R-$band isophotes of NGC 4698, given in steps of 0.4 mag·arcsec^{-2}, with the outermost one corresponding to 21.8 mag·arcsec^{-2} and the central one to 15.8 mag·arcsec^{-2}. In the inset the isophotal map of the inner $10''$ is plotted. North is to the right and east is up. *Right panel:* Observed stellar rotation velocity (after correction for the systemic velocity of $V_\odot = 992 \pm 10$ km s^{-1}) and velocity dispersion as a function of radius along the major (open squares) and minor (filled squares) axes of NGC 4698. The dashed and the dotted lines represent the velocity contribution of the bulge and disk components to the total velocity (thick continuous line) of our model.

to be greater than unity confirming its exceptional property to be elongated along the disk minor axis. Alternative non-parametric decompositions of the NGC 4698 surface-brightness distribution (Moriondo *et al.* 1998) with a disk profile flattening toward the center would produce lower residuals when the modeled surface brightness is subtracted from the observed one. However our standard assumption of an $r^{1/4}$ bulge and an exponential disk does not affect the photometric and kinematical orthogonal decoupling between the bulge and the disk of NGC 4698 (Figure 1).

The major-axis stellar velocity curve is characterized by a central plateau, indeed the net stellar rotation is zero for $|r| \leq 8''$. At larger radii the observed stellar rotation increases from zero to an approximately constant value of about 200 km s^{-1} for $|r| \geq 50''$ up to the farthest observed radius at about $80''$. These velocities are in agreement, within the errors, with those measured by Corsini *et al.* (1999). The stellar velocity dispersion profile has been measured out to $30''$. Its value peaks in the center at 185 km s^{-1}. We measured the minor-axis stellar kinematics out to about $20''$ on both sides of the galaxy. In the nucleus the stellar rotation velocity increases to about 30 km s^{-1} at $|r| \simeq 2''$, decreasing to zero further out. The velocity dispersion profile has a central maximum of 175 km s^{-1}, in agreement with the value measured along the major axis. The stellar rotation velocity curves and velocity-dispersion radial profiles (out only to $28''$ for the spectrum along the major axis) are plotted in Figure 1 (right panel).

In order to demonstrate that the observed velocity curves along the major and minor axes are consistent with the rotation of a bulge and a disk with perpendicular angular momenta, we modeled the observed line-of-sight velocity distribution in the following way: Along the major axis we assumed the bulge rotation velocity was zero, with a constant velocity dispersion of $\sigma_b = 160$ km s^{-1}, while for the velocity curve of the disk we assumed the velocity rising linearly to match the outer points of the plotted curve where the light contribution of the bulge is negligible, with a constant velocity dispersion of $\sigma_d = 130$ km s^{-1}. The resulting velocity curve is obtained by fitting with a Gaussian the sum of the two velocity components corresponding to the bulge and the disk, weighted according to the photometric decomposition. The agreement with the observed points is very good and the flat central part of the observed velocity curve is well reproduced. A similar approach has been applied to reproduce the velocity curve along the minor axis. Constant zero rotation velocity and a constant velocity dispersion $\sigma_d = 130$ km s^{-1} have been assumed for the disk, and again a constant velocity dispersion $\sigma_b = 160$ km s^{-1} for the bulge. The bulge rotation has been maximized in such a way that the resulting velocity curve is, after folding, within the scatter of the data. It reaches a maximum of 35 km s^{-1} in the inner $3''$, decreasing to a local minimum of 15 km s^{-1} at $|r| \simeq 6''$ and then increasing again beyond this radius.

3. Discussion and Conclusion

The observed stellar kinematics can be interpreted as due to an orthogonal kinematical decoupling between the bulge and disk components. Assuming that the intrinsic shape of a bulge is generally triaxial (Bertola, Vietri & Zeilinger 1991), and that the plane of the disk coincides with the plane of the bulge, perpendicular either to the major or to the minor axis, we deduce that the observed configuration indicates that the major axis of the bulge is perpendicular to that of the disk, given that the disk is seen not far from edge-on. The fact that the velocity field of the bulge is characterized by zero velocity along its apparent minor axis, as indicated by the central plateau in the rotation curve along the disk major axis, and by a gradient along its major axis, suggests that the rotation axis of the bulge lies in the plane of the disk.

Our photometric and spectroscopic data, and the ensuing interpretation of the orthogonal decoupling between the bulge and disk of NGC 4698, can be explained if the disk has formed in a distinct process which occurred at some point during the evolution of the galaxy. We suggest that the disk formed at a later stage, due to acquisition of material by a pre-existing triaxial spheroid on its principal plane, perpendicular to the major axis. An example of acquisition on the plane perpendicular to the minor axis could be NGC 7331 (Prada *et al.* 1996), where the bulge has been found counter-rotating with respect to the disk. Up to now, NGC 4698 and NGC 7331 represent the only cases of kinematical evidence that disk galaxies with prominent bulges could initially have been 'undressed spheroids' and their disks accreted gradually over several billion years, as suggested by Binney & May (1986). Recently such kind of processes have been considered within semi-analytical modeling techniques for galaxy formation, where the disks accrete around bare spheroids that formed previously (e.g. Kauffmann 1996; Baugh, Cole & Frenk 1996). In this framework polar-ring elliptical galaxies like NGC 5266 (Varnas *et al.* 1987) and ellipticals with dust lanes along the minor axis (Bertola 1987) could be transient stages towards the formation of spiral systems like NGC 4698.

In order to investigate whether or not the acquisition phenomena giving rise to NGC 4698 and NGC 7331 are general processes of galaxy formation, we need to know how common are such objects. To address this question, we began a photometric and spectroscopic survey of early-type spirals with even only a slight indication of geometrical orthogonal decoupling between bulge and disk. The existence of galaxies with apparently round bulges suggests that the case of the orthogonal geometrical decoupling between the bulge and disk seen in NGC 4698 could indeed be a more general phenomenon. A bulge which appears to have round isophotes on an image becomes intrinsically elongated after the subtraction of an inclined exponential disk. The further spectroscopic analysis of this

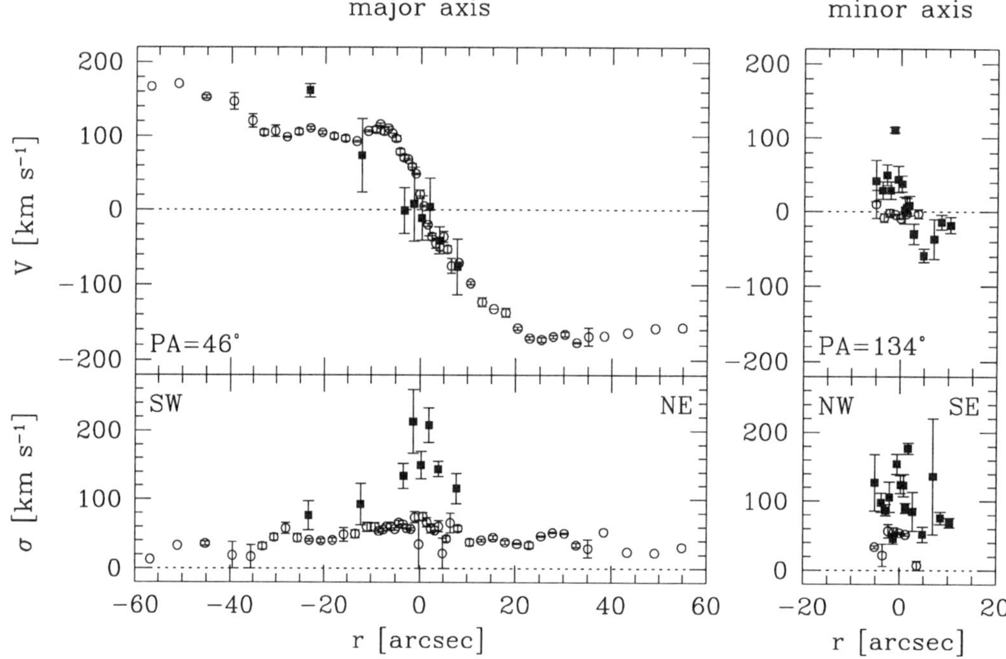

FIGURE 2. The stellar (filled squares) and ionized gas (open circles) kinematics measured along the major (PA = 46°) and minor axis (PA = 134°) of NGC 4672. The systemic velocity is $V_\odot = 3275 \pm 20$ km s^{-1}.

kind of object, with the slit of the spectrograph set along the major axis of their bulges, allows one to determine if the geometrical decoupling is associated with the kinematical decoupling, as it is in NGC 4698.

As a first result of this survey we present the case of NGC 4672 (Sarzi *et al.* 1999, in preparation). This galaxy is very probably a highly-inclined Sa spiral with a prominent $r^{1/4}$ bulge sticking out from the plane of the disk (e.g. Figure 3i by Whitmore *et al.* 1990), rather than a polar-ring galaxy. The stellar velocity field (Figure 3) is characterized by the signature of a central zero-velocity plateau in the rotation curve obtained along the major-axis (PA = 46°) and by a steep velocity gradient observed along the minor axis (PA = 134°), similar to those measured in NGC 4698. In addition, the major-axis gas rotation curve is typical of that of a disk in differential rotation, and not of a ring. The close structural and kinematical resemblance of NGC 4672 to NGC 4698 makes it a good candidate to be a new example of an early-type spiral characterized by a bulge-to-disk orthogonal decoupling.

REFERENCES

BAUGH, C.M., COLE, S., FRENK, C.S. 1996 *MNRAS*, **283**, 1361

BERTOLA, F. 1987, in *Structure and Dynamics of Elliptical Galaxies* (ed. P.T. de Zeeuw) p35. (Reidel)

BERTOLA, F., VIETRI, M., ZEILINGER, W.W. 1991 *ApJ*, **374**, L13

BERTOLA, F., CINZANO, P., CORSINI, E.M., ET AL. 1996 *ApJ*, **458**, L67

BINNEY, J.J., MAY, A. 1986 *MNRAS*, **218**, 743

CORSINI, E.M., PIZZELLA, A., SARZI, M., ET AL. 1999 *A&A*, in press [astro-ph/9809366]

DE VAUCOULEURS, G., ET AL. 1991 *Third Reference Catalogue of Bright Galaxies*. (Springer-Verlag, New York). (RC3)

KAUFFMANN, G. 1996 *MNRAS*, **281**, 487

MORIONDO, G., GIOVANARDI, C., HUNT, L.K. 1998 *A&AS*, **130**, 81

PRADA, F., GUTIERREZ, C.M., PELETIER, R.F., McKEITH, C.D. 1996 *ApJ*, **463**, L9

SANDAGE, A., BEDKE, J. 1994 *The Carnegie Atlas of Galaxies*. (Carnegie Institution, Washington DC). (CAG)

SANDAGE, A., TAMMANN, G.A. 1981 *A Revised Shapley-Ames Catalog of Bright Galaxies*. (Carnegie Institution, Washington DC). (RSA)

VARNAS, S.R., BERTOLA, F., GALLETTA, G., ET AL. 1987 *ApJ*, **313**, 69

WHITMORE, B.C., LUCAS, R.A., McELROY, D.B., ET AL. 1990 *AJ*, **100**, 1489

The Kinematics and the Origin of the Ionized Gas in NGC 4036

By ENRICO M. CORSINI[1], F. BERTOLA[1], M. SARZI[1],
P. CINZANO[1], H.-W. RIX[2] AND W.W. ZEILINGER[3]

[1]Dipartimento di Astronomia, Università di Padova, Vicolo dell'Osservatorio 5, I-35122
Padova, Italy

[2]Steward Observatory, University of Arizona, Tucson AZ-85721, USA

[3]Institut für Astronomie, Universität Wien, Türkenschanzstrasse 17, A-1180 Wien, Austria

The kinematics of stars and ionized gas has been studied near the center of the S0 galaxy
NGC 4036. Dynamical models based both on stellar photometry and kinematics have been
built in order to derive the gravitational potential in which the gas is expected to orbit. The
observed gas rotation curve falls short of the circular velocity curve inferred from these models.
Inside $10''$ the observed gas velocity dispersion is found to be comparable to the predicted circular
velocity, showing that the gas cannot be considered on circular orbits. The understanding of
the observed gas kinematics is improved by models based on the Jeans Equations, which assume
the ionized gas as an ensemble of collisionless cloudlets distributed in a spheroidal and in a disk
component.

1. Introduction

NGC 4036 has been classified $S0_3(8)/Sa$ in RSA (Sandage & Tammann 1981) and S0$^-$
in RC3 (de Vaucouleurs *et al.* 1991). Its total apparent magnitude is $V_T = 10.66$ mag
(RC3). This corresponds to a total luminosity $L_V = 4.2 \cdot 10^{10}$ L$_{V_\odot}$ at the assumed
distance of $d = V_0/H_0 = 30.2$ Mpc, where $V_0 = 1509 \pm 50$ km s^{-1} (RSA) and assuming
$H_0 = 50$ km s^{-1} Mpc^{-1}. At this distance the scale is 146 pc arcsec^{-1}.

We measured the kinematics of stars and ionized gas along the galaxy major axis and
derived their distribution in the nuclear regions by means of ground-based V-band and
HST narrow-band imaging respectively.

2. Modeling the Stellar Kinematics

We apply to the observed stellar kinematics the Jeans modeling technique introduced
by Binney *et al.* (1990), developed by van der Marel *et al.* (1990) and van der Marel
(1991), and extended to two-component galaxies by Cinzano & van der Marel (1994).

The best-fit model to the observed major-axis stellar kinematics is shown in Figure
1. The bulge is an oblate isotropic rotator ($k = 1$) with $(M/L_V)_b = 3.4$ (M/L)$_\odot$. The
exponential disk has $(M/L_V)_d = 3.4$ (M/L)$_\odot$ with the velocity dispersion profile given
by $\sigma_\star(r) = 155\, e^{-r/r_\sigma}$ km s^{-1} with scale-length $r_\sigma = 27.4'' = 4.0$ kpc. The derived
bulge and disk masses are $M_b = 9.8 \cdot 10^{10}$ M$_\odot$ and $M_d = 4.8 \cdot 10^{10}$ M$_\odot$, adding up to
a total (bulge+disk) luminous mass of $M_T = 14.5 \cdot 10^{10}$ M$_\odot$. The disk-to-bulge and
disk-to-total luminosity ratios are $L_b/L_d = 0.58$ and $L_d/L_T = 0.36$.

3. Modeling the Gaseous Kinematics

At small radii both the ionized gas velocity and velocity dispersion are comparable to
stellar velocity and velocity dispersion, for $r \leq 9''$ and $r \leq 5''$ respectively. Moreover a

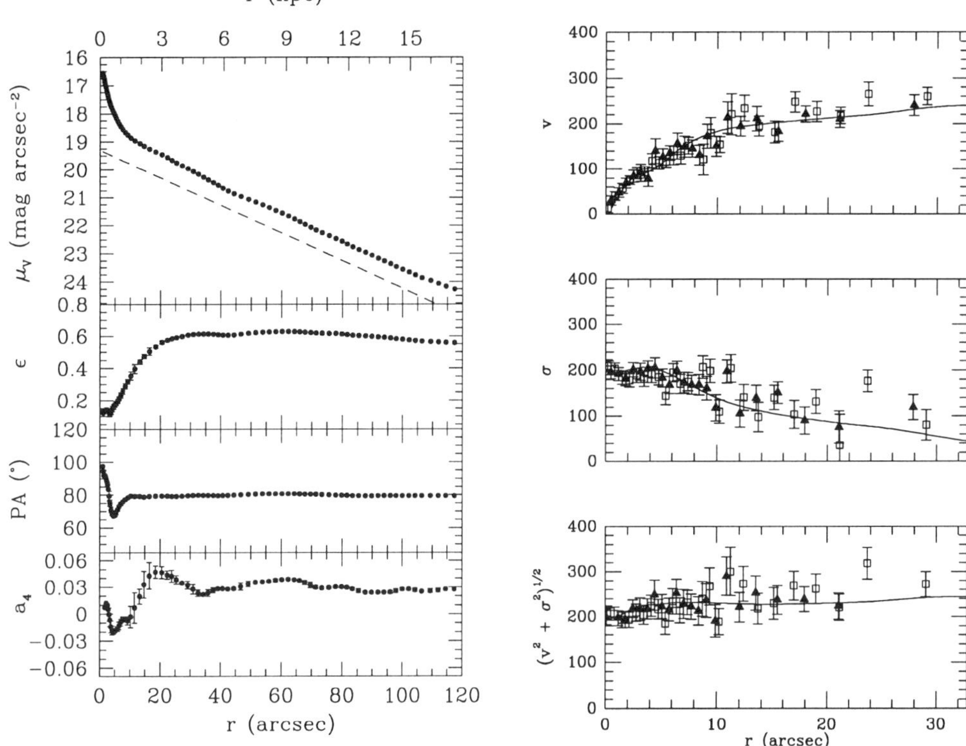

FIGURE 1. *Left panel:* Ground-based V-band surface-brightness, ellipticity, position angle and $\cos 4\theta$ coefficient radial profiles of NGC 4036. The dashed line represents the surface-brightness profile of the exponential disk adopted in our dynamical model. *Right panel:* Comparison of the model predictions to the stellar major-axis kinematics for NGC 4036.

change in the slope of the [O II] $\lambda3726$ intensity radial profile is observed inside $r \simeq 8''$, its gradient appears to be somewhat steeper towards the center. The velocity dispersion and intensity profiles of the ionized gas suggest that it is distributed into two components: a small inner spheroidal component and a disk. We decomposed the [O II] $\lambda3726$ intensity profile as the sum of an $R^{1/4}$ gaseous spheroid and an exponential gaseous disk and the gas spheroid resulted to be the dominating component up to $r \simeq 8''$.

We built up dynamical models for the ionized gas in NGC 4036 (Figure 2). It was assumed to be distributed in a dynamically hot spheroidal and in a dynamically cold disk component and consisting of collisionless individual clumps (cloudlets) which orbit in the total potential. We made two different sets of assumptions based on two different physical scenarios for the gas cloudlets.

Model A: In a first set of models we described the gaseous component consisting of collisionless cloudlets which can be considered in hydrostatic equilibrium. The gaseous spheroid is characterized by a density distribution and flattening different from those of the stars. Its major-axis luminosity profile was assumed to follow an $R^{1/4}$ law. The flattening of the spheroid q was kept as free parameter. To derive the kinematics of the gaseous spheroid and disk we solved the Jeans Equations.

Model B: In a second set of models we assumed that the emission observed in the gaseous spheroid and disk arise from material that was recently shed from stars. Different authors (Bertola *et al.* 1984, 1995b; Fillmore *et al.* 1986; Kormendy & Westpfahl 1989; Mathews 1990) suggested that the gas lost (e.g. in planetary nebulae) by stars was heated by

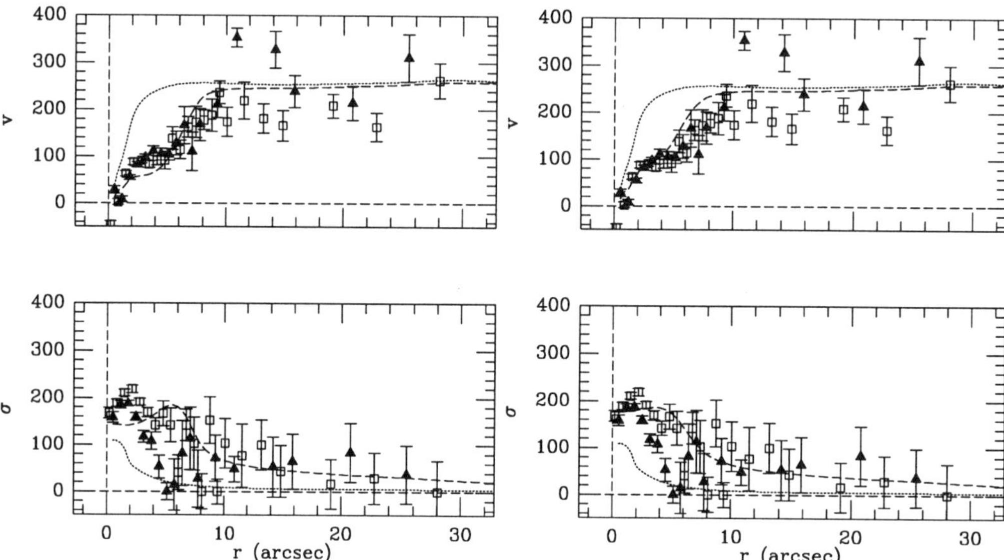

FIGURE 2. Comparison of the predictions (dashed curves) of model A (*left panel*) and model B (*right panel*) to the ionized gas kinematics observed along the major axis of NGC 4036. The dotted curves represent the seeing-convolved circular velocity curve and zero velocity-dispersion profile in the galaxy meridional plane respectively.

shocks to the virial temperature of the galaxy within 10^4 years, a time shorter than the typical dynamical time of the galaxy. Hence in this picture the ionized gas and the stars have the same true kinematics, while their observed kinematics are different due to the line-of-sight integration of their different spatial distribution.

4. Do Drag Forces Affect the Kinematics of the Gaseous Cloudlets?

The discrepancy between model and observations could be explained by taking properly into account the drag interaction between the ionized gas cloudlets of the gaseous spheroid and the hot component of the interstellar medium (Mathews 1990). To have some qualitative insights in understanding the effects of a drag force on the gas kinematics we studied the case of a gaseous nebula moving in the spherical potential generated by an homogeneous mass distribution of density ρ and which, starting onto a circular orbit, is decelerated by a drag force $\mathbf{F}_{drag} = -(k_{drag}v/m)\,\mathbf{v}$, where m and $\mathbf{v}$ are the mass and the velocity of the gaseous cloud and the constant k_{drag} is given following Mathews (1990). We numerically solved the equations of motion of a nebula to study the time-dependence of the radial and tangential velocity components $\dot{r}$ and $r\dot{\psi}$. We fixed the potential assuming a circular velocity of 250 km s^{-1} at $r = 1$ kpc. Following Mathews (1990) we took an equilibrium radius for the gaseous nebula $a_{eq} = 0.37$ pc. It results that $\ddot{\psi} < 0$ and $\ddot{r} > 0$: the clouds spiral towards the galaxy center as expected. Moreover the drag effects are greater on faster starting clouds and therefore negligible for the slowly moving clouds in the very inner region of NGC 4036.

If the nebulae are homogeneously distributed in the gaseous spheroid, only the tangential component $r\dot{\psi}$ of their velocities contributes to the observed velocity. No contribution derives from the radial component $\dot{r}$ of their velocities. In fact for each nebula moving towards the galaxy center, which is also approaching to us, we expect to find along the line-of-sight a receding nebula, which is falling towards the center from the same galactocentric distance with an opposite line-of-sight component of its $\dot{r}$. However the radial components of the cloudlets velocities (typically of 30-40 km s^{-1}) are crucial to explain the velocity dispersion profile and to understand how the difference between the observed velocity dispersions and the model B predictions arises. If the clouds are decelerated by the drag force, their orbits become more radially extended and the velocity ellipsoids acquire a radial anisotropy.

So we expect that (in the region of the gaseous spheroid) including drag effects in our gas modeling should give a velocity dispersion profile steeper than the one predicted by our isotropic model B, and in better agreement with observations.

5. Discussion and Conclusions

The modeling of the stellar and gas kinematics in NGC 4036 shows that the observed velocities of the ionized gas, moving in the gravitational potential determined from the stellar kinematics, cannot be explained without taking the gas velocity dispersion into account. In the inner regions of NGC 4036 the gas is not moving at the circular velocity.

A better match with the observed gas kinematics is found by assuming the ionized gas as made of collisionless clouds in a spheroidal and disk component for which the Jeans Equations can be solved in the gravitational potential of the stars (i.e., model A). A much better agreement is achieved by assuming that the ionized gas emission comes from material which has recently been shed from the bulge stars (i.e., model B). If this gas is heated to the virial temperature of the galaxy (ceasing to produce emission lines) within a time much shorter than the orbital time, it shares the same 'true' kinematics of its parent stars. If this is the case we would observe a different kinematics for ionized gas and stars due only to their different spatial distribution. Except for a complex emission structure inside $3''$, an HST Hα+[N II] image of the nucleus of NGC 4036 confirms the smoothness of the distribution of the emission as we expect for the gas spheroidal component.

This kinematical modeling leaves open the questions about the physical state (e.g. the lifetime of the emitting clouds) and the origin of the dynamically hot gas. We tested the hypothesis that the ionized gas is located in short-lived clouds shed by evolved stars (e.g. Mathews 1990) finding a satisfying agreement with our observational data. These clouds may be ionized by the parent stars, by shocks, or by the UV-flux from hot stars (Bertola *et al.* 1995a). The comparison with the more recent and detailed data on gas by Fisher (1997) opens wide the possibility for further modeling improvement if the drag effects on gaseous cloudlets (due to the diffuse interstellar medium) will be taken into account. These arguments indicate that the dynamically hot gas in NGC 4036 has an internal origin. This does not exclude the possibility for the gaseous disk to be of external origin as discussed for S0's by Bertola *et al.* (1992). Spectra at higher spatial resolution are needed to understand the structure of the gas inside $3''$.

REFERENCES

BERTOLA, F., BETTONI, D., RUSCONI, L., SEDMAK, G. 1984 *AJ*, **89**, 356

BERTOLA, F., BUSON, L.M., ZEILINGER, W.W. 1992 *ApJ*, **401**, L79

BERTOLA, F., BRESSAN, A., BURSTEIN, D., ET AL. 1995 *ApJ*, **438**, 680

BERTOLA, F., CINZANO, P., CORSINI, E.M., RIX, H.-W., ZEILINGER, W.W. 1995 *ApJ*, **448**, L13

BINNEY, J.J., DAVIES, R.L., ILLINGWORTH, G.D. 1990 *ApJ*, **361**, 78

CINZANO, P., VAN DER MAREL, R.P. 1994 *MNRAS*, **270**, 325

DE VAUCOULEURS, G., ET AL. 1991 *Third Reference Catalogue of Bright Galaxies.* (Springer-Verlag, New York). (RC3)

FILLMORE, J.A., BOROSON, T.A., DRESSLER, A. 1986 *ApJ*, **302**, 208

FISHER, D. 1997 *AJ*, **113**, 950

KORMENDY, J., WESTPFAHL, D.J. 1989 *ApJ*, **338**, 772

MATHEWS, W.G. 1990 *ApJ*, **354**, 468

SANDAGE, A., TAMMANN, G.A. 1981 *A Revised Shapley–Ames Catalog of Bright Galaxies.* (Carnegie Institution, Washington DC). (RSA)

VAN DER MAREL, R.P. 1991 *MNRAS*, **253**, 710

VAN DER MAREL, R.P., BINNEY, J.J., DAVIES, R.L. 1990 *MNRAS*, **245**, 582

Optically Thin Thermal Plasma in the Galactic Bulge

By YOSHITOMO MAEDA[1], GORDON GARMIRE[1],
KATSUJI KOYAMA[2,3] AND MASAAKI SAKANO[2]

[1]Dept. of Astronomy and Astrophysics, Pennsylvania State University, 525 Davey Laboratory, University Park PA 16802-6305, U.S.A.

[2]Department of Physics, Graduate School of Science, Kyoto University, Sakyo-ku, Kyoto 606-8502, Japan

[3]CREST, Japan Science and Technology Corporation (JST), 4-1-8 Honmachi, Kawaguchi, Saitama 332-0012, Japan

We present preliminary results of our ASCA observation of the Galactic bulge. We confirm the diffuse (spatially-unresolved) soft X-ray emission in the direction of the bulge. We also detect iron-L and neon-K complex lines in the spectrum. Therefore, the bulge emission undoubtedly originates from an optically thin thermal plasma. The plasma temperature is 0.4 keV. With the results, we present possible implications of the Galactic bulge emission.

1. Introduction

A Galactic Soft X-ray Diffuse Background (SXDB) below ~ 2 keV was discovered by Bowyer, Field & Mack (1968). Four soft X-ray all-sky surveys produced maps of this SXDB (McCammon *et al.* 1983; Marshall & Clark 1984; Garmire *et al.* 1992; Snowden *et al.* 1995, 1997) which show complex features, indicating that the SXDB must be made up of several components. However, Snowden *et al.* (1997) established that the SXDB maps above 0.5 keV are smooth on the south side of the plane, which can be reproduced with only one component: a hot gas in the bulge with a scale height of ~ 1.9 kpc. Thus they named this component as the 'bulge' emission. The typical temperature was estimated to be ~ 0.3 keV.

The ASCA satellite has the capability to observe the SXDB with a reasonable energy resolution (Tanaka *et al.* 1994), which allows an improved study of line emission. We present here results of our initial analysis of the ASCA spectrum and discuss the bulge emission.

2. Observation and Data-Reduction

ASCA observations of the bulge emission were performed on April 17th, 1998. The central position of the field was placed at $(l,b) = (0°.0, -11°.5)$ because this field was consistent with the bulge model by Snowden *et al.* (1997). The total exposure time was 19 ksec. ASCA carries two Gas Imaging Spectrometers (GIS2 and GIS3) and two Solid-state Imaging Spectrometers (SIS0 and SIS1) at the foci of nested thin foil mirrors (Tanaka *et al.* 1994). The SIS achieve higher energy resolution (2% at 6keV) and higher sensitivity in the soft X-ray bands than the GIS. Therefore we will focus on a report of the SIS results. A GIS study of the bulge emission is given in Sakano *et al.* (1999). The SIS was operated in the 1CCD (4sec exposure) Faint mode. For details of the instrumentation, one can refer to Tanaka, Inoue & Holt (1994), Burke *et al.* (1994), Ohashi *et al.* (1996) and Serlemitsos *et al.* (1995).

3. Analysis and Results

In order to test possible contamination from discrete sources in the field of view, we searched for sources in the 0.7–2 keV band. We found that there are no significant detections of discrete sources at more than 3σ confidence limit of $\sim3\times10^{-14}$ ergs s^{-1} cm^{-2} corresponding to an intrinsic luminosity of 3×10^{32} ergs s^{-1} at 8.5 kpc. Thus we extracted the diffuse X-ray spectrum from the entire region.

Figure 1 shows the SIS spectrum of the diffuse emission. We detected significant X-ray emission up to ~4 keV. Giacconi *et al.* (1962) discovered the Cosmic X-ray Background (CXB) which is nearly isotropic. The CXB spectrum was nearly described with a power-law model below 10 keV. The spectrum above ~2 keV was consistent with the CXB power-law model but a significant excess appeared below ~2 keV. Therefore we concluded that the origin of the hard X-rays is extragalactic while that of the soft X-rays is galactic, which we assign to the bulge emission.

In order to test the emission mechanism of the soft excess, we fitted it with two different models; a thin thermal plasma model and a power-law model. Due to the limited statistics, we assumed the solar abundance and fixed absorption column of 1×10^{21} H cm^{-2} which was derived from a Galactic HI column density map (Dickey & Lockman 1990).

The thin thermal plasma model reproduced the spectral shape quite well, while the power-law model showed large residuals around 1 keV and was rejected. The residuals seen in the power-law model are due to the iron L and neon K complex lines. Thus we found that the soft excess, that is the bulge emission, is undoubtedly thin thermal. The temperature was found to be about 0.4 keV. The hot gas density is estimated to be 10^{-3} electron cm^{-3} assuming the geometry given by Snowden *et al.* (1997).

4. Discussion

The origin of the thin thermal bulge emission has been discussed in several papers (Snowden *et al.* 1997 and reference therein). While not a unique interpretation, hot gas filling the bulge is a strong candidate for explaining the thermal x-ray emission.

One of the great results of the Einstein X-ray satellite is the discovery of the hot halo around galaxies such as elliptical galaxies and starburst galaxies (see Fabbiano in this volume.). These hot gases are often tracers of the gravitational potential or of AGN or starburst activities. By comparison with these galaxies, we present possible implications of the bulge hot gas.

4.1. *Comparison with Gravity*

If a hot gas is gravitationally bounded, stellar motions and the hot gas should be in energy equipartition. Figure 2 shows a relation between the velocity dispersion and the gas temperature. For comparison, we plotted the data points for cluster of galaxies and elliptical galaxies (Hatsukade 1989; Matsumoto *et al.* 1997). We adopted a velocity dispersion of the SiO maser for the bulge (Izumiura *et al.* 1995). The cluster of galaxies and a part of elliptical galaxies ride on the equipartition line. Therefore these hot gases are good tracers of their gravitational potential. However, the bulge gas is far away from the equipartition line, i.e., the hot gas in the bulge is not bounded by the gravity.

4.2. *Implication of Heating Sources*

In order to heat a gas to 0.4 keV by shocks, a particle velocity must be more than 300 km s^{-1}. Massive stars or supernova remnants release energetic particles whose velocity is high enough. However these sources are not found in the bulge. Therefore we infer that the heating sources should be located not at the bulge but at the Galactic center or disk. Since the hot gas is not bounded by gravity, the bulge hot gas can be interpreted as the left-overs of past AGN or starburst activity in the Galactic Center region.

5. Summary

ASCA detected the diffuse X-ray emission in the direction of the bulge. By comparison with the CXB spectrum, we confirm a significant excess in the soft X-ray band, which is due to the bulge emission. The iron-L and neon-K complex lines are detected in the soft excess. Therefore the bulge emission is undoubtly thin thermal emission. The temperature is 0.4 keV. By comparison with the hot gases observed in clusters of galaxies and elliptical galaxies, we find that the hot gas in the bulge is not a tracer of the gravitational potential.

We thank all the members of the ASCA team for their support in this study. Our analysis was performed using the FTOOLS and XANADU packages, which were provided by GSFC/NASA. Y. M. and M. S. are financially supported by the Japan Society for the Promotion of Science.

REFERENCES

BOWER, S., FIELD, G.B., MACK, J.E. 1968 *Nature*, **217**, 32

BURKE, B., ET AL. 1994 *IEEE Trans. Nucl. Sci.*, **41**, 375

DICKEY, J.M., LOCKMAN, F.J. 1990 *ARA&A*, **28**, 215

GARMIRE, G.P., ET AL. 1992 *ApJ*, **399**, 694

GIACCONI, R., ET AL. 1962 *Phys. Rev. Lett.*, **9**, 439

HATSUKADE, I. 1989, Ph. D. Thesis, Osaka University

IZUMIURA, H., ET AL. 1995 *ApJ*, **453**, 837

MARSHALL, F.J., CLARK, G.W. 1984 *ApJ*, **287**, 633

MATSUMOTO, H., ET AL. 1997 *ApJ*, **482**, 133

McCAMMON, D., ET AL. 1983 *ApJ*, **269**, 107

OHASHI, T., ET AL. 1996 *PASJ*, **48**, 157

SAKANO, M., ET AL. 1999 *Adv. Space Res.*, in press

SERLEMITSOS, P.J., ET AL. 1995 *PASJ*, **47**, 105

SNOWDEN, S.L., ET AL. 1995 *ApJ*, **454**, 643

SNOWDEN, S.L., ET AL. 1997 *ApJ*, **485**, 125

TANAKA, Y., INOUE, H., HOLT, S.S. 1994 *PASJ*, **46**, L37

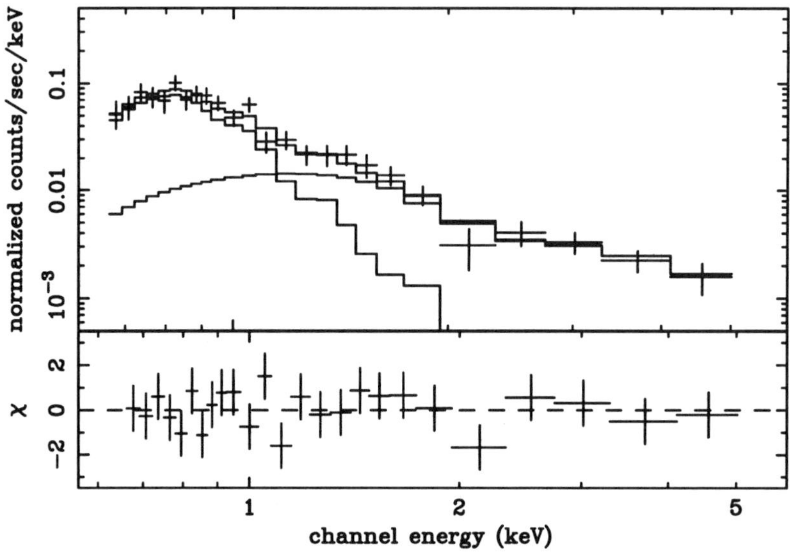

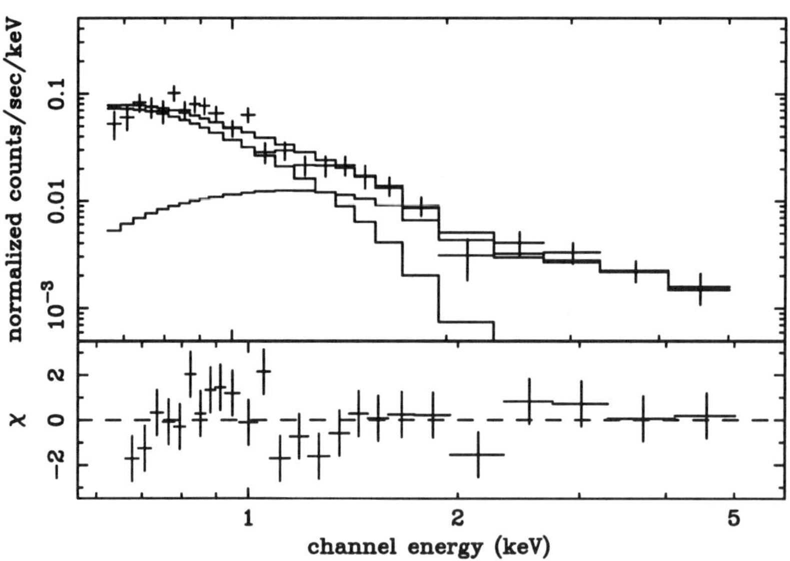

FIGURE 1. SIS spectrum of the diffuse X-ray emission in the direction of the bulge. The thin-thermal plasma and power-law models with the CXB model are shown in the top and bottom panels, respectively. The residuals between data and models are shown in the bottom box of each panel. The solid lines are the best-fit parameters. The spectrum above 2 keV is consistent with the CXB model while the excess due to the bulge emission is seen below 2 keV. The thin thermal plasma model for the soft excess was accepted (Red χ^2/d.o.f.=0.718/22) while the power-law model was rejected (Red χ^2/d.o.f.=1.32/22).

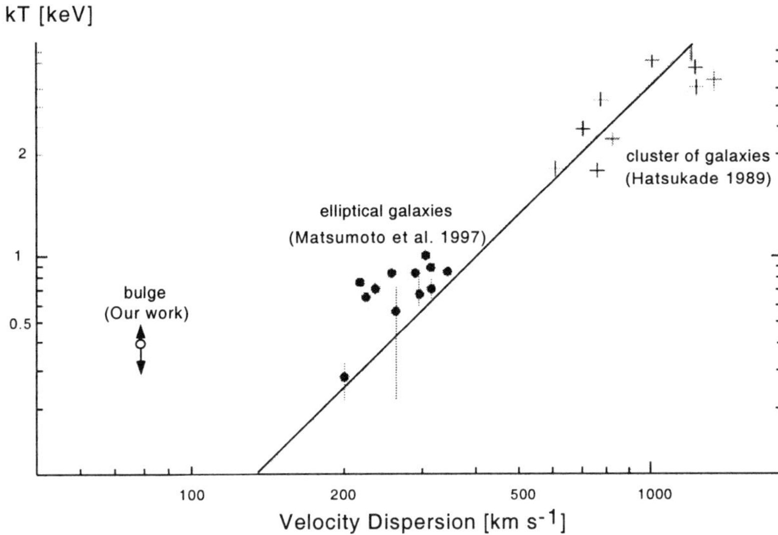

FIGURE 2. The velocity dispersion vs. the gas temperature. The solid line is the case for the energy equipartition (see Matsumoto *et al.* 1997 for detail). Crosses and filled circles denote clusters of galaxies and elliptical galaxies, respectively.

X-ray Properties of Bulges

By GIUSEPPINA FABBIANO

Harvard-Smithsonian Center for Astrophysics, 60 Garden St., Cambridge MA 02138, USA

Integrated X-ray observations of early-type spirals and detailed observations of the bulge of M31 suggest a complex picture of the X-ray emission of bulges. There is a baseline population of point-like X-ray sources, most likely accretion binaries, which is seen to dominate the emission of the bulge of M31. There are also spectral results suggesting an additional gaseous emission component in some X-ray bright galaxies. Future deep observations with the *Chandra X-ray Observatory* (CXO) will allow us to shed light on the nature of the X-ray emission.

1. Introduction

The nature of the X-ray emission of early-type, prominent-bulge spirals has been the subject of an on-going controversy, which has sought to establish if and how much of this emission can be ascribed to thermal emission of an optically thin hot gaseous medium. This is an important issue, because if it can be established that the X-ray emission is dominated by gravity-bound gaseous halos, the X-ray data may be used to measure the mass of these galaxies (see review in Fabbiano 1989).

In what follows, I give a summary of the work on this subject, and point out future opportunities.

2. A Brief History of X-ray Studies of Early-type Spirals

With the clear exception of M31, most of the bulges of early-type spirals could not be studied in detail with X-ray observatories, starting with the *Einstein Observatory*, in the early '80s, and including all the facilities in orbit and operational at this time. The available sensitivity and resolution allow only integrated flux measurements in most cases. X-ray photometry and spectra of the integrated galaxy emission have been derived for a subset of galaxies.

The picture that emerged from early *Einstein* work on the emission of early-type spirals is somewhat controversial. An early claim (Forman, Jones & Tucker 1985) would have Sa galaxies dominated in X-rays by the emission of hot tenuous halos. However, later work provided contrary evidence, suggesting instead that X-ray binaries may be responsible for the bulk of the $\sim 0.2 - 4$ keV X-ray emission of early-type spirals. In particular, the few Sa galaxies in Forman, Jones & Tucker 1985 tended to have X-ray to optical flux ratios in the lower envelope of values found for the E and S0 sample. Moreover, these values were consistent with the bulk of those found in spirals of all morphological types (Fabbiano & Trinchieri 1985, hereafter FT). These X-ray to optical flux ratios were consistent with those of the bulge of M31, which is dominated in the *Einstein* band by a collection of point-like individual sources (Trinchieri & Fabbiano 1985; see review in Fabbiano 1989; more recent ROSAT work can be found in Primini *et al.* 1993 and Supper *et al.* 1997). Furthermore, the analysis of a sample of 51 spiral galaxies (Sa-Irr) (Fabbiano, Gioia & Trinchieri 1988, hereafter FGT) reported similar, linear, X-ray – B-band correlations in both bulge- and disk-dominated systems. These linear correlations suggested that X-ray binaries (which can be assumed to represent a fraction of the normal stellar content of a galaxy) would be responsible for the X-ray emission of all spiral systems.

More recent work, however, has raised again the possibility of a gaseous emission component in early-type spirals. The spectral analysis of Kim, Fabbiano & Trinchieri 1992 showed that the average X-ray spectrum of Sa galaxies in the *Einstein* band is softer than that of later types, suggesting a resemblance with that of E and S0s and perhaps the possibility of some hot ISM contributing to the X-ray emission. The ROSAT detailed observations of one such system, NGC 4594 - the Sombrero galaxy -, proved inconclusive for establishing the presence of a substantial hot gaseous halo (Fabbiano & Juda 1997), but an emission line gaseous component may be suggested by the ASCA spectrum (Terashima *et al.* 1994). Similarly, an extended soft gaseous component has been suggested in NGC 4736 (Cui *et al.* 1997).

3. Recent Results

A recent analysis (Shapley, Fabbiano & Eskridge 1999; Fabbiano & Shapley 1999) of the sample of normal galaxies in the *Einstein Observatory* X-ray Catalog and Atlas of Fabbiano, Kim & Trinchieri (1992) supports the possibility of gaseous emission components dominating the X-ray emission of the most X-ray luminous S0/a-Sab galaxies. Contrary to previous reports based on much smaller samples (FT, FGT), the X-ray – B correlation is steeper than linear ($L_X \sim L_B^{1.4}$). Comparisons with the analogous correlation observed in E and S0s, and with the distribution of AGNs in the $L_X - L_B$ plot, suggest that hot ISM in X-ray brighter galaxies (as in E and S0s), rather than nuclear activity, is a more viable explanation.

This work also suggests the possibility that the radio continuum emission of S0/a-Sab galaxies may be dominated by an active nuclear component. A new result of this analysis is that in S0/a-Sab galaxies the strongest radio continuum – infrared (IR) link is with the mid- (rather than far-) IR; instead, the radio continuum – far-IR (FIR) correlation dominates in later morphological types. Nuclear activity may be responsible for the radio – mid-IR correlation. The close link with the mid-IR argues in itself for this possibility, since mid-IR emission is enhanced in AGN (Elvis *et al.* 1994). Moreover, the correlation is also consistent with the distribution of bright AGNs in the same diagram.

The 6cm – 12μm correlation of S0/a-Sab galaxies is steep ($L_{6cm} \propto L_{12\mu}^{1.45\pm0.23}$). This power-law slope is difficult to explain. Linear correlations would be expected both in the picture of shock-related nuclear emission (Ho *et al.* 1989), and in the more traditional AGN scenario, where the IR emission may be due to re-irradiation of the nuclear continuum. The observed non-linearity may be related to the presence of hot gaseous halos in the more luminous galaxies, which may result in depletion of small-size dust grains (and thus a decrease of the 12μm flux). A similar mechanism may operate in E and S0 galaxies, where a steep radio – IR correlation has also been reported (Eskridge, Fabbiano & Kim 1995).

4. Conclusions

We are at a turning point in X-ray astronomy. In the next year we will be able to observe the X-ray sky with the *Chandra X-ray Observatory* (formerly AXAF) and with XMM. With these facilities we will be able to address and explore some of the issues raised by the presently available X-ray observations. In particular, with *Chandra* we will be able to obtain deep, high resolution spatial/spectral images of nearby bulges, and derive a direct measurement of the different X-ray emission components. XMM will help greatly in the assembling of large, well defined samples of galaxies for more in depth statistical studies.

I acknowledge support from the *Chandra X-ray Center* (CXC).

REFERENCES

CUI, W., FELDKHUN, D., BRAUN, R. 1997 *ApJ*, **477**, 693

ELVIS, M., WILKES, B.J., McDOWELL, J.C., GREEN, R.F., BECHTOLD, J., WILLNER, S.P., OEY, M.S., POLOMSKI, E., CUTRI, R. 1994 *ApJS*, **95**, 1

ESKRIDGE, P.B., FABBIANO, G., KIM, D-W. 1995 *ApJS*, **97**, 141

FABBIANO, G. 1989 *ARAA*, **27**, 87

FABBIANO, G., JUDA, J.Z. 1997 *ApJ*, **476**, 666

FABBIANO, G., SHAPLEY, A. 1999 *ApJ*, submitted

FABBIANO, G., TRINCHIERI, G. 1985 *ApJ*, **296**, 430 (FT)

FABBIANO, G., GIOIA, I.M, TRINCHIERI, G. 1988 *ApJ*, **324**, 749 (FGT)

FABBIANO, G., KIM, D-W., TRINCHIERI, G. 1992 *ApJS*, **80**, 531 (FKT)

FORMAN, W., JONES, C., TUCKER, W. 1985 *ApJ*, **293**, 102

HO, P.T., TURNER, J.L., FAZIO, G., WILLNER, S.P. 1989 *ApJ*, **344**, 135

KIM, D-W., FABBIANO, G., TRINCHIERI, G. 1992 *ApJ*, **393**, 134

PRIMINI, F., FORMAN, W., JONES, C. 1993 *ApJ*, **410**, 615

SHAPLEY, A., FABBIANO, G., ESKRIDGE, P.B. 1999 *ApJ*, submitted

SUPPER, R., HASINGER, G., PIETSCH, W., TRUEMPER, J., JAIN, A., MAGNIER, E.A., LEWIN, W.H.G., VAN PARADIJS, J. 1997 *A&A*, **317**, 328

TERASHIMA, Y., SERLEMITSOS, P.J., KUNIEDA, H., IWASAWA, K. 1994, in *New Horizon of X-Ray Astronomy - First results from ASCA* (ed. F. Makino & T. Ohashi), p523. (Tokyo: Universal Academy Press)

TRINCHIERI, G., FABBIANO, G. 1985 *ApJ*, **296**, 447

The Host Galaxies of Radio-Loud AGN

By C. MEGAN URRY[1], RICCARDO SCARPA[1],
MATTHEW O'DOWD[1], MAURO GIAVALISCO[1],
RENATO FALOMO[2], JOSEPH E. PESCE[3],
AND
ALDO TREVES[4]

[1] Space Telescope Science Institute, 3700 San Martin Dr., Baltimore MD 21218, USA
[2] Osservatorio Astronomico di Padova, Vicolo Dell'osservatorio 5, 35122 Padova, Italy
[3] Eureka Scientific, 657 Cricklewood Dr., State College PA 16803, USA
[4] University of Insubria, via Lucini 3, 22100 Como, Italy

AGN are known to lie in galaxies, and both galaxies and AGN evolve similarly over cosmic time (e.g., Silk & Rees 1998). This suggests a close connection between the nuclear phenomena associated with black holes and the formation and evolution of ordinary galaxies. The host galaxies of AGN are a direct probe of the AGN-galaxy connection. Among AGN, BL Lac objects are know to reside mostly, if not systematically, in elliptical galaxies. BL Lac can therefore probe (massive) spheroids to large redshifts. Results from an HST WFPC2 survey of ~ 100 BL Lac objects are here presented.

1. Introduction: The Range of Radio-Loud AGN

While AGN are clearly unified through orientation (Antonucci 1993; Urry & Padovani 1995), important intrinsic differences remain. For example, extended radio lobes form only when the radio power exceeds a threshold that increases with galaxy luminosity (Ledlow & Owen 1996, Bicknell 1995). Powerful FRII radio galaxies (defined by their lobe morphologies; Fanaroff & Riley 1974) correspond to the most luminous quasars, while lower luminosity FRI radio galaxies correspond to BL Lac objects (Urry & Padovani 1995).

At any given redshift z, the full range of luminosity needs to be explored in host galaxy studies, to separate trends in host galaxy properties with nuclear AGN luminosity from a possible redshift dependence.

2. Our HST Survey of 100 AGN

We observed over 100 BL Lac objects with the HST WFPC2, mostly in the F702W filter. These low-luminosity AGN span the redshift range $0 \lesssim z \lesssim 1.4$, and so can be compared directly to the many quasars observed with HST over the same range. Host galaxies were detected in almost all BL Lacs with $z < 0.5$, but in only about 20% of those at higher redshift.

HST adds critical morphological information, since galaxy radii between a few tenths and a few arcseconds from the nucleus can be investigated. In nearly all cases an elliptical morphology ($r^{1/4}$ law) is statistically preferred over an exponential disk. Both statistical and systematic errors were carefully assessed and in general the latter dominate (Urry *et al.* 1999a,b).

3. Results: Luminous Elliptical Host Galaxies

BL Lac host galaxies are luminous ellipticals with average K-corrected absolute magnitude $M_R \sim -23.8$ mag, i.e., ~ 1 mag brighter than L_*, comparable to FRI radio galaxies

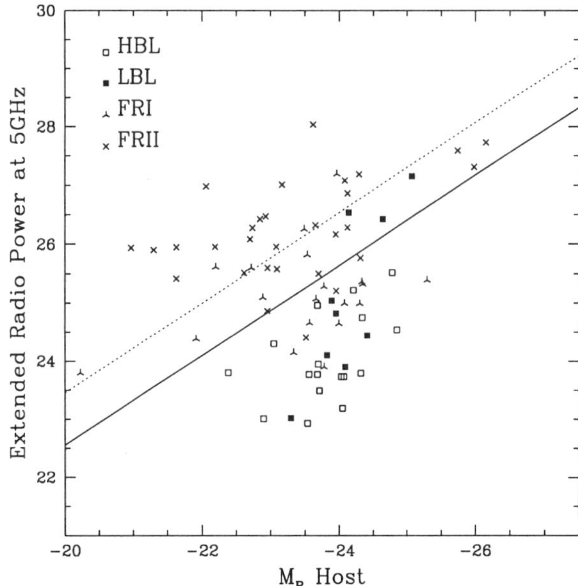

FIGURE 1. Extended radio power versus host galaxy luminosity for radio galaxies from the complete 2 Jy sample (Wall and Peacock 1985; Morganti *et al.* 1993), and BL Lac objects. The dotted and solid lines are two different models for the FRI/FRII dividing line; these two models span the assumed parameter space in the work of Bicknell (1995). Basically, the division of the two morphological types of radio galaxy is a function of the increasing jet-power needed to overcome the 'stopping' power of increasingly massive galaxies. The BL Lac host galaxies and jet-powers are completely consistent with those of FRI radio galaxies. The combination of relatively low radio power and massive host galaxy means that the jets are decelerated before they form extended radio lobes.

and $\lesssim 0.5$ mag fainter than the brightest cluster galaxies. Their average effective radius is $r_e \sim 10$ kpc. We find no dependence of host galaxy properties on the physical state of the BL Lac jets (HBL or LBL, defined by Padovani & Giommi 1995; see also Figures 1, 2 and 3). A number of BL Lacs have close companions and/or appear to be in poor groups or clusters, consistent with unification with FRIs. With somewhat limited multicolor information for a few objects, we see only old stellar populations though we are not sensitive to very recent (blue) star formation.

Surprisingly, the $\mu_e - r_e$ relation for BL Lac hosts (Figure 4) is statistically indistinguishable from what is found for non-active spheroids. This result indicates their structure is unaffected by the strong nuclear activity.

REFERENCES

ANTONUCCI, R.R.J. 1993 *ARA&A*, **31**, 473

BICKNELL, G. 1995 *ApJS*, **101**, 29

FANAROFF, B.L., RILEY, J.M. 1974 *MNRAS*, **167**, 31

LEDLOW, M.J., OWEN, F.N. 1996 *AJ*, **112**, 9

MCLEOD, K.K., RIEKE, G.H. 1995 *ApJ*, **454**, L77

MORGANTI, R., KILLEEN N.E.B., TADHUNTER, C.N. 1993 *MNRAS*, **263**, 1023

PADOVANI, P., GIOMMI, P. 1995 *MNRAS*, **277**, 1477

SILK, J., REES, M.J. 1998 *A&A*, **331**, L1

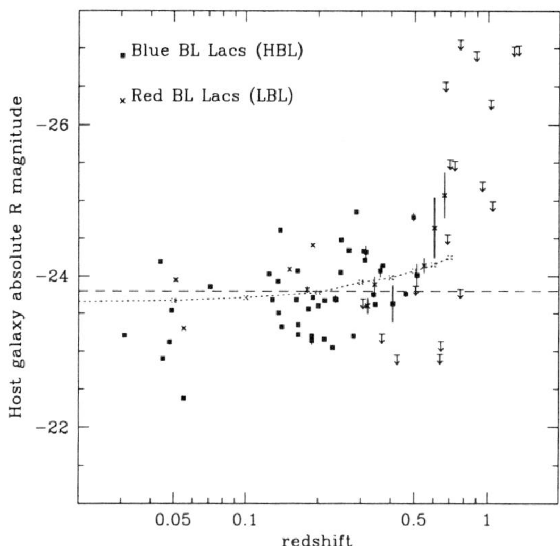

FIGURE 2. Host galaxy magnitude (extinction- and K-corrected) versus redshift for BL Lacs with known redshift. The average absolute magnitude is $M_R \sim -23.8$ mag (dashed line), with relatively small scatter. All $z < 0.3$ objects were resolved, while for $z > 0.5$, longer integration times are needed to detect host galaxies of similar magnitude. The dotted line gives the expected variation in magnitude for a passively evolving elliptical galaxy.

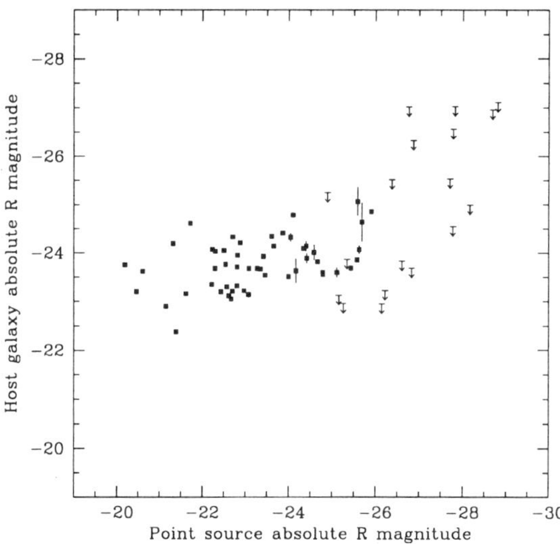

FIGURE 3. Nuclear versus host galaxy luminosity. A correlation was reported for high-luminosity quasars (McLeod & Rieke 1995; cf. Taylor *et al.* 1996) but we see no significant correlation in our lower luminosity sample.

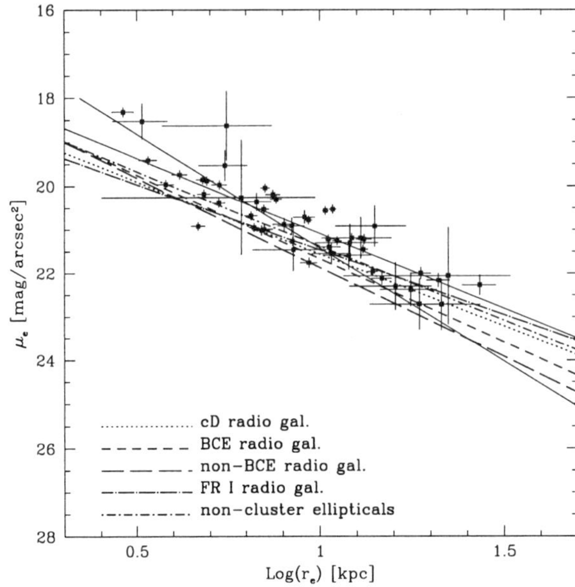

FIGURE 4. The surface brightness vs. effective radius relation for BL Lacs follows the well-known trend for elliptical galaxies and is very similar to that of brightest cluster galaxies. The minimum and maximum acceptable slope for BL Lac (solid lines) is plotted and compared with the relation found for several types of AGNs.

TAYLOR, G.L., DUNLOP, J.S., HUGHES, D.S., ROBSON E.I. 1996 *MNRAS*, **283**, 930

URRY C.M., FALOMO R., SCARPA R., PESCE J.E., TREVES A., GIAVALISCO M. 1999a *ApJ*, **512**, in press

URRY, C.M., SCARPA R., FALOMO R., O'DOWD, M., GIAVALISCO M., PESCE J.E., TREVES A. 1999b, in preparation

WALL, J.V., PEACOCK, J.A. 1985 *MNRAS*, **216**, 173

The Centers of Radio-Loud Early-Type Galaxies with HST

By GIJS A. VERDOES KLEIJN[1], S. A. BAUM[1]

AND

P. T. de ZEEUW[2]

[1]Space Telescope Science Institute, 3700 San Martin Drive, Baltimore, MD 21218, USA

[2]Leiden Observatory, Postbus 9513, Leiden, 2300 RA, The Netherlands

We briefly discuss the properties of radio-loud spheroids, and present the first results from a HST/WFPC2 imaging survey of a sample of nearby Fanaroff-Riley-I nuclei.

1. Introduction: Radio Emission from Spheroids

Radio emission is observed from the centers of both active spiral bulges and E/S0 galaxies. There are distinct differences in the properties of the central radio emission from these classes of galaxies (Slee *et al.* 1994; Sadler *et al.* 1995). Spirals sometimes contain compact radio cores, possibly not related to starburst activity, but a large fraction of the emission originates from an extended region of several hundred parsecs. In early-type galaxies the emission is always completely dominated by the unresolved core. The spectral index α of the core emission (with $S \sim \nu^{\alpha}$ for flux density S and frequency ν) is typically around −1 for spirals and around 0.3 for ellipticals. These differences appear to hold for bulges and early-type galaxies of the same luminosity. Thus, rather surprisingly, it seems that radio cores 'know' what kind of host they reside in.

There is also a difference in radio emission between low- and high-luminosity ellipticals. Only in high-luminosity ellipticals do we see radio-jets on the scales of hundreds of kiloparsecs, i.e., galaxies classified as FRI or FRII (Fanaroff & Riley 1974, types I and II, respectively). As noted by Sadler (1997) this threshold roughly coincides with the break which marks differences in structural properties such as stellar rotation and central cusp slope (e.g., Faber *et al.* 1997). Spheroids with different dynamical properties also have different radio properties.

It is now generally believed that the energy of the radio-jets and core of FR galaxies is generated by accretion of matter onto a central supermassive black hole (SMBH; e.g., Rees 1984). With the rapidly increasing number of detections of SMBHs in centers of both 'normal' and active galaxies (e.g., van der Marel 1998; Richstone *et al.* 1998, for reviews) it appears plausible that this process is also responsible for the radio emission in lower luminosity spheroids. The three main ingredients for this process are fuel, an accretion disk and a black hole. How then do the properties of these ingredients differ from galaxy to galaxy to produce the different forms of radio emission and how is this connected to the global properties and formation history of the host galaxy?

FRI galaxies have spheroids with large total luminosities, so they are expected to have large black hole masses ($\sim 10^{8-9} M_{\odot}$) (Kormendy & Richstone 1995). A few FRIs have now been observed with HST. Its high spatial resolution proved necessary to (i) resolve small nuclear disks of gas and dust (the signs of the fuel supply) discovered in some of these galaxies, and (ii) to sample the kinematics of the central emission-line gas to determine for instance black hole masses. To investigate the morphology, physical conditions and kinematics of the nuclear dust and gas distributions and to analyse the central stellar populations and dynamics of radio-loud nuclei with HST in a systematic way, we constructed a radio flux-limited nearby sample of 20 galaxies (recession velocity < 7000 km s^{-1}). FRI type radio jets were detected in 19 of these (Xu *et al.* 1999 and references therein). Observations with WFPC2 in V and I and in narrow bands centered on the Hα+[NII] emission lines will be completed in February 1999. Here we describe some preliminary results of the WFPC2 survey (see Verdoes Kleijn *et al.* 1999 for the analysis of the complete set of WFPC2 observations).

2. Central Properties of Nearby FRIs

We have detected dust in 16 of the 18 galaxies observed to date. This fraction agrees with previous findings for radio-loud galaxies, and is more than 2.5 times larger than found for normal early-type galaxies (Van Dokkum & Franx 1995). There is a large variety in dust morphology, ranging from irregular dust patches and filaments to nuclear dust disks hundreds of parsecs in size (Figure 1). In 12 out of 14 galaxies the orientation of the dust is consistent with being roughly perpendicular to the radio jets, as commonly seen in radio-loud galaxies at all redshifts (Kotanyi & Ekers 1979; De Koff *et al.* 1999). The orientation is off by more than $40°$ in the remaining 2 galaxies. Dust masses inferred from extinction range between 10^4 M$_\odot$ and 10^5 M$_\odot$. All observed galaxies have detectable nuclear emission-line gas. In addition low surface-brightness emission is commonly associated with the dust. Total emission luminosities vary between 10^{39} ergs^{-1} and 2×10^{40} ergs^{-1} ($H_0 = 75$ km s^{-1} Mpc^{-1}). Quite often determination of the luminosity profile in the inner few arcseconds is hindered by dust. However, the luminosity profile of sample galaxies without central dust obscuration invariably flattens off to a shallow core, consistent with the large luminosities. Isophotes are either 'boxy' or almost perfect ellipses. The ellipticity of the isophotes increases with radius in the inner $15''$ of nearly all galaxies, with the increase sometimes confined to a well-defined radial range. The position angle of the isophotes twists by more than $10°$ in 3 cases.

To illustrate the variety of central properties we briefly discuss four galaxies which are representative of the sample (see Figure 1 and Table 1). NGC 315 contains an inclined disk which is aligned with the major axis of the galaxy and has an apparent major axis of $\sim$800 pc. The patches of dust at $\sim 2.5''$ to the right of the disk and a little protuberance at the left end of the disk indicate that not all the dust has settled yet. A small disk of emission gas forms the inner part of the dust disk. If the dust disk is intrinsically circular it is inclined by $76°$. The larger attenuation of stellar light on the lower side of the dust disk suggests this side of the disk is closest to the observer. The two-sided radio-jet is perpendicular to the dust disk. The radio emission from the side of the jet extending upwards from the disk is many times stronger, suggesting this side of the jet is beamed towards us. The disk in NGC 383 has a diameter of 1.1 kpc, is almost face-on, and shows a counter clock-wise spiral structure which becomes flocculent at the edge. The difference in obscuration of starlight over the disk suggests the right side is nearer to the observer. The one-sided radio jet extends up suggesting again we only see the beamed sight of the jet. No dust is detected in NGC 2892. The jet is two-sided with comparable radio luminosities from both sides. NGC 4335 displays large dust lanes, $\sim 15''$ in extent, sweeping across the nucleus. They seem to settle down towards the center in a plane perpendicular to the two-sided radio jet.

3. Next Steps

Our isophotal results sofar suggest that many of the host galaxies are quite similar. They are typical high-luminosity ellipticals with an increased amount of gas and dust in

(a): NGC 315   **(b)**: NGC 383

(c): NGC 2892   **(d)**: NGC 4335

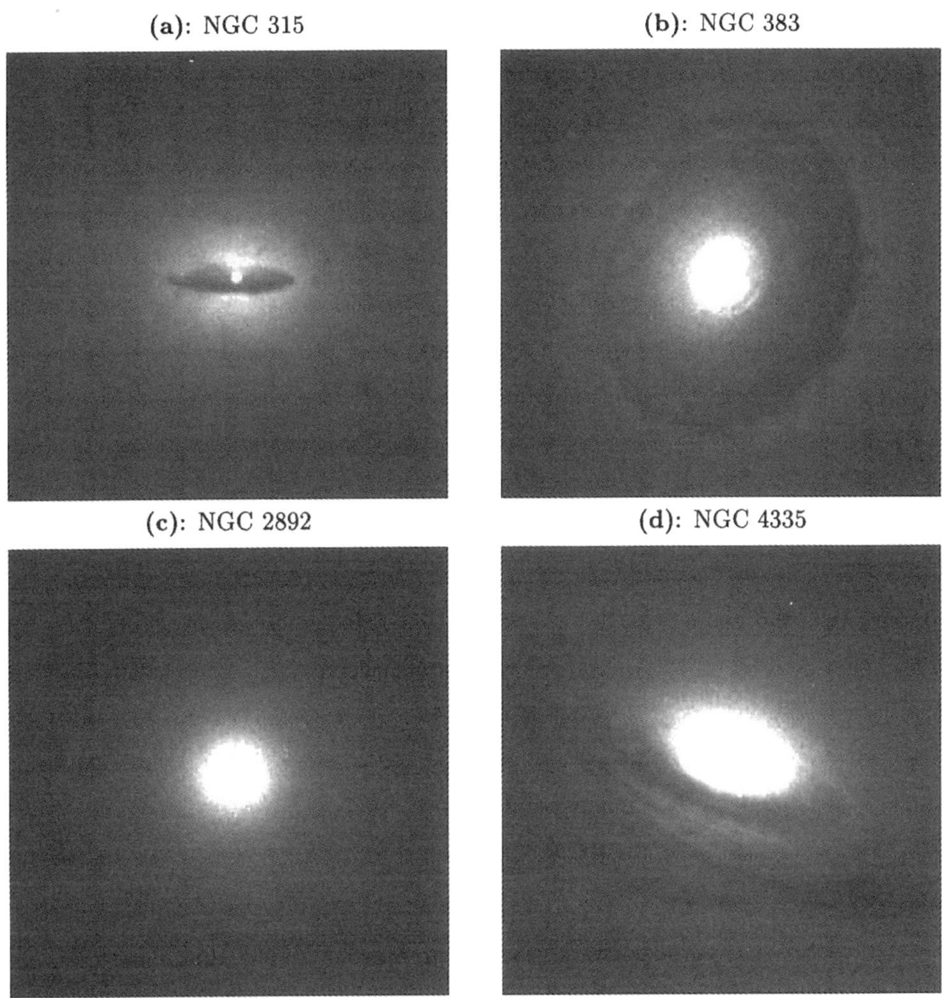

FIGURE 1. $10''\times10''$ V band images of the centers of 4 sample galaxies. All images are rotated such that the apparent orientation of the large-scale radio jet is vertical.

Name	Type	D (Mpc)	M_p (mag)	L_{radio} (log WHz^{-1})
(1)	(2)	(3)	(4)	(5)
NGC 0315	E	67.9	-21.8	24.19
NGC 0383	S0	65.2	-20.6	24.64
NGC 2892	E	90.8	-20.4	23.62
NGC 4335	E	61.5	-20.3	23.20

TABLE 1. Properties of the galaxies: Name, Hubble classification, distance, absolute photographic magnitude and total radio luminosity at 1400 MHz.

various morphologies and anomalous variations in the central ellipticity profile. It will be interesting to determine the full range of central stellar dynamics for the complete sample. A narrow range would indicate a tight relation between the processes that shape the spheroid and those that produce the radio jets. This could indicate that the typical properties of the SMBH (and possibly also of the fuel and accretion) which lead to the formation of a FRI radio jet are present in only a specific class of galaxies. On the other hand this might indicate that the ingredients themselves affect the central dynamics. A clear example of this is provided by recent studies on the influence of a SMBH on the central stellar dynamics (e.g., Merritt & Quinlan 1998). A related question is whether the central structure of FRI nuclei differs from the nuclei of 'normal' ellipticals with similar luminosities. For instance, do the latter nuclei show the same anomalies in ellipticity?

We will also look for correlations between the onset of activity and the infall of fuel by comparing the morphology of dust and gas to the properties of the radio jets from parsec to kiloparsec scales. We will determine the relative orientations in the disk-jet systems, check for the presence of warps and put constraints on beaming of the radio emission. In Cycle 8 we have been awarded STIS spectroscopic observations of the centers of 18 of our 20 galaxies. The remaining 2 are in the archive. We will model the kinematics of the emission lines to determine the mass of the SMBH. The various line ratios will constrain the ionization mechanism (e.g., shocks and/or photo-ionization). This systematic effort should significantly clarify the nature of active galactic nuclei.

REFERENCES

DE KOFF, S., ET AL. 1999 *ApJ*, submitted

FANAROFF, B.L., RILEY, F.M. 1974 *MNRAS*, **167**, 31

KORMENDY, J., RICHSTONE, D. 1995 *ARA&A*, **33**, 581

FABER, S.M., ET AL. 1989 *ApJS*, **69**, 763

KOTANYI, C.G., EKERS, R.D. 1979 *A&A*, **73**, L1

MERRITT, D., QUINLAN, G.D. 1998 *ApJ*, **498**, 625

REES, M. 1984 *ARA&A*, **22**, 471

RICHSTONE, D., ET AL. 1998 *Nature*, **395**, 14

SADLER, E.M. 1997, in *The Nature of Elliptical Galaxies* (ed. M. Arnaboldi, G.S. Da Costa & P. Saha), ASP Conf. Ser. **115**, p411. (ASP)

SADLER, E.M., SLEE, O.B., REYNOLDS, J.E., ROY A.L. 1995 *MNRAS*, **276**, 1373

SLEE, O.B., SADLER E.M., REYNOLDS J.E., EKERS R.D. 1994 *MNRAS*, **269**, 928

VAN DER MAREL, R.P. 1998, *IAU Symp.* **186**, p102. (Kluwer)

VAN DOKKUM, P.G., FRANX, M. 1995 *AJ*, **110**, 2027

VERDOES KLEIJN, G.A., BAUM, S.A., DE ZEEUW, P.T. 1999, in preparation

XU, C., ET AL. 1999, in preparation

Central UV Spikes in Two Galactic Spheroids†

By MICHELE CAPPELLARI[2], F. BERTOLA[2],
D. BURSTEIN[3], L. BUSON[4], LAURA GREGGIO[5,6],
AND
A. RENZINI[7,8]

[2]Dipartimento Astronomia, Università di Padova, Vicolo Osservatorio 5, I-35122 Padova, Italy

[3]Department of Physics & Astronomy, Arizona State University, Tempe AZ 85287-1504, USA

[4]Osservatorio di Capodimonte, Via Moiariello 16, I-80131 Napoli, Italy

[5]Osservatorio di Bologna, Via Ranzani 1, I-40127 Bologna, Italy

[6]Universitäts Sternwarte, Scheinerstr. 1, D-81679 München, Germany

[7]Dipartimento di Astronomia, Università di Bologna, Via Zamboni 33, I-40126 Bologna, Italy

[8]European Southern Observatory, Karl-Schwarzschildstr. 2, D-85748 Garching, Germany

FOS spectra and FOC photometry of two centrally located, UV-bright spikes in the elliptical galaxy NGC 4552 and the bulge-dominated early spiral NGC 2681, are presented. These spectra reveal that such point-like UV sources detected by means of HST within a relatively large fraction ($\sim 15\%$) of spheroids can be related to radically different phenomena. While the UV unresolved emission in NGC 4552 represents a transient event likely induced by an accretion event onto a supermassive black hole, the spike seen at the center of NGC 2681 is not variable and it is stellar in nature.

1. Introduction

HST UV images of nearby galaxies presented by Maoz *et al.* (1996) and Barth *et al.* (1998), as well as analogous space-borne optical images of early-type galaxies discussed by Lauer *et al.* (1995) and Carollo *et al.* (1997) have shown that about 15% of imaged galaxies show evidence of unresolved central spikes.

In the following we discuss two 'prototype' galactic spheroids, NGC 2681 and NGC 4552, that we properly monitored with HST—which host UV-bright, unresolved spikes at their center. While the early-spiral (Sa) galaxy NGC 2681 shows a *nonvariable* unresolved cusp, the UV spike which became visible at the center of the Virgo Elliptical NGC 4552 is a UV flare caught in mid-action, presumably related to a transient accretion event onto a central supermassive black hole (Renzini *et al.* 1995; Cappellari *et al.* 1998).

Although radically different phenomenologies are involved, the appearance of either nuclei—recently imaged in the UV (FOC/96 F342W) by means of the refurbished HST— is quite similar. Nevertheless, basic pieces of information can still be extracted from photometric profiles alone which represent a potential diagnostics to disentangle the above scenarios. For instance, the UV-bright unresolved spike observed at the center of NGC 2681 *does not* vary and matches a pure Nuker-law profile of the power-law type (Cappellari *et al.* 1999). On the contrary, in order to model the flaring UV spike at the center of NGC 4552 one has to add to the observed galaxy profile the contribution of an unresolved central point source, whose intensity is allowed to vary (see Figure 1).

† Based on observations with the NASA/ESA Hubble Space Telescope, obtained at the Space Telescope Science Institute, which is operated by AURA, Inc., under NASA Contract NAS 5-26555.

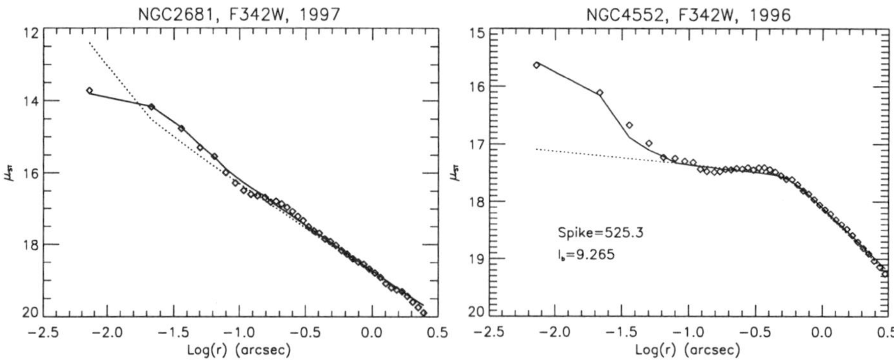

FIGURE 1. The inner surface brightness profiles (arbitrary magnitudes) of NGC 2681 (left) and NGC 4552 (right) vs. log r in the FOC/96 F342W waveband. Diamonds represent the observed profiles, the dashed lines represent models of the true galaxy profiles and finally the solid line shows the above models after convolving with the proper PSF. NGC 2681 is a typical power-law bulge, while NGC 4552 is a classic giant elliptical with core, but one needs to add a central point-like source to the Nuker-law in order to reproduce the observed profile.

2. Observations and Reductions

FOC UV observations of NGC 4552 obtained in 1991, 1993 and 1996 are described in detail by Cappellari *et al.* (1998). These data include a single FOC/96 F342W frame obtained on July 19, 1991 and subsequent images obtained on November 27-28, 1993 in four consecutive UV passbands (FOC/96 F175W, F220W, F275W, F342W). We observed NGC 4552 for a third time on May 24, 1996 with COSTAR-Corrected HST making use of a comparable set of UV filters as in 1993 (FOC/96 F175W, F275W and F342W). Initial FOC images of NGC 2681 were obtained by our group on November 4-5, 1993 in the FOC F175W, F220W, F275W, and F342W filters, pre-COSTAR. As with NGC 4552, we also obtained a set of post-COSTAR UV images on February 1, 1997 of NGC 2681 with the same FOC filter set (apart from F220W) as used in 1993. All FOC images have been re-calibrated in a self-consistent manner, including all required correction factors for PSF and sensitivity differences (zoom/non-zoomed modes and COSTAR) as well as nonlinearity effects. In addition to the FOC images obtained in 1996, we were also able to obtain FOS spectra of both galaxies. The FOS peak-up procedure was used to locate the $0.''2$ square aperture on the nucleus of each galaxy (as confirmed via the multiple peak-up output). FOS gratings G270H, G650L and G780H were used for each galaxy. The nuclear spectra of NGC 4552 and NGC 2681 were obtained on May 24, 1996 and on February 2, 1997, respectively.

3. Results

The ultraviolet-bright source in NGC 4552 was first detected in 1991, it increased in luminosity by a factor of $\sim$ 4.5 by 1993, and then declined a factor of $\sim$ 2.0 by 1996. On the contrary the 1993 and 1997 UV FOC observations of NGC 2681 are consistent with no variation at all.

The overall nuclear FOS spectra of NGC 2681 and NGC 4552, together with the IUE and optical underlying spectra normalized to the visual region, are shown in Figure 2. In the case of NGC 2681 the match of the two spectra is quite striking, thus implying that the UV continuum flux of NGC 2681 is simply dominated by its stellar population, essentially identical in its innermost regions and in the whole $10''\times20''$ IUE aperture. On the other hand, FOS spectroscopy of NGC 4552 reveals a strong UV continuum

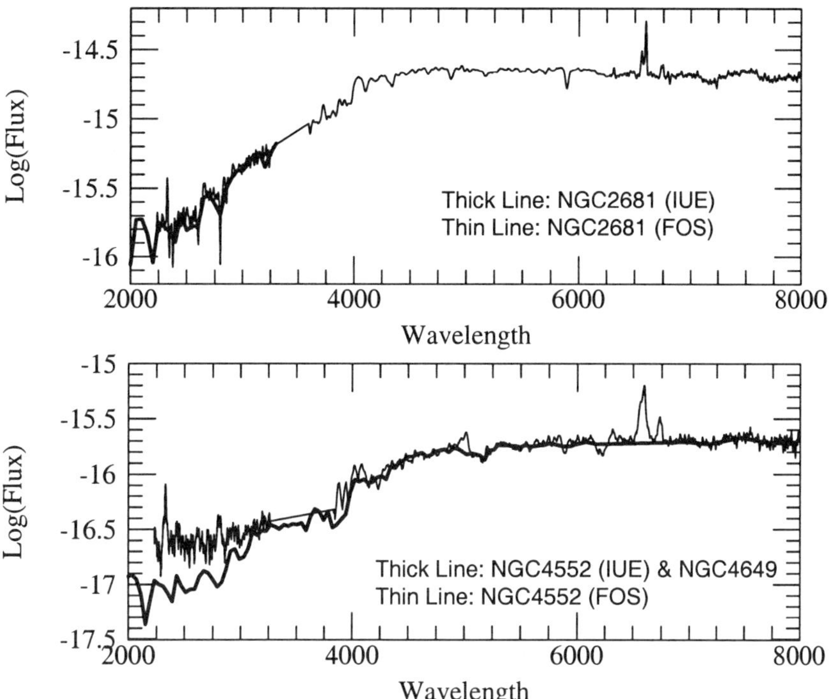

FIGURE 2. *Upper panel:* The overall 1997 FOS spectrum of NGC 2681 within the $0''\!.2 \times 0''\!.2$ aperture centered on the spike (thin line), is superimposed to the IUE spectrum of the same galaxy within a $10'' \times 20''$ aperture (Burstein *et al.* 1988), properly normalized to the visual region. *Lower panel:* The 1996 FOS spectrum of NGC 4552 centered on the spike (thin line), is superimposed to a scaled combination of the IUE spectrum of the same galaxy matched to ground-based optical spectrum of NGC 4649, a giant elliptical whose SED is virtually the same as that of NGC 4552 (thick line). The spectra have been normalized to the visual region. The NGC 4552 spectrum appears quite different owing to a continuum UV excess shortward of $\lambda \sim 3000$ Å. This UV excess is absent in NGC 2681.

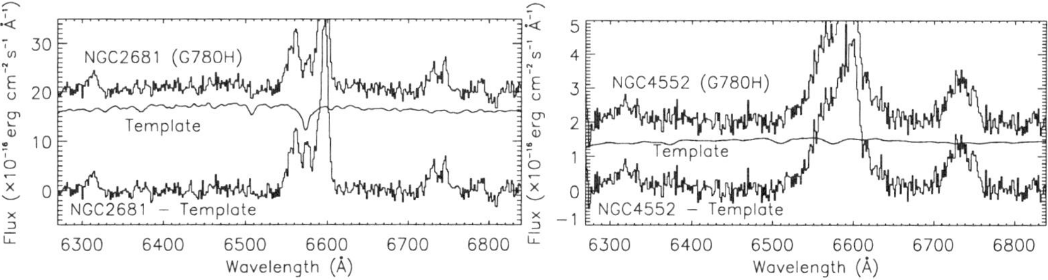

FIGURE 3. The FOS G780H spectrum of NGC 2681 (left panel) and NGC 4552 (right panel), showing the region of the [O I], [N II], Hα, and [S II] emission lines. In both panels a starlight template (obtained as in Ho *et al.* 1997) has been subtracted from the upper plot spectrum to obtain the continuum-free spectrum of the lower plot.

over the spectrum of the underlying galaxy, along with several emission lines in both the UV and the optical ranges. The SED of the spike alone—obtained by subtracting the V-mag normalized IUE spectrum of the galaxy from the FOS spectrum—indicates a temperature of $T \sim 15000$ K for the spike in 1996, if a thermal origin for the UV flux is assumed.

The FOS G780H spectra of NGC 2681 and NGC 4552 including the [O I], [N II], Hα,

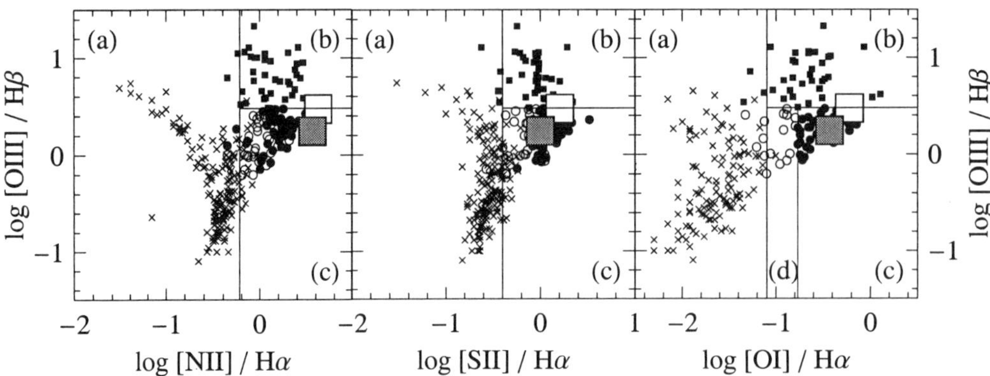

FIGURE 4. The location of the NGC 2681 (large gray square) and NGC 4552 (large open square) nucleus (as derived from the narrow line emission components measured on the FOS spectra) on the diagnostic diagrams used by Ho *et al.* (1997). The corresponding errors are of the size of the smaller simbols. The other symbols represent the nuclei included in the Ho *et al.* sample (crosses = H II nuclei, filled squares = Seyfert nuclei, filled circles = LINERs, open circles = transition objects). The vertical and horizontal lines delineates the boundary adopted by Ho *et al.* between (a) H II nuclei, (b) Seyfert galaxies, (c) LINERS and (d) Transition objects.

and [S II] emission lines are presented in Figure 3. In NGC 2681 all lines are well fitted by a single gaussian component with FWHM of ~ 470 km s^{-1}. In the case of NGC 4552, however, both permitted and forbidden lines are best modelled with a combination of broad and narrow components, with FWHM of ~ 3000 km s^{-1} and ~ 700 km s^{-1}, respectively. The 1996 broad Hα luminosity of this mini-AGN is $\sim 5.6 \times 10^{37}$ erg s^{-1}, about a factor of two less than that of the nucleus of NGC 4395, heretofore considered to be the faintest known AGN (Filippenko *et al.* 1993).

The FOS spectroscopy indicates also a significant similarity between the two nuclei, namely their emission line ratios and related gas diagnostics and UV-source classification. A comparison of the emission line ratios of the narrow components for both the NGC 4552 and NGC 2681 spikes with the distribution of Seyfert galaxies, LINERS and H II regions in the diagnostic emission line diagrams of Ho *et al.* (1997) is given in Figure 4. As is evident, the line ratios definitively place both spikes among extreme AGNs. The ratios for NGC 4552 fall just on the borderline between Seyferts and LINERs, while those measured for NGC 2681 indicate that this nucleus can be classified as a LINER.

REFERENCES

BARTH, A.J., HO, L.C., FILIPPENKO, A.V., SARGENT, W.L.W. 1998 *ApJ*, **496**, 133

CAPPELLARI, M., RENZINI, A., GREGGIO, L., DI SEREGO ALIGHIERI, S., BUSON, L.M., BURSTEIN, D., BERTOLA, F. 1999 *ApJ*, **519**, in press (astro-ph/9807063)

CAPPELLARI, M., BERTOLA, F., BURSTEIN, D., BUSON, L.M., GREGGIO, L., RENZINI, A. 1999, in preparation

CAROLLO, C.M., STIAVELLI, M., DE ZEEUW, P.T., MACK, J. 1997 *AJ*, **114**, 2366

FILIPPENKO, A.V., HO, L.C., SARGENT, W.L.W. 1993 *ApJ*, **410**, L75

HO, L.C., FILIPPENKO, A.V., SARGENT, W.L.W. 1997 *ApJS*, **112**, 315

LAUER, T.R., ET AL. 1995 *AJ*, **110**, 2622

MAOZ, D., FILIPPENKO, A.V., HO, L.C., MACCHETTO, F.D., RIX, H-W., SCHNEIDER, D.P. 1996 *ApJS*, **107**, 215

RENZINI, A., GREGGIO, L., DI SEREGO ALIGHIERI, S., CAPPELLARI, M., BURSTEIN, D., BERTOLA, F. 1995 *Nature*, **378**, 39

Part 6

CONFERENCE SUMMARY

Where Do We Stand?

By ROSEMARY F. G. WYSE

Department of Physics & Astronomy, The Johns Hopkins University, 3400 N. Charles Street,
Baltimore MD 21218, USA

I review the understanding of bulges that emerged from the lively discussions and presentations during the meeting, and emphasize areas for future work. The evidence is for a diversity of 'bulges', and of formation mechanisms.

1. What is a Bulge?

Classical bulges are centrally-concentrated, high surface density, three-dimensional stellar systems. Their high density could arise either because significant gaseous dissipation occurred during their formation, or could simply reflect formation at very high redshift (or some combination of these two, depending on the density). For illustration, equating the mean mass density within the luminous parts of a galaxy (assumed to have circular velocity v_c and radius r_c) with the cosmic mean mass density at a given redshift, z_f, gives (e.g. Peebles 1989)

$$z_f \sim 30 \, \frac{1}{f_c \Omega^{1/3}} \, (\frac{v_c}{250 \text{km/s}})^{2/3} \, (\frac{10 \text{kpc}}{r_c})^{2/3}, \tag{1.1}$$

where f_c is the collapse factor of the proto-galaxy, being at least the factor 2 of dissipationless collapse, and probably higher so that bulges, as observed, are self-gravitating, meaning that they have collapsed relative to their dark halos.

The majority view at the meeting, consistent with the observations, is that indeed proto-bulges radiated away binding energy, but also at least their stars formed at relatively high redshift. One must always be careful to distinguish between the epoch at which the stars now in a bulge formed, and the epoch of formation of the bulge system itself (as emphasized by Pfenniger, this volume). Of course if the bulge formed with significant dissipation, meaning gas physics dominated, then the star formation and bulge formation probably occurred together.

The small length-scale of bulges, combined with their modest rotation velocity, leads to a low value of their angular momentum per unit mass. Indeed, in the Milky Way Galaxy, the angular momentum distribution of the bulge is similar to that of the slowly-rotating stellar halo, and different from that of the disk, strongly suggestive of a bulge–halo connection, perhaps via gas ejection from halo star-forming regions (e.g. Wyse & Gilmore 1992). One can appeal to bulges forming from the low angular momentum regions of the proto-galaxy, a variant on the Eggen, Lynden-Bell & Sandage (1967) 'monolithic collapse' scenario, explored further by van den Bosch (1998 and this volume). Or one can posit angular momentum transport prior to the formation of the bulge, taking angular momentum away from the central regions, and depositing it in the outer regions. Such transport of angular momentum could perhaps occur during hierarchical merging, by dynamical friction and gravitational torques, although one must be careful not to end up with too small a disk due to over-efficient angular momentum re-arrangement (e.g. Zurek, Quinn & Salmon 1988; Navarro & Benz 1991; Navarro & Steinmetz 1997). More modest amounts of angular momentum transport may be achieved by some viscosity in the early disk (e.g. Zhang & Wyse 1999).

A recurring theme of the meeting was that large bulges (of early-type disk galaxies?)

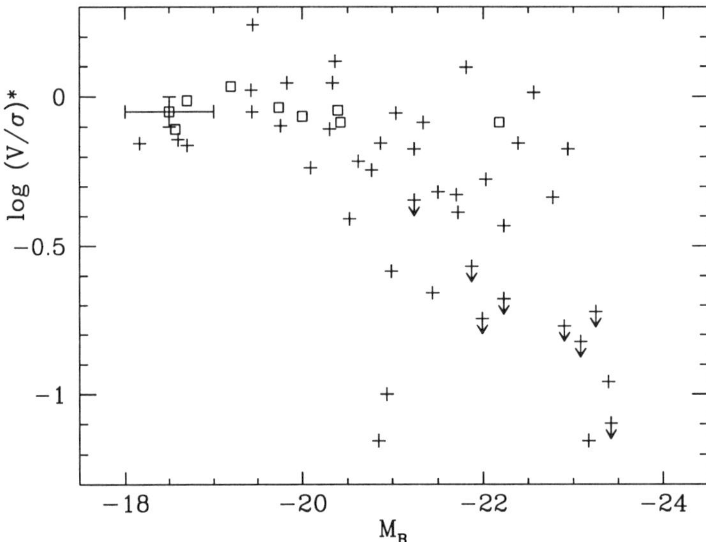

FIGURE 1. The level of rotational support as measured by $(V/\sigma)^*$, which has the value unity for an isotropic oblate rotator, against absolute magnitude for elliptical galaxies (crosses) and bulges of early-type spirals (open squares); data from Davies *et al.* 1983. The bulge of the Milky Way, with kinematic quantities and flattening estimated in a similar manner as for the external galaxies, is indicated by the point with error bars.

are related to ellipticals while small bulges (intermediate–late-type disk galaxies?) are more closely allied to disks. We need to be very clear about the observational selection criteria used in the definition of samples, and how this could bias our conclusions. As we will see below, the Milky Way bulge shows characteristics of *both* early- and late-type bulges, and will feature in *both* bulge–elliptical connections and bulge–disk connections.

1.1. *The Elliptical–Bulge Connection*

There has been remarkably little new kinematic data for representative samples of bulges (as opposed to detailed study of particular individual bulges, chosen for their unusual characteristics) since the pioneering work of the 1970s and 1980s. As demonstrated by Davies *et al.* (1983), the bulges of early-type spirals are like ellipticals of equal luminosity in terms of rotational support, and are consistent with being isotropic oblate rotators, i.e., with having an isotropic stellar velocity dispersion tensor, and being flattened by rotation about their minor axis. This sample was biased towards early-type spirals to facilitate bulge–disk decomposition, by observing edge-on systems with a prominent bulge. The bulge of the Milky Way Galaxy can be observed to match the techniques employed in the study of the bulges of external galaxies, and also then has stellar kinematics consistent with being an isotropic rotator (Ibata & Gilmore 1995a,b; Minniti 1996), as shown in Figure 1 here.

The trend apparent in Figure 1, and discussed more fully in Davies *et al.* (1983), is that the level of rotational support in ellipticals increases as the luminosity of the elliptical decreases. The surface brightness of ellipticals also increases with decreasing luminosity, at least down to the luminosity of M32 (the dwarf spheroidal galaxies are another matter), as noted by Kormendy (1977), Wirth & Gallagher (1984) and many subsequent papers. These two relations are consistent with an increasing level of importance of dissipation in ellipticals with decreasing galaxy luminosity (Wyse & Jones 1984).

Further, the bulges of S0-Sc disk galaxies follow the general trend of the Kormendy (1977) relations, in that smaller bulges are denser (de Jong 1996; Carlberg, this volume; see Figure 3 below for details). Thus one interpretation of Figure 1 is then that (some) bulges too formed with significant dissipation.

As discussed by several speakers, the bulges of S0-Sc disk galaxies have approximately the same Mg2 – velocity dispersion relation as do ellipticals (Jablonka *et al.* 1996; Idiart *et al.* 1997; see Renzini, this volume), although the actual physics behind this correlation is not uniquely constrained. The properties of line-strength gradients in ellipticals of a range of luminosities are consistent with lower luminosity ellipticals forming with more dissipation than the more luminous ellipticals (Carollo, Danziger & Buson 1993). Again these results are suggestive that bulges are similar to low-luminosity ellipticals, and that gas dissipation was important.

The detailed interpretation of the line-strength data in terms of the actual age and metallicity distributions of the stars is extremely complex and as yet no definitive statements can be made. There is a clear need for more data, including radial gradients, and for more models (see Trager *et al.*, this volume). The broad-band colors of (some) bulges are consistent with those of the stellar populations in early-type galaxies in the Coma cluster (Peletier & Davies, this volume). We still need better models to interpret even broad-band colors.

1.2. *The Disk–Bulge Connection*

The surface-brightness profiles of bulges in later-type disk galaxies are better fit by an exponential law than by the steeper de Vaucouleurs profile, which in turn is a better fit for the bulges of early-type disk galaxies (Andredakis, Peletier & Balcells 1995; de Jong 1996). The sizes of bulges are statistically related to those of the disks in which they are embedded, and indeed the (exponential) scale-lengths of bulges are around one-tenth that of their disks; this correlation is better for late-type spirals than for early types (Courteau, de Jong & Broeils 1996). The projected starlight of the bulge of the Milky Way can be reasonably well-approximated by exponentials (vertically and in the plane); the Milky Way then fits within the scatter of the correlation of the external galaxies.

The optical colors of bulges are approximately the same as those of the inner disk, for the range of Hubble types S0-Sd (Balcells & Peletier 1994; de Jong 1996), but as ever the decomposition of the light curves is difficult, as is correction for dust. This correlation implies similar stellar populations in bulges and their disks, as may be expected if bulges form from their disks (see Pfenniger, this volume). Thus, should there be a variation of mean stellar age from disk to disk, as may be expected from the range of colors observed, and indeed from observations of gas fraction etc., together with models of star formation in disks, one would expect a corresponding range in the mean stellar age of the different bulges. However, Peletier & Davies (this volume) find only a narrow range in bulge ages for their sample, based on optical–IR colors. More data are clearly needed.

Figure 1 demonstrated the similarity in their kinematics between bulges and ellipticals of the same luminosity; Figure 2 (taken from Franx 1993) illustrates some of the complexity of bulge kinematics, and emphasizes the need to be aware of the selection criteria – not all bulges are the same. The left-hand panel shows that in terms of the ratio of stellar velocity dispersion to true circular velocity (not the rotational streaming velocity), bulges scatter below ellipticals. Further, the right-hand panel shows that bulges of late-type disk galaxies have values of this ratio similar to that typical of inner disks (from Bottema 1993). The Milky Way bulge in this plot is quite typical ($\sigma/V_c \sim 0.5$, $B/T \sim 0.25$).

Complexity in the relationship between surface brightness and scale-length for bulges

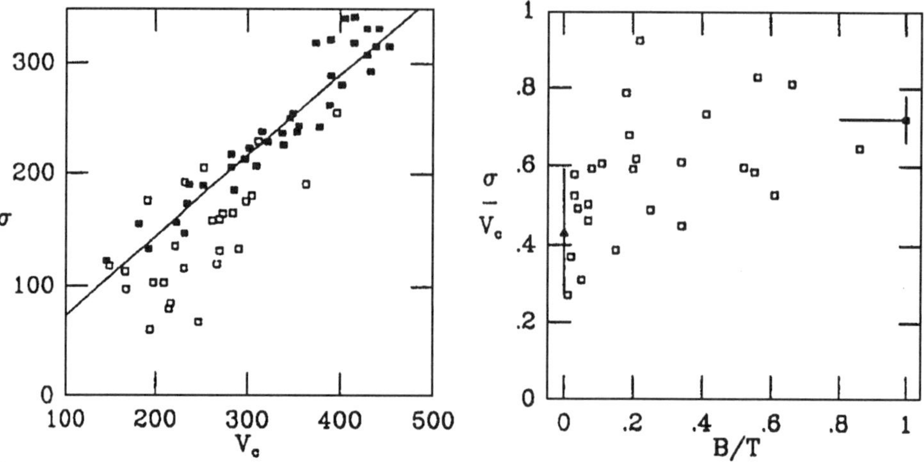

FIGURE 2. *a*) The central velocity dispersion of stellar tracers, σ, against dark halo circular velocity, v_c. Open symbols represent bulges; closed symbols represent ellipticals. Circular velocities for the ellipticals are derived from models, as described by Franx (1993). (*b*) The ratio of velocity dispersion in the bulge to dark halo circular velocity, σ/v_c, taken from Franx (1993), plotted as a function of bulge-to-total luminosity (B/T) ratio, for the entire range of Hubble Type. The triangle at left is valid for the inner regions of pure disks, the square at right for ellipticals. Note that systems with low B/T have kinematics almost equal to those of inner disks.

is illustrated in Figure 3, based on WFPC2 data from Carollo (1999). The plot shows that while the large, $R^{1/4}$-law bulges follow the same scaling as ellipticals, the smaller, exponential-profile bulges are offset to lower surface brightnesses and occupy the extension to smaller scalelengths (by about a factor of ten, as noted above) of the locus of late-type disks. This strengthens the disk–bulge connection for these small bulges. However, Carollo (1999) finds both $R^{1/4}$ and exponential bulges in apparently very similar disks, so some additional parameter is important.

Association of 'peanut' bulges with bars, which are essentially a disk phenomenon, was made in several contributions, using both gas and stellar kinematics (Kuijken, Bureau, this volume). However, the pronounced 'peanut' in the early COBE images of the Milky Way was apparently largely an artefact of patchy dust, and the amplitude of such a morphology in the bulge of the Milky Way is not reliably established (Binney, Gerhard & Spergel 1997). As emphasized by Pfenniger (this volume), the kinematical and dynamical effects of bars are 3-dimensional; they can scatter stars by resonances, and/or themselves go unstable, fatten and dissolve, leading to a bulge. Which process dominates? There is a wealth of fascinating physics to explore. The modellers need to make more contact with observations, including predictions for direct comparison with the stellar kinematics, ages of stars, surface brightness profiles etc.

M33 has neither a bulge nor a bar, but does have a central nucleus, and of course a substantial disk. Such systems need to be discussed in this context. The central nucleus of the Milky Way contains a black hole and star clusters of mass fraction well below the 1% or so estimated to destroy a bar (Norman, Sellwood & Hasan 1996), if we associate all the $10^{10} M_\odot$ of the bulge with the bar. Indeed it is somewhat of a curiosity that the

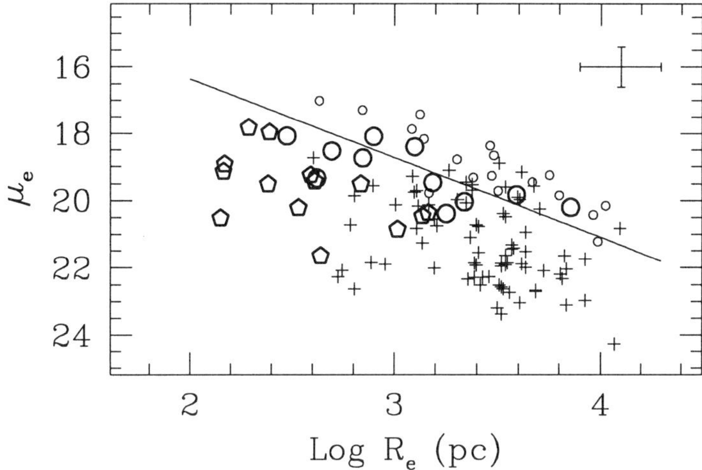

FIGURE 3. The mean V-band surface brightness μ_e within the half-light radius R_e, as a function of $\log R_e$ (in pc). The WFPC2 measurements are shown with pentagons for the exponential bulges and large circles for the 'classical' $R^{1/4}$-law bulges. Comparison data from the literature are shown for the $R^{1/4}$ bulges from Bender *et al.* (1992; small circles) and Scd-Sm disks from Burstein *et al.* (1997; crosses). The solid line is the best fit to the elliptical galaxy sequence (data from Bender *et al.* 1992 and from Burstein *et al.* 1987). The typical 1-σ error bar for the WFPC2 masurements is shown in the upper-right corner.

Milky Way does not fit the relationship between black hole mass and bulge mass found by Magorrian *et al.* (1998).

2. When do Bulges Form?

The fossil evidence from Local Group galaxies constrains the ages of the stars presently in bulges, which could be rather different from the age of the morphological system.

The overwhelming evidence (contributions by Gilmore, Frogel, Renzini, Rich) for the Milky Way bulge is that its stars are old, except for a very small scaleheight young component – and since all components of the Galaxy have their peak surface brightnesses in the center,† this is as likely to be associated with the disk. Further, as discussed by Rich (this volume) the dominant stellar population even in the nuclear regions is apparently old.

The situation in external bulges in the local universe is more uncertain, but is consistent with stars in large bulges being 'old', which means forming perhaps 10Gyr ago.

Direct studies of morphology at high redshift require HST and are at present based on small samples and must be treated with caution if attempting to draw general conclusions. The Hubble Deep Field (HDF) has provided much of the field galaxy sample (as opposed to members of galaxy clusters). Recently, Abraham *et al.* (1999a) analysed the spatially-resolved colors of galaxies of known redshift in the HDF. In contrast to the case of cluster ellipticals discussed by Renzini (this volume), they find that almost half of their (small)

---

† Some of the decompositions of the COBE data have modelled the disk with a hole in the central regions, which I believe points to continuing uncertainty in the interpretation of those data in terms of the parameterizations of the different components along the line-of-sight.

sample of field ellipticals at intermediate redshift ($0.4 < z < 1$) show evidence for a range of stellar ages. The color gradients in the galaxies for which they could derive a reasonable bulge–disk decomposition are consistent with the mean stellar ages of the bulges being older than those of their disks. These authors argue that this presents difficulties for secular evolution models, but again one must remember the possible selection biases. Abraham *et al.* (1999b) further find a significant deficit of barred galaxies for redshifts above 0.5; as those authors note, more data for a wider range of rest-frame colors and redshifts are needed confirm this result, and then to decide on a robust interpretation. As discussed by Lilly (these proceedings), there is strong evidence from SCUBA data for the existence of compact galaxies with high star formation rates at high redshift, consistent with proto-spheroids forming in a starburst.

The age distribution of inner disks is of obvious importance for constraining scenarios of disk–bulge formation. Unfortunately, we do not know this well, even in the Milky Way. Indeed, we do not have a good understanding of the star formation history even at the solar neighborhood. We do know that out to a few kpc from the Sun there are stars in the thin and thick disks that are as old as the globular clusters (Edvardsson *et al.* 1993; Gilmore, Wyse & Jones 1995). The stellar color–magnitude data, the chemical abundances and the white dwarf luminosity function data are all broadly consistent with a local (solar neighborhood) star formation rate that has been approximately constant, back to ~ 12Gyr (e.g. Noh & Scalo 1990; Rocha-Pinto & Maciel 1997). Most models of star formation in disks predict that the central regions should evolve faster, and hence the mean stellar age should be older in the inner disk than in the outer disk. Thus perhaps indeed predominantly-old bulges can be formed recently, from old stars in the central parts of disks. But one really has to be careful to avoid a significant age range in the bulge, reflecting the continuing star formation in the disk up to the time of bulge formation.

Simulations of hierarchical clustering galaxy formation predict 'bulges' to form stars at redshift of $z \sim 2$ (peak) even if assembled later (Frenk, oral presentation at this meeting; Baugh, Cole, Frenk & Lacey 1998). In these scenarios, bulges (and ellipticals) form from mergers between pre-existing disk galaxies, and consist of a mix of the disk stars, plus, in some versions, new star formation in the central regions resulting from the disk gas being driven there during the merger. Disks are then (re-)accreted around these bulges. Thus bulges in galaxies with relatively big disks (i.e. Scs) should be the oldest bulges, and bulges with small disks should be the youngest (Baugh, Cole & Frenk 1996; Kauffmann 1996). This is not obviously consistent with the observations presented at the meeting.

A preliminary attempt to make detailed predictions and see if the 'bulges' in these models fit the observed scaling between size and luminosity was presented at the meeting by Lacey; the models did not include dissipation and failed to produce small enough bulges. This is further evidence that the high phase space densities of bulges require dissipation (cf. Wyse 1998).

3. What are the Timescales of Bulge Formation?

The finest time-resolution in studies of stellar populations is available from study of the patterns of elemental abundances in individual stars, as discussed in this volume by Renzini; the elemental signature of a short duration of star formation is a pattern of enhanced α-elements as produced by Type II supernovae alone only. The bulge of the Milky Way is surprisingly under-studied and we really do need more data for field stars; for the extant small sample, different α-elements show different patterns (McWilliam & Rich 1994), unexplained within the context of solely Type II supernovae yields (e.g.

Worthey 1998) or in comparison with the element ratios of stars in the stellar halo. It is worth noting however that the elemental abundances seen in the bulge field stars and in the bulge (or thick-disk?) globular clusters are consistent with a normal massive-star stellar IMF (cf. Wyse & Gilmore 1992), as also seen via star counts for the lower mass stars in Baade's window (Holtzman *et al.* 1998).

Color-magnitude diagrams of old populations can only constrain the duration of star formation to be less than many Gyr, due to the crowding of the isochrones (reflecting the long main-sequence lifetimes of low-mass stars). Further, one needs to know the metallicity distributions, and crucially for the Milky Way bulge, the distance distribution, since foreground disk stars are a difficult contaminant.

As mentioned above, hierarchical-clustering and merging scenarios predict a many Gyr spread in ages of bulge stars, but we need a better quantification of 'many'. And again, a significant age spread is predicted in the simpler secular evolution models, although the restriction to only one early disk–bar–bulge episode would minimise it.

The shortest durations of star formation are predicted by the starburst models wherein pre-assembled gas forms stars on only a few free-fall times, but the physics of the assembly of the gas will also play a role (Carlberg, this volume, who favors wind-regulated accretion of gas-rich satellites; Elmegreen, this volume, who favors unregulated, monolithic collapse). That very high star formation rates happened in some systems at high redshift is supported by the SCUBA observations (Lilly, this volume), but important aspects of the model obviously need to be worked out (e.g. is there or isn't there a dominant supernova-driven wind?)

4. Constraints from Physical Properties

4.1. *Angular Momentum Distributions*

The hierarchical-clustering and merging scenario predicts misalignment in the angular momentum vector of different shells of material around a peak. This may be expected to translate into some persistent misalignment between disk and bulge, and even counter-rotating components. While examples of such systems exist (see Bertola *et al.*, this volume), these would appear to be the exception rather than the rule (see Kuijken, this volume).

Quantification of the specific angular momentum distributions of disks and bulges is obviously desirable, but the observational determinations are dependent on not only detailed kinematic data, but also the decomposition of the light profile (and M/L). Note that in the Milky Way the determination of the kinematic properties of the bulge – and in particular any gradients – requires very careful treatment of contamination by the disk (see Ibata & Gilmore 1995a,b; Tiede & Terndrup 1997 for details). The extant theoretical predictions of angular momentum distributions of bulge and of disk are also not sufficient.

4.2. *Central Star Clusters and Bars*

'Secular evolution' models for forming bulges from inner disks naively predict an anti-correlation between significant central mass concentrations and bars, since in these models the clusters destroy the bar. There is a particular need to determine how many cycles of bar formation/dissolution are expected theoretically, and how many are allowed by the observations. The uniform old age of bulges, including that of the Milky Way, suggested by most of the evidence presented at this meeting (but again remember possible selection effects) argues strongly for only one such episode, and as noted by Gilmore (this volume), the disk must still continue into the central regions. The relative frequency of bars, exponential versus $R^{1/4}$ bulges, central star clusters etc. is as yet poorly quantified. The initial results of an HST WFPC2 and NICMOS imaging survey of nearby spiral galaxies (Carollo 1999) have revealed some of the complexity of the inner regions of these systems, finding a high fraction of photometrically-distinct compact sources sitting at the galactic centers. These 'nuclei' have surface brightnesses and radii ranging from those typical of the old Milky Way globular clusters to those of the young star clusters found in interacting galaxies (e.g. Whitmore *et al.* 1993; Whitmore & Schweizer 1995), with typical half-light radii of a few pc up to $\approx$ 20pc. Many of the nuclei are embedded in bulge-less disks or in bulge-like structures whose light distribution is too dusty/star-forming to be meaningfully modelled. Every exponential bulge was found to contain a nucleus, and further the luminosity of the nucleus was consistent with its being sufficiently massive to have destroyed a bar of the same mass as the (exponential) bulge. Are these nuclei the central mass concentrations of the models?

The $V - H$ color distribution of the exponential bulges is rather broad, and peaks at $V - H \sim 0.96$, significantly bluer, by about 0.4 mag, than the value typical of the $R^{1/4}$ bulges (Carollo *et al.* 1999). If this bluer color can be ascribed to a younger age, this would indicate that exponential bulges are the preferred mode for bulges forming more recently. The relatively massive central clusters found in these exponential bulges could theoretically prevent subsequent bar formation, and removing the possibility of successive cycles of bar formation – gas inflow – formation of central object – bar dissolution mechanism (as was discussed also by Rix in his oral presentation at this meeting).

4.3. *Chemical Abundance Distributions*

The K-giants in the Milky Way bulge have a very broad metallicity distribution, both in Baade's window (at a projected Galactocentric distance of $\approx$500pc; Rich 1988) and at projected distances of several kpc (Ibata & Gilmore 1995a,b). The breadth of the metallicity distribution in the bulge contrasts with that narrow distribution observed in the disk at the solar neighborhood. The lack of metal-poor stars in the local disk conflicts with predictions of the 'simple model' of chemical evolution, and is the famous 'G-dwarf problem'. One hastens to add that the fact that the Milky Way bulge has a broad distribution, and indeed fits the predictions of the 'simple model' of chemical evolution, does not mean that any or all of the many assumptions of the 'simple model' are valid; another example of a stellar system with a metallicity distribution that is well-fit by the simple model (albeit with a reduced yield) is the stellar halo of the Milky Way. The G-dwarf problem has many solutions, the most popular of which is to postulate gas inflows (e.g. Tinsley 1980). The width of a metallicity distribution is related to the ratio of inflow time to star formation time, and perhaps the wider metallicity distribution in the bulge can be interpreted in terms of very rapid star formation, occuring too fast for inflow to affect the metallicity structure.

The M-giants studied by Frogel (this volume) in the inner 100pc or so of the bulge do appear to have a narrow metallicity distribution, but this may reflect the bias inherent in the sample selection by such a late spectral type; data for K-giants are desirable, both because they are a more representative evolutionary phase of low-mass stars, and because their spectra are easier to interpret and use to determine metallicities, than are M-giants.

From the width of the giant branch in color-magnitude diagrams, the M31 bulge is inferred to have a rather broad metallicity distribution in its outer parts, but a narrow metallicity distribution interior to 1kpc (Renzini, Rich, this volume). Perhaps this variation in width also reflects a variation of the ratio of star formation rate to gas inflow rate, this time a variation with radius within the bulge. At face value, the opposite trend – one with a broader metallicity distribution in the inner, more dense parts – may be expected in models where the local star formation rate is determined by a non-linear function of gas density, but the flow rate is given by the inverse of the dynamical time (proportional to the square root of density), so that the ratio of star formation time to flow time decreases with increasing density.

The stellar populations of the resolved bulges in the Local Group are not compatible with their formation via accretion and assimilation of satellites and or globulars like those remaining today – the bulges are too metal-rich, and have too narrow an age distribution. However, perhaps some part of the metal-poor tail in the Milky Way bulge could be due to accretion of the dense, metal-poor and old globular clusters. Note that for stellar satellite systems with a realistic density profile, a significant fraction of the stars will be tidally removed far out in the halo, and only a fraction will make it into the center (Syer & White 1998; see also Kuijken, this volume). Kuijken (this volume) notes that the timescale of satellite accretion is rather long, so that any bulge-building by this means should be on-going. This raises a further difficulty, in that the old, metal-rich bulge stars are unlike those in typical satellites. A graphic illustration of the difference in stellar populations between the bulge of the Milky Way and the Sagittarius dwarf, one of the more massive satellite galaxies of the Milky Way, is shown in Figure 4.

4.4. *Chemical Abundance Gradients*

A strong chemical abundance gradient is a signature of slow, dissipative collapse. Such gradients are weakened, but not erased, by any subsequent mergers (e.g. White 1980; Barnes & Hernquist 1992). There are no clear predictions for secular evolution models (but are needed).

Observationally, there are weak or minimal amplitude gradients in mean metallicity in resolved bulges (Milky Way Galaxy – Gilmore, Frogel, this volume; M31 – Frogel, Rich, Renzini, this volume). As mentioned, the interpretation of absorption line-strengths remains ambiguous, and we need more data and models.

5. Summary

Bulges are diverse in their properties, and probably in their formation mechanisms, or at least in the dominant physics at the epochs of star formation and/or assembly. Perhaps the differences are just a matter of degree, since, for example, even 'monolithic collapse' involves fragmentation, with subsequent star formation in the fragments. A centrally-concentrated profile appears to match 'maximum entropy' arguments (Tremaine, Henon & Lynden-Bell 1986) for the end-point of violent relaxation of a cold, clumpy system, independently of the details of the evolution to that end-point.

The overall trends of the observations are that small bulges, of late-type disk galaxies, show a strong connection to their disk, while big bulges, of early-type disk galaxies, are more like the low-luminsity extension of the elliptical galaxy sequence. The bulge of the Milky Way appears to straddle these two generalities, having an affinity for its disk in terms of structure, but having the old, metal-rich population associated with 'spheroids'.

What does this mean? Even the casual reader should have noted the not-infrequent occurrence of the sentiment 'more data and models are needed' in the text above. We are at the stage of requiring robust quantitative results from both theory and observations.

More specifically, for the Milky Way, we require good HST color-magnitude diagrams for more lines-of-sight towards the Milky Way bulge, following the work of Feltzing & Gilmore (1999) in establishing the association of a younger stellar population with

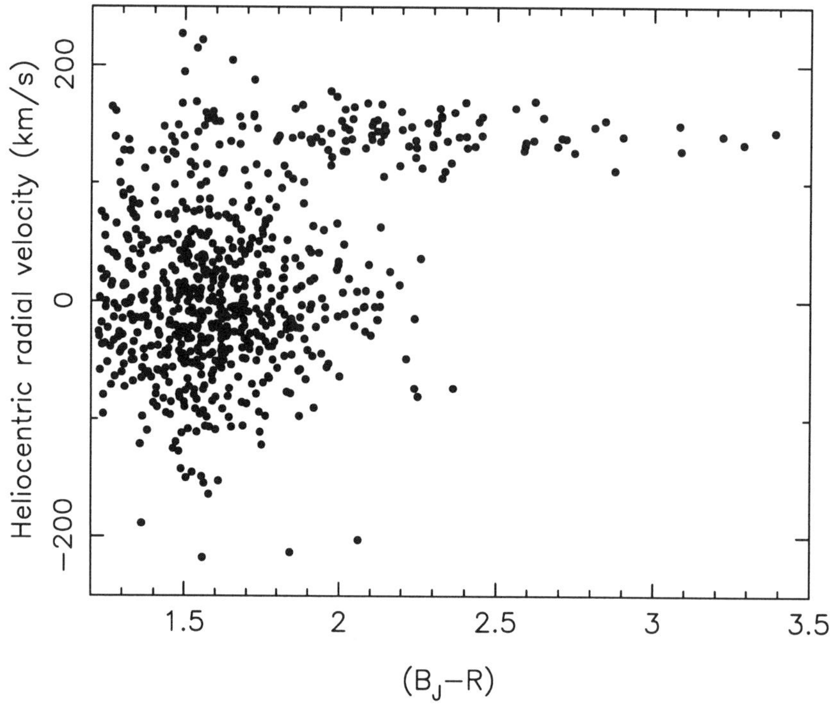

FIGURE 4.   Heliocentric radial velocities of the sample of K-giant stars observed by Ibata, Gilmore & Irwin (1994) in lines-of-sight towards the Milky Way bulge ($\ell = -5°$, $b = -12°$, $-15°$, $-20°$). Note the narrow velocity-dispersion subsample centered at around 150km/s. These stars are members of the Sagittarius dwarf spheroidal galaxy, which was discovered from these data. The reddest stars are exclusively members of this galaxy, illustrating the real difference in stellar populations between the bulge and this satellite galaxy. The bulge cannot have formed from a simple merger of satellites like the Sagittarius dwarf.

foreground disk. We also require good reddening maps and metallicity data to aid the interpretation of these color-magnitude diagrams. The inner disk of the Milky Way is remarkably under-studied, and again age and metallicity distributions – and stellar kinematics – are obviously crucial in determining the similarity or otherwise of inner disk and bulge. Further, we need to understand the relationship between the 'bulge' globular clusters and the bulge field population; present models of globular-cluster formation appeal to pre-enrichment to provide the uniform enrichment within a given cluster, so it is not obvious that the enrichment signatures of cluster stars and field stars should be the same. Elemental abundances for statistically-significant samples of unbiased tracers of the field in a variety of lines-of-sight are required to understand the history of star formation.

A combination of HST and ground-based (to probe both small- and large-scale structure) broad-band optical and IR colors, and surface brightness profiles, are still lacking for large samples, including the whole range of spiral Hubble types. These data should allow a robust quantification of the correlations between morphologies. Basic kinematic data, including gradients, should be obtained for a representative sample of bulges and disks. While we may lack the means at present for a unique interpretation of absorption line-strength data, the straightforward test for continuity in the line strengths from bulges to their disks is meaningful.

The redshift of statistically-significant samples of galaxies is being continually pushed back (at what point will this pose a real problem for CDM?) and HST and the next generation of telescopes should provide robust morphological classifications. We will no doubt see evolution, but need to have the model predictions to be able to distinguish the underlying physics behind the evolution.

'Secular-evolution' models are at their early stages of development, but several key questions may be posed. While it may be reasonable to comment that a correlation between bulge scale-length and disk scale-length points to a connection between bulge and disk, can the models 'post'-dict the factor of ten that is observed? Can they predict the frequency with which one should see barred spirals today, even ones with big bulges? Are all bars the same? Are there too many bars and/or central concentrations observed for the models of bar dissolution? Or is the dominant mechanism of bulge-building in this scenario actually scattering of disk stars through resonant coupling, rather than bar dissolution? How can this be compatible with uniformly old bulges? But are exponential bulges (apart from the Milky Way bulge) composed of old stars?

Cold-dark-matter dominated cosmologies gained popularity partially because of their robust predictive power, a requirement for a good theory, in terms of the large-scale structure formed by the dissipationless dark halos, (e.g. Davis, Efstathiou, Frenk & White 1992). The predictions for the luminous components, the galaxies as we observe them, have not yet achieved the same level of maturity. Advocates of merging and hierarchical clustering should quantify further the ages of stars now in bulges, and the epoch of assembly into bulges. What is predicted for the age spread within a typical bulge like the Milky Way? What fraction of bulges should have angular momentum vector misaligned with their disk? Should colors of bulge and disk be correlated?

If bulges form in a 'star-burst', what is the role of a supernova-driven wind? In this context, the X-ray properties of bulges, including the Milky Way, should constrain the ability of the bulge potential well to retain hot gas.

Where do we stand? – inspired to get to work!

I acknowledge supported from NASA ATP grant NAG5-3928.

REFERENCES

ABRAHAM, R.G., ELLIS, R.S., FABIAN, A.C., TANVIR, N., GLAZEBROOK, K. 1999a *MNRAS*, **303**, 641

ABRAHAM, R.G., MERRIFIELD, M., ELLIS, R.S., TANVIR, N., BRINCHMANN, J. 1999b, preprint (astro-ph/9811476)

ANDREDAKIS, Y.C., PELETIER, R.F., BALCELLS, M. 1995 *MNRAS*, **275**, 874

BALCELLS, M., PELETIER, R. 1994 *AJ*, **107**, 135

BARNES J., HERNQUIST L. 1992 *ARA&A*, **30**, 705

BAUGH, C.M., COLE, S., FRENK, C.S. 1996 *MNRAS*, **283**, 136

BAUGH, C.M., COLE, S., FRENK, C.S., LACEY, C. 1998 *ApJ*, **498**, 504

BENDER R., BURSTEIN D., FABER S.M. 1992 *ApJ*, **399**, 462

BENDER R., BURSTEIN D., FABER S.M. 1993 *ApJ*, **411**, 153

BINNEY J., GERHARD O., SPERGEL D. 1997 *MNRAS*, **288**, 365

BOTTEMA, R. 1993 *A&A*, **275**, 16

BURSTEIN, D., BENDER, R., FABER, S.M., NOLTHENIUS, R. 1997 *AJ*, **114**, 1365

BURSTEIN, D., DAVIES, R., DRESSLER, A., FABER, S.M., STONE, R., LYNDEN-BELL, D., TERLEVICH, R., WEGNER, G. 1987 *ApJS*, **64**, 601

CAROLLO, C.M., 1999 *ApJ*, **523**, in press

CAROLLO, C.M., DANZIGER, J.J., BUSON, L. 1993 *MNRAS*, **265**, 553

CAROLLO, C.M., STIAVELLI, M., SEIGER, M., DE ZEEUW, P.T., DEJONGHE, H. 1999, in preparation

COURTEAU, S., DE JONG, R., BROEILS A. 1996 *ApJ*, **457**, L73

DAVIES, R.L., EFSTATHIOU, G.P., FALL, S.M., ILLINGWORTH, G., SCHECHTER P. 1983 *ApJ*, **266**, 41

DAVIS, M., EFSTATHIOU, G.P., FRENK, C.S., WHITE, S.D.M. 1992 *Nature*, **356**, 489

DE JONG R. 1996 *A&A*, **313**, 45

EDVARDSSON, B., ANDERSEN, J., GUSTAFSSON, B., LAMBERT, D.L., NISSEN, P., TOMKIN, J. 1993 *A&A*, **275**, 101

EGGEN, O., LYNDEN-BELL, D., SANDAGE, A. 1962 *ApJ*, **136**, 748

FELTZING, S., GILMORE, G. 1999 *A&A*, in press

FRANX M. 1993 in *Galactic Bulges* (ed. H. Dejonghe & H. Habing), IAU Symp. **153**, p243. (Kluwer)

GILMORE, G., WYSE, R.F.G., JONES, J.B. 1995 *AJ*, **109**, 1095

HOLTZMAN, J.A., ET AL. 1998 *AJ*, **115**, 1946

IBATA R., GILMORE, G. 1995a *MNRAS*, **275**, 591

IBATA R., GILMORE, G. 1995b *MNRAS*, **275**, 605

IBATA, R., GILMORE, G., IRWIN, M. 1994 *Nature*, **370**, 194

IDIART, T., DE FREITAS PACHECO, A., COSTA, R. 1996 *AJ*, **112**, 2541

JABLONKA, P., MARTIN, P., ARIMOTO N. 1996 *AJ*, **112**, 1415

KAUFFMANN G. 1996 *MNRAS*, **281**, 487

KENT, S., DAME, T.M., FAZIO, G. 1991 *ApJ*, **378**, 131

KORMENDY, J. 1977 *ApJ*, **218**, 333

MCWILLIAM, A., RICH, R.M. 1994 *ApJS*, **91**, 749

MAGORRIAN, J., ET AL. 1998 *AJ*, **115**, 2285

MINNITI, D. 1996 *ApJ*, **459**, 175

NAVARRO, J., BENZ, W. 1991 *ApJ*, **380**, 320

NAVARRO, J., STEINMETZ, M. 1997 *ApJ*, **478**, 13

NOH, J., SCALO, J. 1990 *ApJ*, **352**, 605

NORMAN, C., SELLWOOD, J., HASAN, H. 1996 *ApJ*, **462**, 114

PEEBLES, P.J.E. 1989, in *The Epoch of Galaxy Formation* (ed. C.S. Frenk *et al.*), p1. (Kluwer)

PELETIER, R., BALCELLS, M. 1996 *AJ*, **111**, 2238

RICH, R.M. 1988 *AJ*, **95**, 828

ROCHA-PINTO, H.J., MACIEL, W.J. 1997 *MNRAS*, **289**, 882

SYER, D., WHITE, S.D.M. 1998 *MNRAS*, **293**, 337

TIEDE, G.P., TERNDRUP, D.M. 1997 *AJ*, **113**, 321

TINSLEY, B.M. 1980 *Fundam. Cosmic Phys.*, **5**, 287

TREMAINE, S., HENON, M., LYNDEN-BELL, D. 1986 *MNRAS*, **219**, 285

VAN DEN BOSCH, F. 1998 *ApJ*, **507**, 601

WHITE, S.D.M. 1980 *MNRAS*, **191**, P1

WHITMORE, B.C, SCHWEIZER, F., LEITHERER, C., BORNE, K., ROBERT, C. 1993 *AJ*, **106**, 1354

WHITMORE B.C., SCHWEIZER, F. 1995 *AJ*, **109**, 960

WIRTH, A., GALLAGHER, J.S. III 1984 *ApJ*, **282**, 85

WORTHEY, G. 1998 *PASP*, **110**, 888

WYSE, R.F.G. 1998 *MNRAS*, **293**, 429

WYSE R.F.G., GILMORE G. 1992 *AJ*, **104**, 144

WYSE R.F.G., JONES, B.J.T. 1984 *ApJ*, **286**, 88

ZHANG, B., WYSE, R.F.G. 1999 *MNRAS*, submitted

ZUREK, W., QUINN, P., SALMON, J. 1988 *ApJ*, **330**, 519

Index